FAILURES IN EARTHWORKS

FAILURES IN EARTHWORKS

Proceedings of the symposium on Failures in Earthworks, organized by the Institution of Civil Engineers and held in London, 6–7 March 1985

WITHDRAWN

ⱦ| THOMAS TELFORD, LONDON

Organizing Committee: Dr R.T. Murray (Chairman), Dr R. Jewell, Dr C.J.F.P. Jones, E.A. Snedker, S. Thorburn

First published 1985

British Library Cataloguing in Publication Data
Symposium on Failures in earthworks (1985: London)
Failures in earthworks: Proceedings of the symposium on
   Failures in earthworks
   1. Earthwork
   I. Title
   624.1'5   TA715

ISBN: 0 7277 0243 2

Published for the Institution of Civil Engineers by Thomas Telford Ltd, P.O. Box 101, 26–34 Old Street, London EC1P 1JH

Printed in Great Britain by Billing and Sons Ltd, Worcester

Contents

Technical Notes

1. Excavated slopes in fissured tills

A. McGOWN, BSc, PhD, MICE, MIHT, FGS, Reader in Civil
Engineering, University of Strathclyde

SYNOPSIS. In Central Scotland a number of the excavated
slopes for roadworks failed although conventional soil
testing methods and analytical techniques were employed in
their design. The paper describes detailed investigations
of both failed and stable slopes which revealed that the
till was fissured in a systematic manner and it discusses the
techniques developed to identify and allow for the fissures
in future designs.

INTRODUCTION
1. Large ice-moulded mounds of glacial lodgement till, known
as drumlins, are common over much of Central Scotland. They
are typically asymmetrical, with their long axis aligned in
the known direction of ice flow over this area. They are
distributed over the low lying ground of Central Ayrshire, on
either side and beneath the post glacial deposits of the
Clyde Basin and across the Midland Valley to Stirling and
Edinburgh. In these areas there is approximately one
drumlin per square kilometre with two or three drumlins
often impinging on one another. The presence of these
large mounds of stiff till has governed the pattern of
development of the communication systems in the area, with
all the principal roads, railways and canal systems routed
around the mounds rather than through them.
2. In the late 1950's and early 1960's a new network of
urban and rural roads was planned. These new roads were
designed to be high speed routes with restrictions placed
on both horizontal and vertical alignments. This meant
routing through drumlins with deep cuttings inevitably
required. As there were few such cuttings previously and no
records available of major difficulties with them, it was
reasonably assumed that cuttings in these stiff tills would
present few design problems. Thus, after 100 mm diameter
open drive sampling and generally 37.5 mm diameter by 75 mm
high triaxial testing, a large number of cuttings through
drumlins were assessed using circular slip analyses and
judged to be safe at slopes of between 1 in 1.5 and 1 in
2.0. They were then constructed from the late 1960's onwards,
however, a number of them failed at varying periods after

construction, some only weeks after completion of the
excavations and installation of surface drainage systems.
Typically only one side of the cutting failed whereas the
other side remained, and still does remain, stable. The
usual remedial measure was to excavate the failed soil and
replace it with rock fill.

3. A study of the behaviour of cutting slopes in drumlins
has been underway at the University of Strathclyde for a
number of years and a pattern of behaviour has emerged
which is shown to directly depend upon the presence and
orientation of fissures and bedding features within the tills
forming the drumlins. In this paper, the findings of the
study are reported and it is shown how they can be applied
to the design of cutting slopes in these materials.

FORMATION AND SHAPE OF DRUMLINS

4. Many theories have been put forward over the last
century to explain the formation of drumlins by glaciers.
They are either based on the till being directly deposited
in this shape by the ice or reshaped by the ice after it has
been deposited. It is, however, likely that drumlins are
formed by a range of somewhat different processes. This
is evidenced by the difference in character of the drumlins
that may be encountered in any specific area. These can be
summarised as follows:

(a) Individual drumlins may consist of a variety of
materials in addition to till, including sand layers,
stratified silts and clay and pockets of gravel.

(b) Some drumlins may have rock cores or large blocks or
rafts of rock within them, whilst adjacent drumlins may be
composed entirely of till.

(c) In most cases the drumlins are aligned parallel to the
last known ice flow direction, whilst others are aligned
in directions at variance to this.

5. Often as part of investigations into the formation
of drumlins, their shape has been the subject of study.
For example, Wright (1912) noted two principal groups of
drumlins in Donegal, Ireland; a) the most common type, which
is cigar shaped and b) a much less common group, which may
be oval, triangular or crescentic in shape. He postulated
that the oval drumlins were the result of the ice flow
direction continually changing and that the triangular
and crescentic drumlins occurred where the ice flow
direction changed most rapidly. Later Chorley (1959)
suggested that a mathematical model of the plan shape of the
cigar-like drumlins was a lemniscate loop. He and many others
employing very similar techniques, have since proven that
most drumlins in any area have a very regular shape. As
part of the study at the University of Strathclyde, the
shape of the drumlins east of Kilmarnock, in Central Glasgow,
to the north of Falkirk and around Stirling were recorded.

This showed that the drumlins were regular in overall shape with heights above the surrounding ground ranging from 5 to 20 m, averaging 8 m. Their long axis varied in length from 200 to 900 m with an average of 620 m. The ratio of the lengths of their long axis to their maximum widths varied from 0.5 to 3.0 with an average of 1.9. Based on this study, a normalised plan shape for the drumlins in Central Scotland was produced and used to correlate the position of sample points in various drumlins, as indicated later in this paper.

NATURE OF THE TILLS STUDIED

6. A detailed study of the basic engineering properties of the various tills in Central Scotland has previously been reported by McGown et al (1975). These showed that although there are slight variations between the weathered materials of the upper 2 or 3 m of the tills from the various areas, the unweathered tills cannot be distinguished in terms of their basic engineering properties. Thus they may all be represented by the gradings shown in Fig. 1 and the other properties indicated in Table 1.

7. Visual examinations of lumps of fresh till extracted from borehole sample tubes or open excavations would suggest that the till is an intact stoney clay soil. Only after the failure of a cutting slope at Hurlford in Ayrshire was the first serious investigation undertaken to identify and record discontinuities in these tills, McGown et al (1974). The technique employed in this investigation, and in the many others that have followed, was the cavity technique. Most of the recording methods and terminology adopted were those suggested by Fookes and Denness (1969) but these have been modified and developed with time, McGown et al (1980).

8. At each site studied, at least three cavities were dug into the till at points which were unaffected by surface failures. One of the cavities was usually located near the original ground level with the others lower down the slope and some 25-50 m apart. For each discontinuity exposed in the cavities, measurements were taken of its strike and dip, dimensions, nature of any surface coatings, surface asperity (roughness) and the spatial location of its centroid with respect to a fixed set of axes. From the data on the location of the centroid, the spacing of the various sets (groups) of discontinuities were identified.

9. Two distinctive types of discontinuities have been found:

(a) A single set of low dip angle features, generally planar, of low asperity (1 to 4 mm) with a large majority coated in a thin layer of sand, silt or clay. The remainder of the low dip angle features are clean and thought to be stress relief features formed during the melting of the ice or removal of the soil overburden when forming the slope. The

3

Table 1. Typical moisture contents, plasticity characteristics and bulk densities of tills in Central Scotland

Soil Property	Unweathered Till		Weathered Till		Silty Discontinuities		Clayey Discontinuities	
	Average	Range	Average	Range	Average	Range	Average	Range
Moisture Content (%)	12.5	8-20	20.5	8-42	–	–	–	–
Liquid Limit (%)	27	24-44	33	23-47	39	36-42	49	47-51
Plasticity Index (%)	13	8-19	15	6-21	23	21-26	29	27-31
Bulk Density (t/m³)	2.26	2.05-2.45	2.03	1.7-2.35	–	–	–	–

coated features are considered to be bedding features formed
when the basal ice locally melted and allowed accumulation
of fines on the underlying till. The grading of these
coating materials are as given in Fig. 1. The plasticity
characteristics of the silty and clayey coatings are also
significantly different from the till matrix as shown in
Table 1.

(b) Two or four sets of near vertical features which are
generally planar, of medium to high asperity (2 to 8 mm),
clean below 2 to 3 m and stained by weathering above this
level. They are all considered to be stress relief
features formed when the over-riding glacier slowed down
and melted, so relieving the translational forces imposed
by the ice.

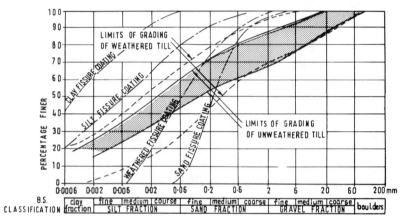

FIG. 1. Particle size distribution of unweathered and
weathered till and the discontinuities within them.

The size and spacing of the discontinuities in the tills are
not found to be sensitive to their orientation. The
majority of the features range in area from 1,600 to
40,000 mm^2 and their average spacings vary from approximately
30 mm near the original ground surface to 140 mm at 10 m
depth.

10. In order to build up an overall view of the pattern of
discontinuities in the drumlins, the data from a number of
investigations carried out at several sites were correlated
using the normalised drumlin shape referred to earlier. Fig.
2 shows the simplified discontinuity pattern diagrams from
these sites located in their relative positions with respect
to the long axis of the normalised drumlin (LL). From this
it can be seen that the low dip angle bedding features are
lying approximately parallel to the original ground surface
of the drumlins whilst the near vertical discontinuities are
forming conjugate sets about the direction of the long axis
of the drumlin, the approximate direction of ice flow. From
the orientation and spacing data, a very idealised model of

5

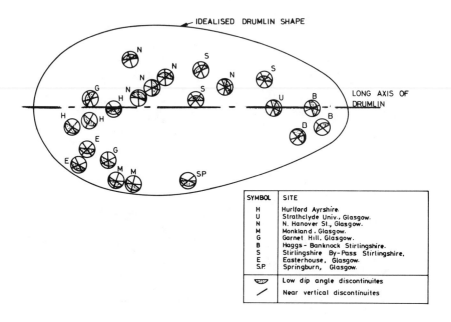

SYMBOL	SITE
H	Hurlford Ayrshire.
U	Strathclyde Univ., Glasgow.
N	N. Hanover St., Glasgow.
M	Monkland . Glasgow.
G	Garnet Hill. Glasgow.
B	Haggs - Banknock Stirlingshire.
S	Stirlingshire By-Pass Stirlingshire.
E	Easterhouse, Glasgow.
S.P.	Springburn, Glasgow.

⟨⟩	Low dip angle discontinuites
/	Near vertical discontinuites

FIG. 2.   The measured discontinuity orientations in the
drumlins of Central Scotland

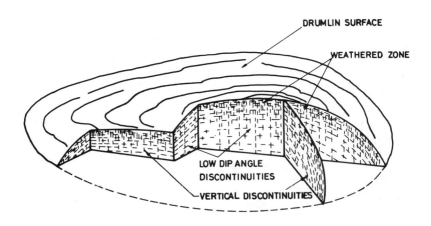

FIG. 3.   An idealised model of the distribution and
orientation of discontinuities in drumlins of
Central Scotland

the spatial distribution of discontinuities in drumlins may be
constructed as shown in Fig. 3. Of course no single drumlin
is likely to be structured exactly in the manner suggested by
the figure but it does give a broad indication of the pattern
of discontinuities to be expected.

MEASURED SHEAR STRENGTHS

11. An extensive programme of undrained and drained shear
strength testing has been undertaken in the tills of the area.
The most common type of testing carried out has been undrained
triaxial testing using 75 mm x 37.5 mm dia. test specimens
extracted from open drive 102 mm dia. sample tubes. As part
of the research programme at the University of Strathclyde,
small hand vane tests on the soil matrix, small undrained
shear box tests on discontinuities and 102 and 230 mm dia.
undrained triaxial tests on the soil mass, were conducted on
samples from large block and tube samples. Further, small
drained shear box tests on discontinuities and remoulded soil
and drained triaxial tests on 37.5 and 102 mm dia. specimens,
were also carried out.

12. The main finding of the undrained shear strength
testing was that the measured undrained cohesion of the tills
varied considerably, with values in the unweathered till
ranging from 60 to 600 kN/m². This was judged to be due to
three principal factors; the local variability of the soil
matrix, the depth of sampling and the nature and orientation
of the discontinuities in the soil mass. Of these three
factors, the influence of the discontinuities was the most
important and this was well illustrated in two ways; firstly,
by comparing the influence of sample size on the measured
vertical strength of the till and secondly, the effect of
the orientation of the sample from which the test specimen
is taken, as indicated in Fig. 4 for the N. Hanover St. site
in Glasgow.

13. The much more limited amount of drained triaxial
testing was principally concerned with the determination of
the peak drained strength of the till mass although some small
drained shear box tests on oriented specimens were carried
out to determine both peak and residual effective stress
parameters for both the matrix and the discontinuities.
Additionally, tests on remoulded samples were carried out to
confirm the residual strength parameters. From all of
these tests the average effective stress parameters identified
were as shown in Table 2.

BEHAVIOUR OF CUTTING SLOPES IN THE DRUMUNISED TILLS

14. With the recognition of the presence of discontinuities
in the tills, the relationships between the original ground
slope, cutting slope and discontinuity orientations were
established for cuttings which had failed and those which had
remained stable. As shown in Fig. 5, the slopes that failed
were invariably found to have low dip angle bedding features,
(i.e. coated discontinuities), with strikes and dips close to

7

the strike and dip of the cutting slope, whereas those that
remained stable did not.

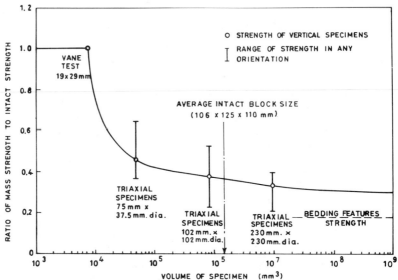

FIG. 4. The relationship between undrained shear strength
and volume of specimen tested for till from North
Hanover St., Glasgow.

Table 2. Typical effective stress strength
Parameters for the Tills

Soil		Peak Strength		Residual Strength*
		Cohesion (C') (kN/m²)	Angle of Friction (Ø'°)	Angle of Friction (Ø'°r)
Soil Mass		0 - 60	28 - 33	25 - 27
Clean Vertical Fissures		-	25 - 38+	not measured
Bedding Features	Sandy	-	38 - 40	not measured
	Silty	-	20 - 24	14 - 18
	Clayey	0 - 35	17 - 19	12 - 14

Notes   *Assuming cohesion C'r = 0
+ Greatly influenced by the presence of stones

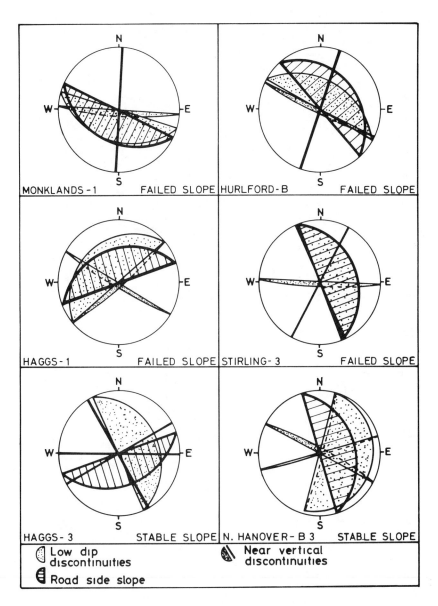

FIG. 5. Simplified discontinuity patterns measured at failed and stable cutting slopes

15. The stability of the various slopes was analysed using
the Bishop Method as employed within the Genesys "Slip Circle -
1" program. The soil strength parameters appropriate to this
type of analysis are those of the soil mass. Using the measured
peak strength values from vertically oriented triaxial test
specimens, for both total and effective stress conditions, the
factors of safety calculated for all the slopes was much
greater than unity. Using effective stress parameters
approaching residual values for the soil mass, reduced factors
of safety were of course calculated but still this did not
predict failure of the slopes which had failed.

16. Having regard to the presence of the discontinuities
and the translational nature of failures that had occured in
the tills, use was also made of the analysis suggested by
Barton (1971) for rock slopes containing discontinuities,
Fig. 6. The data used in this method are the dip of the
bedding features at the strike of the cutting slope, (i.e. β
their orientation with respect to the cutting), the peak
and residual effective stress strength parameters of the
bedding features and the pore water pressures within the
soil. Factors of safety less than unity were consistently
obtained using these data with the Barton method for the failed
slopes and the critical importance of the presence of silty or
clayey bedding features at the correct orientation was
highlighted.

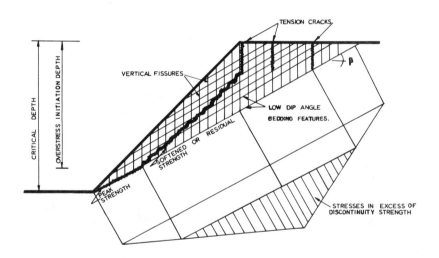

FIG. 6. Translational slide mechanism adopted in Barton
(1971) analysis.

THE LOANS BY-PASS

17. The Loans By-Pass forms part A78 route and runs
parallel to the Clyde Coast just to the east of Troon. The
line of the By-Pass as initially proposed lay along the
inland boundary of the raised beach deposits of sand and
gravel that cover that part of the coastline. Just north of
the village of Loans (Site A) and 1.5 km further north

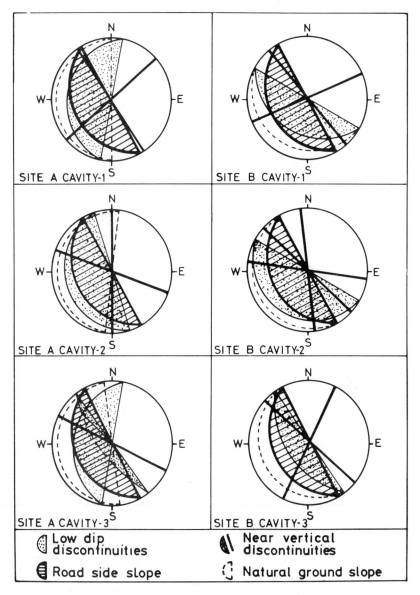

FIG. 7. Simplified discontinuity patterns measured at
Loans By-Pass

(Site B), the line and level of the proposed Road required cuttings into the shoulders of two drumlins which project out into the raised beach. In view of the stability problems with cuttings in other drumlins in the area, it was decided to carrty out discontinuity studies at the sites of the proposed cuttings.

18. At each drumlin three trial pits, each 4 m deep, were dug across the line of the proposed road. The pits were carefully shored and fenced off to make them safe to work in. Cavities were dug in the uphill side of the pits at depths of 3 - 4 m. As in previous studies, the orientation, distribution and size of the discontinuities were measured and their surface properties described.

19. As can be seen in Fig. 7, the discontinuities exposed in the cavities fitted closely the pattern of observations obtained in other drumlins in the area. Firstly, there are three distinct sets of discontinuities; two near vertical sets and one low dip angle set. The majority of the low dip angle discontinuities were found to be coated with silty materials whereas the near vertical sets were found to be relatively clean, although all three showed evidence of weathering. This is not surprising bearing in mind the relatively shallow depth of the trial pits.

20. Using the effective stress strength parameters given in Table 2 for the soil mass and silty bedding features, both Bishop and Barton slope stability analyses were run for the slopes. Once again stability was predicted by the Bishop analysis and failure by the Barton analysis. Based on this the Local Authority Engineers responsible for the design decided to avoid possible earthworks failures and so changed the alignment of the road to reduce the extent of the cuttings in length and height. This modified route has recently been completed and no problems were experienced with the reduced excavations.

CONCLUSIONS

21. The investigations reported in this paper have clearly identified a relatively consistent pattern of discontinuities in the drumlinised tills of Central Scotland. It has shown the close correlation between this pattern and the occurence of slope failures in cuttings through the materials. As Henkel (1982) suggested, this relationship between the geology of deposits and their engineering behaviour should always be sought, but as he also pointed out, the unexpected should always be expected, particularly in glacial materials. So that the idealised pattern of discontinuities suggested in this paper should not be taken as a rule, only as a guide, the presence of sand lenses, laminated silt and clay beds and other features within drumlins should therefore be expected.

22. The paper has highlighted the need to identify appropriate soil parameters and methods of analysis for slope stability problems. The value of collecting, collating and back-analysing failures in order to assist in this, has once

again been proven.

23. Lastly, the practicality of carrying out soil discontinuity studies has been proven, at least for the 3 to 10 m depths of cutting slopes encountered in the drumlins of Central Scotland. Using the data collected from these it has been shown possible to identify potential slope failures and to provide sufficient information for the designer to avoid them, or make special provision for them.

REFERENCES

1. BARTON N. Progressive failure of excavated rock slopes. Proc. 13th Sym. on Rock Mech., Univ. Illinois, Urbana, 1971, 139-170.
2. CHORLEY R.J. The shape of drumlins, Jour. Glaciol. 3, 1959, 339-344.
3. FOOKES P.G. and DENNESS B. Observational studies on fissure patterns in cretace ous sediments of S.E. England. Geotechnique, 19, 1969, 453-477.
4. HENKEL D.J. Geology, geomorphology and geotechnics. Geotechnique 32, 1982, 3, 175-194.
5. McGOWN A., ANDERSON W.F. and RADWAN A.M. Geotechnical properties of the tills in West Central Scotland. Sym. Engg. Behaviour of Glacial Materials, Univ., Birmingham, 1975, 89-99.
6. McGOWN A., MARSLAND A., RADWAN A.M. and GABR A.W.A. Recording and interpreting soil macrofabric data. Geotechnique 30, 1980, 4.417-447.
7. MCGOWN A., SALI A. and RADWAN A.M. Failure patterns and slope failure in boulder clay at Hurlford, Ayrshire. Quart. Jour. Engg., Gol. 7, 1974, 1-26.
8. WRIGHT W.B. Drumlin topography of South Donegal. Geol. Magazine, 1912, 9.4.

2. Earthworks in Hong Kong—their failure and stabilisation

A. J. VAIL, JP, BSc, FICE, FIWES, FIE(M), FHKIE, Senior Resident Partner, and A. A. BEATTIE, BA, FICE, FHKIE, MASCE, Associate, Binnie & Partners, Hong Kong

SYNOPSIS. Slope failures and slope stability studies in Hong Kong have featured prominently in recent literature. The remedial works carried out to restore these unstable slopes and the measures adopted to prevent the failures have received less attention. This paper briefly reviews the history of earthwork failures in Hong Kong, differentiates between the behaviour of natural and man made slopes and discusses possible mechanisms of failure in the light of new information. Stabilising methods are briefly described with examples of their application.

INTRODUCTION

1. Hong Kong lies on the south coast of China in the Pearl River estuary. Seventy five percent of the land area is mountainous, the remaining area being relatively flat or gently sloping coastal land which has been intensely developed.

2. Between November and April, the weather is relatively dry and cold. Between May and October the hot and humid southwest monsoons prevail and 80% of the territory's annual rain falls. Precipitation exceeding 100 mm in an hour has been recorded with peak intensities of over 150 mm per hour.

3. The geology of Hong Kong has been described in detail in Ref. 1 and the weathering profile in Ref. 2.

4. Initial development of the Territory took place on the relatively flat land bordering Victoria Harbour on Hong Kong Island. Within a few years this development had reached the lower slopes and the peaks. Cut slopes and filled embankments were formed for roads,buildings and gardens. Many platforms were supported by thin rubble retaining walls, some of considerable height, and some backed by stabilised soil.

5. Development continued at a modest pace until the outbreak of war in 1940. After the war, development recommenced and accelerated during the 1950's as Hong Kong's population increased rapidly with the influx of immigrants from China. In 1953 the Government instituted an ambitious public housing programme which placed heavy demands on land resources.

6.   The available low lying natural and reclaimed sites were soon exhausted and platforms were then formed in the surrounding hillsides by cutting ridges and filling valleys. Increasing wealth of the population provoked a demand for private housing and since the early 1960's there has been increasingly intense development of the steep hillsides.

7.   In 1977, following a series of disastrous failures, the Government introduced stringent geotechnical regulations.   As a consequence the awareness of the geotechnical problems of Hong Kong improved dramatically, until today the state of the art in the territory is as good as anywhere in the world.

HISTORY OF FAILURES
Natural slopes

8.   While aerial photographic evidence suggests that there have been large failures of natural slopes in Hong Kong in the past, recent failures generally have been less than about 50 m^2 and not much more than 1 m deep.   Once the natural vegetation has been disturbed, failure scars enlarge by erosion.   There have been few reports of loss of life or significant economic loss as a result of failures of virgin natural slopes, except where boulders have dislodged or where the slopes have been undermined by excavation.

9.   In or about 1965, a developer cut into a hillside above Kotewall Road to form a platform for a building.   He then abandoned the site which remained unsupported for seven years until, in June 1972, during a period of exceptional rainfall, the hillside above collapsed carrying Po Shan Road with it.   The resulting landslides destroyed a 12 storey block of flats and killed 67 people (Ref. 3), (Ref. 4).

Man made earthworks

10.   Man made earthworks have had a history of failure for the past 100 years and many of these have caused loss of life and extensive economic loss.

11.   The first recorded major earthwork failure, in May 1889, occurred when 672 mm of rain fell in 24 hours (302 mm in 4 hours).   This caused the collapse of a building platform near the Peak.   Recent aerial photographs together with oblique photographs of the time, suggest that the slope was about 40 m high and was formed at an angle of about 35°.   The failure debris travelled 600 m from an elevation of 460 m to 130 m.   This however was not the first slope failure since newspaper reports indicate that during the same storm a new retaining wall, constructed to stabilise a slope which had a history of failures, also collapsed.

12.   In at least 15 out of the next 25 years, there were landslides and wall failures in the most heavily developed areas, some resulting in loss of life.

13.   In 1925 a vertical series of three retaining walls at Po Hing Fong collapsed onto a row of houses killing 125 people.   Water had been seen seeping from the lower wall

during previous wet seasons. At the time of the collapse a
site above the third wall was being redeveloped. Although
the existing walls had shown signs of distress prior to the
start of this redevelopment it was decided not to re-build
them because they had been standing for 30 years and there
were no signs of recent movement. During the rainstorm which
led to the failure, a total of 237.5 mm of rain fell in
approximately 8 hours, the collapse taking place at the end
of the period. During the storm, water had been discharging
onto the site and collecting in foundation excavations on the
platform which otherwise was covered with concrete paving.
The cause of the failure was found to be "inadequate design
and inadequate drainage of the footings of the wall".
Subsequent investigation revealed that many similar walls in
the area were in very poor condition.

14. In most storms, damage by boulder movements was also
reported. In 1926 a boulder weighing approximately 20 tonnes
crashed onto one of the main pumping stations causing
considerable damage. Subsequent proceedings in the
Legislative Council suggested a beginning of public awareness
to the problems of slope stability as one member requested
that a Committee be formed to "search for, examine and report
on the vulnerable parts of the Colony contiguous to houses
and roads which are liable to storm damage".

15. Whilst earthwork failures continued intermittently,
the next major failure occurred in 1966 when 108 mm fell in
one hour. A fill slope about 40 m high standing at about 34°
to the horizontal failed, the debris travelling 100 m and
dropping 200 m. During this storm there were 181 landslide
and retaining wall collapses on Hong Kong Island and Kowloon.

16. In 1972, an embankment in Sau Mau Ping failed. The
embankment was approximately 40 m high and built at a slope
of approximately 34°, the debris from the failure travelling
between 100 and 160 m over land that sloped at 1 in 25. The
slope was formed of decomposed granite specified to be placed
and compacted in 1 m layers. A Commission of Inquiry (Ref.
5) found the cause of failure to have been "Softening of the
fill material caused by infiltration of rainwater through the
sloping face as a result of an exceptionally long and intense
rainstorm". These failures led to a programme to identify
and study areas in Hong Kong which were landslide prone
(Ref. 6).

17. Heavy rain in 1976 led to four more landslides in the
Sau Mau Ping area, each 2 to 3 m deep (Refs. 4 and 7). The
slip material in all four slopes liquefied on failure.
Investigations concluded that failure was caused by rain
infiltrating the slopes which had been formed of inadequately
compacted decomposed granite. It was thought that, at the
failure surfaces, there existed relatively impervious layers
of material. When vertical infiltration of rainfall reached
these layers, flow down the slope developed and the resulting
positive pore pressures caused the failures. The report on
these investigations highlighted the dangers of placing loose

fill in embankments particularly where that fill could become saturated.

18. Following the investigation of the 1976 failures, very stringent geotechnical controls on construction were introduced by Government. A massive earthwork stabilisation programme was also instigated. Inevitably, there have still been minor landslides but these have occurred away from the residential and industrial areas which have received priority treatment under this stabilisation programme.

Construction failures

19. The Po Shan Road landslide could be regarded as a construction failure as it was triggered by an inadequately supported excavation into the hillside. Most construction failures, however, receive little attention unless they cause loss of life or disruption to the public.

20. In dry conditions quite large cut slopes as steep as 60° can be formed in both the decomposed granites and volcanics. This can and did induce an unwarranted sense of security in the construction industry and precautions to protect slopes against the effects of rainfall were in the past often left too late, particularly at the beginning of the wet season when some of the most intense rain falls.

21. Other temporary slopes have failed because the structure of the parent rock has been ignored and wedge failures have taken place along relict joints forming a plane of weakness in the soil mass.

22. Recently there have been several failures of temporary works (Ref. 8). For example, during excavation for a basement, sheet piles were allowed to cantilever for 4 m before any raking shores were installed. Two layers of shoring were erected and excavation was carried down to a third level but before this could be installed the support system failed. Investigations revealed that deflection of the piles (80 mm) had allowed movement of a 400 mm water main which lay within 1 m of the piles (although record drawings showed the distance to be 4 m). This 80 mm movement was sufficient to pull a joint on an unrecorded 150 mm hydrant connection crossing the adjacent highway. When the 150 mm main burst, the sheet piles were subjected to loads for which they were not designed.

23. In other instances, failures of temporary works during construction have been caused by site staff removing supporting members to facilitate construction.

24. At one site an old masonry culvert under a loose fill embankment collapsed when H piles were driven within 1 metre of it. In subsequent heavy rain high pore pressures in the vicinity of the collapse caused the slope to fail.

MODES OF FAILURE

25. Apart from construction failures caused by carelessness or poor supervision, nearly all failures have

been associated with heavy rain or leakage of water from services. The mechanics of these failures have been the subject of debate for many years and this continues despite recent advances and intensive research by Government.

26. Analysis of many Hong Kong slopes, using parameters derived from triaxial testing of samples saturated by back pressure, yields factors of safety which are often less than or close to unity. Although some of these slopes fail, others remain stable for many years as was shown in the CHASE study (Ref. 9). Only rarely are there obvious factors to indicate why one slope should fail and another not. Whilst the method of back saturating the samples is open to criticism, since the stress path followed in the test does not follow that in the field (Ref. 10), the parameters derived lead in general to designs which are safe, although it must be recognised that in some cases they may be conservative.

27. Predicting the pore pressure conditions is difficult for the engineer analysing slope stability in Hong Kong. Calculations are often based upon the maximum observed piezometric level in a wet season upon which is superimposed the calculated rise due to a descending wetting band (Ref. 6). The slope is analysed assuming failure to be caused by a rising ground water level. Different factors of safety are considered to be acceptable for different rainfall conditions(Ref. 11).

28. However, many slopes fail above the level which would be affected by ground water unless a perched water table has formed on a discontinuity or zone of low permeability. Such a feature is rarely observed. Consequently, the destruction of suction pressures, due to increasing saturation, has been suggested as one possible reason for these failures.

29. Recent work (Ref. 12) indicates that the incidence of major landslides is closely related to intensity of rainfall. Thus the model of rainfall infiltration used in assessing slope stability must be modified. To this end an examination of rainfall conditions prior to the Sau Mau Ping failures is useful.

30. Rainfall records, at stations nearest to the slopes which failed, for the 1966, 1972 and 1976 storms show maximum hourly intensity to have been approximately 100 mm in 1972 (Fig. 1 & 2). In that year only Slope A (Fig. 3) failed. The maximum hourly intensity during the 1976 storm was approximately 70 mm per hour but four slopes failed within 20 minutes. In 1976 there was continuous rain for 32 hours prior to the failure whilst in 1972 there was only 10 hours. This was of generally lower intensity than that which fell in 1976, however, during the antecedent 3 days there had been heavy rainfall. During the 1966 storm there had been 27 hours of continuous rain but of much lower intensity than in either 1972 or 1976 and the maximum occurred early in the storm.

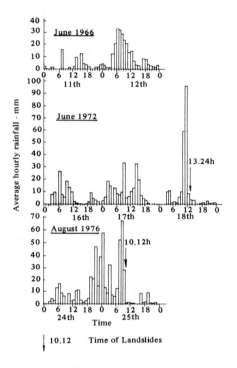

Fig. 1   Average of rainfalls at HK airport and
Tates Cairn for storms in 1966, 1972, 1976

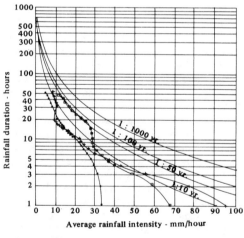

Fig. 2   Comparison of average rainfall intensity
and duration for 1966, 1972 and 1976 storms

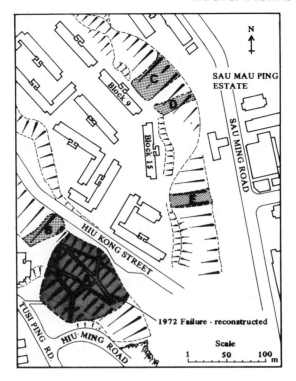

Fig. 3　Location of failures at Sau Mau Ping
1972 and 1976.

31. Whilst, in 1972, the mode of failure was not examined in detail, the 1976 investigation suggested that infiltration into the slopes was inhibited by layers of impervious material in the slope on which perched water tables formed. However all four slopes were formed of fill probably obtained from different sources and built at slightly different times. The depths of all failures were similar and if the suggested mode is correct impervious layers must have existed in all the slopes at the same depths. This would require a remarkable coincidence.

32. An alternative hypothesis is that, prior to intense rainfall, (in excess of 50 mm per hour) infiltration created a band of nearly saturated soil parallel to the surface of the slopes. As a result of the subsequent intense rain a band of soil at the surface became fully saturated. As this fully saturated band migrated down, the displacement of pore air by the advancing wetting front was inhibited or totally prevented by an "air-lock". At this stage flow changed from being vertical to downslope, the resulting positive pore water pressures causing failure.

33. The sensitivity of such a mechanism to soil properties can be demonstrated by considering the stability of an

21

infinite slope with flow parallel to the surface the factor of safety being given by:

$$F_s = \frac{c'}{Z \, Sin \, \beta Cos \beta} + \frac{(\gamma - \gamma W)}{\gamma} \frac{tan \, \phi'}{tan \, \beta} \ldots\ldots\ldots\ldots(1)$$

where Z is the depth measured vertically to the potential failure surface and β is the slope angle. At Sau Mau Ping the bulk density of the soil was 16 $K_N m^{-3}$; the slope angle was 34° and consolidated undrained triaxial testing of the fill yielded a ϕ' of 30°. Substituting these values into equation (1) for $F_s = 1$ we get:

$$Z = 0.2c' \text{ metres, (where c' is expressed in Kpa.)} \ldots.(2)$$

Triaxial testing of the decomposed granite yielded a c' of 12 Kpa which is consistent with the observed failure depths of between 2 and 3 m. Furthermore, using data collected during the investigation it can be shown that 30 hrs of continuous rain would have been required to develop a wetting band of this depth. In 1972 with only 10 hrs continuous rain the wetting band would not have developed to this depth and the slopes did not fail.

34. The importance of compacting fill is demonstrated by substituting typical values for compacted fill into equation (1) i.e. a ϕ' of 39° and a bulk density of 19.3 $K_N m^{-3}$. Then, for $F_s = 1$, we get:-

$$Z = 0.27c' \text{ metres, (where c' is expressed in (Kpa.)} ..(3)$$

If c' does not increase as a result of compaction, a wetting band only 35% deeper than that necessary for uncompacted soil is required to generate a failure. However as compaction can reduce the permeability by at least three orders of magnitude the duration of antecedent rainfall needed to cause failure of a compacted slope would be 1000 times greater. Therefore this mode of failure can not develop in a compacted slope.

35. Thus high intensity rainfall triggers a failure only if the wetting band has advanced to a critical depth dictated by the soil shear strength properties. This is consistent with the observation of Brand et al (ref. 12) that only the immediately antecedent rainfall is important in causing failure.

36. The loose fill slopes which have failed during heavy rainfall have been extremely destructive as they have formed flowslides. Natural and cut slopes in Hong Kong do not liquefy on failure and therefore the debris travels shorter distances and poses less of a threat to the public. For this reason the Government identified fill slopes as having priority for landslide preventive treatment.

37. Where rock lies close to the surface, small scale slips occur in continuous heavy but not necessarily intense rainfall and the conclusion must be that in these cases the conventional model of failure caused by a rising ground water level is probably applicable.

38.   Leaking services can contribute significantly to
rising ground water levels.   In intense rainfall old sewers
become surcharged and under these conditions leak into the
surrounding soil.   For this reason, in carrying out
stabilising works, services have received close attention.

39.   Rock slopes have not been considered in detail in this
paper but for completeness it should be recorded that many
have failed, often during heavy rain.

REMEDIAL MEASURES

40.   Whilst some earthworks in Hong Kong have failed
progressively and relatively slowly, in most cases failure
has been without warning.   As a consequence, since 1976
Government policy has required that the stability of all
existing slopes and retaining walls, posing a potential
hazard to the public, be assessed analytically and if found
lacking, be improved.   In most cases this is done using limit
equilibrium methods of analysis, the structures being
required to possess minimum factors of safety in accordance
with Government established criteria (Ref. 11).   In
implementing this policy many techniques have been adopted to
stabilise potentially unstable earthworks.

Slope reconstruction

41.   To prevent flow slides of the type which had occurred
at Sau Mau Ping it was decided in 1976 that a 3 m skin of
fill compacted to 95% of B.S. density would be provided on
the surface of all embankments where the fill was shown to
have densities lower than the critical value (Ref. 13).   Such

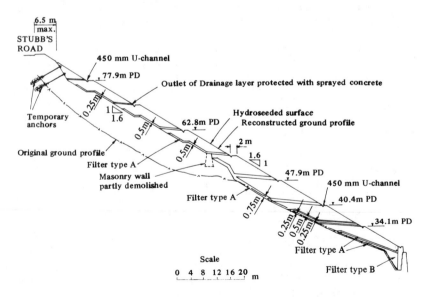

Fig. 4   Typical slope reconstruction profile (Ref. 14)

23

a skin would not liquefy on shearing and would prevent infiltration of water into the slope. The 3 m depth was selected as being the minimum required for the mechanical plant needed for construction. Furthermore it was decided that, in general, fill slopes should possess a factor of safety of 1.4 against slope failure. Most embankments had been built to a standard gradient of 1 in 1.5. Reconstructed slopes have been regraded to 1 in 1.7 with surface drainage channels on berms at 7.5 m vertical centres. Drainage layers of various geometrical forms have been incorporated in and behind the compacted fill layers to prevent groundwater being dammed behind the compacted face. Figure 4 shows a section of one typical reconstructed slope.

42. Where services pass close to the crest of a slope they have been diverted or have been re-installed in a watertight duct drained to surface channels.

43. Whilst most slopes have been stabilised in this way, the surface of a small number of slopes have been compacted using a dynamic compaction technique (Ref. 14) in an attempt to find a cheaper and quicker solution to the problem.

Rock anchors

44. Where rock lies relatively close to the surface, rock anchors have been used to stabilise existing retaining walls and cut slopes. Anchors have not been used to stabilise fill slopes except as a temporary measure. Building regulations in Hong Kong require the total length of anchors to be within the lot boundary. Thus where rock is very deep or structures are built near the lot boundary, permanent anchors are precluded although temporary anchors are allowed, providing these are destressed on completion of the work (Ref. 16). To overcome these problems anchors used to stabilise a cut slope at Trafalgar Court were fixed in a tunnel (Fig. 5). At the same site anchors were also fixed in deadmen within caissons (hand dug piers) for the same reasons.

Retaining walls

45. Many different types of wall have been used to support fills and cuttings:

46. Gabions: Gabions have been used for construction of both small and fairly large walls (Ref. 8). The uncertain life of the gabion boxes in tropical conditions has caused the Building Authority to question their use for long life structures close to buildings. They have been used however for limited life structures and their performance is being monitored.

47. Concrete block: Difficulty of access in Hong Kong often leads to older manual construction methods being retained in a slightly different form. An in-situ concrete block retaining wall constructed simultaneously with the placing of backfill was used recently to repair a failure on an important but remote access road (Fig. 6). The work was carried out relatively cheaply by a small contractor with

limited plant and relied on techniques familiar to his small
labour force.

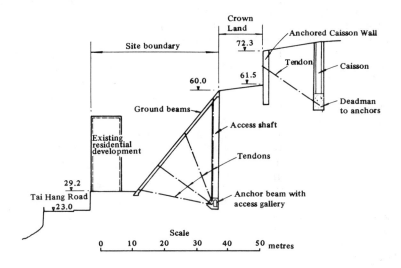

Fig.5   Slope stabilisation work at Trafalgar Court

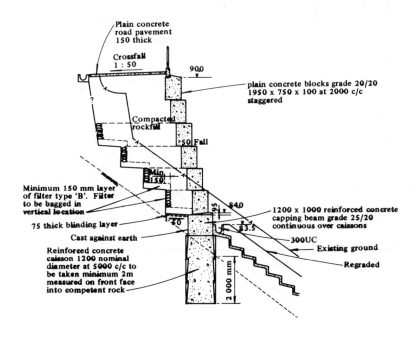

Fig. 6   Concrete block wall

25

48. Caisson walls: Caisson walls are common in Hong Kong where the technology is well understood. Non contiguous walls are now favoured, with drains installed between caissons (Fig. 7) to minimise the effect of these walls on ground water levels in adjacent slopes (Ref. 17).

49. Crib Walls: Crib walls have been used for construction on slopes where access is difficult. They require a relatively large excavation for construction but are aesthetically more pleasing than concrete.

Drainage

50. Where the ground water level is high, stability has been improved by installing various forms of drainage system. Whilst initially much use was made of small diameter drilled raking drains, attempts have been made to provide more reliable systems. For example that installed at Yau Tong, (Fig. 8) used vertical drains formed in hand dug caissons to intercept ground water flows behind the slope.

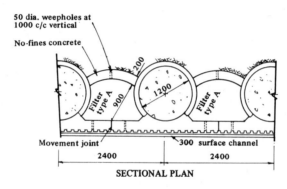

Fig. 7 Typical cantilever caisson wall

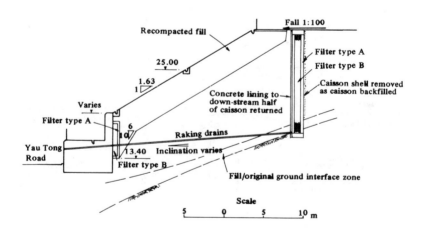

Fig. 8 Drainage works at Yau Tong

Erosion protection

51. The residual soils of Hong Kong are subject to severe erosion during heavy rain, particularly when disturbed. To prevent this and to improve the appearance of earthworks, extensive use is made of hydroseeding to establish a grass cover over both cut and fill slopes. Recently erosion protection on one project has been improved by planting shrubs suited to the area being hydroseeded.

52. Where slopes are too steep to support grass growth or on temporary slopes a cement mortar plaster, sprayed concrete, no fines concrete and dressed stone and concrete lattice grids enclosing grass have been used to resist erosion.

Conclusion

53. The history of earthwork failures in Hong Kong and the present programme of rehabilitation is one of many problems which become increasingly important as a country develops and the standard of living of the population improves. In the early stages of development, capital costs of individual projects are minimised. In doing so, risks are accepted, consciously or unconsciously, which the population later demands be reduced to conform with improving conditions of life. A lack of understanding of the factors causing landslides initially caused Hong Kong to accept them as an inevitable consequence of heavy rain. Recent study has shown this not to be the case and the efforts of the Hong Kong Government to reduce the risks appear to have been successful. However it must be said that had the knowledge gained in other parts of the world been applied when it first became available the marginal cost of adopting the improved standards would have been much lower than the present cost of stabilisation. Undoubtedly, earthworks, similar to those now being stabilised in Hong Kong, are being built in other countries where the consequences of failures, when compared with immediate social and economic problems, are relatively unimportant. They will however assume increasing importance in the future when reconstruction programmes similar to those being conducted in Hong Kong are implemented.

References

1. ALLEN, P.M. & STEPHENS, E.A. (1971), Report on the Geological Survey of Hong Kong. Hong Kong Government Press, 1971.
2. RUXTON, B.P. & BERRY, L. Weathering of Granite and associated erosional features in Hong Kong. Bulletin of the Geological Society of America, Vol 68, 1957, pp1263-1291
3. VAIL A.J. Final Report of the Commission of Enquiry into the Rainstorm Disasters 1972. Appendix V. A Report on the Po Shan Landslide, Government Printer, Hong Kong 1972.
4. VAIL A.J. Two landslide disasters in Hong Kong, Proceedings of Fourth International Symposium on

Landslides, Toronto, Vol 1 1984 pp. 717-722.

5. GOVERNMENT OF HONG KONG Interim Report of the Commission of Inquiry into the Rainstorm Disasters 1972, Government Printer, Hong Kong 1972.

6. BEATTIE A.A. & ATTEWILL E.J.S. A landslide study in the Hong Kong residual soil. Proceedings of the Fifth South East Asian Conference on Soil Engineering Bangkok, 1977 pp. 117-188.

7. MORGENSTERN, N.R. Mobile Soil and Rock flows, Geotechnical Engineering, Vol 9, 1975 pp.123-141.

8. MALONE, A.W. Moderator's report on retaining walls and basements. Proceedings of the Seventh South East Asian Geotechnical Conference, Hong Kong, 1982, Vol 2, pp.275-280.

9. BRAND, E.W. & HUDSON, R.R. CHASE - An empirical approach to the design of cut slopes in Hong Kong soils. Proceedings of the Seventh Southeast Asian Geotechnical Conference, Hong Kong, 1982, Vol 1, pp.1-16.

10. BRAND, E.W. Some thoughts on rain-induced slope failure. Proceedings of the Tenth International Conference on Soil Mechanics and Foundation Engineering, Stockholm, 1981, Vol. 3, pp 373-376.

11. GEOTECHNICAL CONTROL OFFICE Geotechnical Manual for Slopes, Hong Kong Government Printer, 1984.

12. BRAND, E.W., PREMCHITT, J & PHILLIPSON, H.B. Relationship between rainfall and landslides in Hong Kong. Proceedings of the Fourth International Symposium on Landslides, Toronto, Vol 1, pp.377 & 384.

13. BEATTIE, A.A. & LOVEGROVE G.W., Contribution to the session Slopes and Foundations, Proceedings of the Ninth International Conference on Soil Mechanics and Foundation Engineering, Tokyo, Vol 3, pp.427-428.

14. RODIN, S. Insitu compaction of loose fill slopes. Hong Kong Engineer, Vol 9, No. 6, pp47-50.

15. BOWLER, R.A. & PHILLIPSON, N.B. Landslip preventive measures - a review of construction. Hong Kong Engineer, Vol 10, 1983, No. 10, pp.13-31.

16. VAIL, A.J. & HOLMES D.G., A retaining wall in Hong Kong. Proceedings of the Seventh Southeast Asian Geotechnical Conference, Hong Kong, Vol 1, pp.557-570.

17. BEATTIE, A.A. & MAK, B.L. Design and construction of an anchored - buttress caisson retaining wall, Proceedings of the Seventh South East Asian Geotechnical Conference, Hong Kong, Vol 1, 1982 pp.465-480.

3. A57 Snake Pass, remedial work to slip near Alport Bridge

A. D. LEADBEATER, MICE, MIHT, DipTE, Assistant County Surveyor, Derbyshire County Council

SYNOPSIS. This paper describes the cause, effect and solutions to a long term landslip which threatened to close a 120 metre length of the Principal Road A57 (Snake Pass) between Manchester and Sheffield. The various alternative solutions are discussed, whether or not the road could be closed during construction; the derivation of the slip circles and soil parameters considered in the final design and how that information was translated into the adopted solution of large diameter bored piles with a reinforced concrete top slab. The construction method adopted is described in some detail.

INTRODUCTION

1. <u>History</u>. The Snake Pass section of Principal Road A57 carries the shortest route between Manchester and Sheffield and crosses the southern end of the Pennines via the valleys of the River Ashopton and Holden Clough. The summit of the road is at level 494.00m (1620 ft). This route has been a track since Roman times (Doctors Gate Road), was then a pack horse route and finally in 1810 a turnpike road was built linking Sheffield and Manchester cut into the side of the valley. This road has been widened since that date to take modern traffic. The valley itself has been subject to extensive glaciation which has led to several lengths of very unstable road. Over the last fifteen years seven major slips have occurred on the road and several different solutions have been adopted including drainage works, toe loading, counterweight slabs and minor road realignment. The slip near Alport Bridge has been treated in a structural manner and essentially consists of pinning the slip in position.

2. <u>Alternative Routes</u>. No through roads leave Snake Pass between the Ladybower Reservoirs (A6013 Bamford Road) and Glossop (A624 Hayfield Road) and, therefore, alternative routes between Sheffield and Manchester are the A628 across the Woodhead Pass and A6013; A623; A6 across Tideswell Moor. These routes add at least 25 kms to the Sheffield–Manchester route, and for farms, smallholdings and the Snake Inn etc sited on Snake Pass add considerable greater diversion lengths. The value of the route as a tourist attraction is

Failures in earthworks. Thomas Telford Ltd, London, 1985

29

also important. The site of this slip, together with others on the route is shown on Fig. 1.

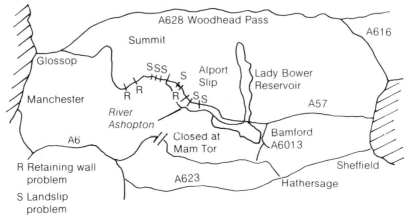

Fig 1. Road layout and slip site

3. <u>Geology</u>. The landslip is sited in a situation where, for a short length, the road is cut on a ledge on a steep hillside, approximately two-thirds up the side of the hill. The River Ashopton is meandering in the adjacent valley and slowly cutting into the hill on which the road is built. The river carries little water in the summer but in the winter during heavy rain and at the time of snow melt becomes a fierce, raging torrent, full of large, moving boulders which quickly erodes the base of the hillside. The hillside itself consists of rough beds of black, brown or grey shale or shaley clay occasionally incorporating gritstone and mudstone fragments with traces of iron stains. The borehole nearest the river shows traces of gritstone throughout its depth which suggests that this area has been re-made during the last ice-age. The permanent water table is some 12 metres below the road and slopes gently down to river level. However, an impermeable layer exists some 6 metres above that level which is water bearing.

4. <u>Behaviour of slip</u>. The road has been unstable in this area for a very long time and the turnpike company, prior to 1900, carried out stabilisation works consisting of stone buttresses inset in the lower side slope. Early this century some additional drainage was installed to drain the water above the impermeable layer some 6 metres below the road. Slow continuous movement has rendered these works non-effective by breaking both the buttresses and the drains. In recent times the road has settled slowly, each winter up to a differential depth of 50 - 75mm generally, and the road has been re-surfaced each spring to take out the differential levels. The average depth of tarmac fill is 1.5 metres below existing road level and this demonstrates an important con-clusion for maintenance expenditure "it is cheaper to top up a slip every year (in most cases) than to repair the slip totally". The 1.5 metres of tarmac filling represents the

maintenance cost of keeping the road open for a period of
50 - 60 years and at approximately £2,000 a time compares
very favourably with carrying out permanent works before they
are necessary. The total cost of the remedial works at
Alport slip was £400,000 and whilst this is an extreme case,
I believe the point about patching is well made and proved.
1976 in Derbyshire was an extremely dry year and, indeed,
rainfall figures at the Killhill Weather Station some three
miles from the site of Alport slip showed that for the first
time since 1901 (when records commenced) no rain fell for
four consecutive months and these were the first months when
no rainfall was recorded on a monthly basis. Autumn 1976
and Winter 1977 were extremely wet culminating in a massive
snow melt. Alport slip moved downwards and was patched and
within a two week period, four movements and patchings
occurred and it became quite clear that past methods of slip
control would no longer work. The road was reduced to single
way and soil surveys and remedial design works commenced.
The total settlement recorded over this period was in the
order of 500mm and cracks opened up on the hillside below
the road to a width of 150mm with depths measured up to
9 metres.

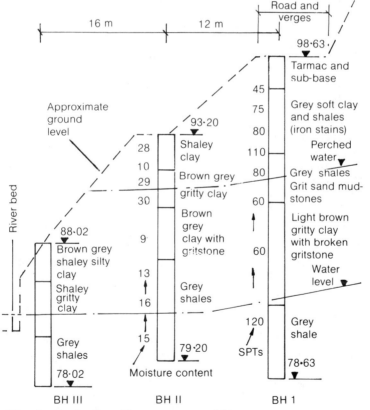

Fig 2. Typical soil survey results

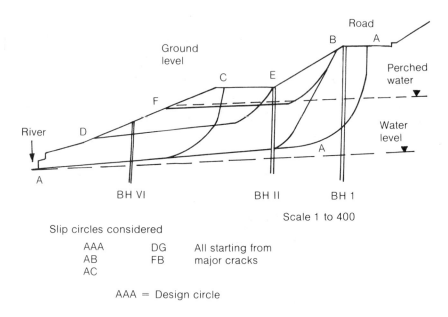

Slip circles considered

AAA	DG	All starting from
AB	FB	major cracks
AC		

AAA = Design circle

Fig 3. Slip circles used in analysis

STABILITY CHECKS AND ALTERNATIVE SOLUTIONS

5.   Borehole Results.   In slip analysis it is vital to
obtain the very best results from an experienced Company
with proven experience in recovering cores from difficult
materials (glacially disturbed and materials subject to movement must be amongst the most difficult to produce and inter-
pret). The whole analysis and remedial design will be solely
dependent on the ability of the site foreman and his men and
in some cases Engineers must recommend a limited number of
tenderers rather than the normal "free for all". The
recovery of cores is vital as in many cases the location of
the slip zone can be seen from the cores recovered. In
these situations, the installation of slip indicators and
piezometers is a must, however, care must be taken in interpreting the results from these devices. Slip indicators may
indicate the position of a subsidiary rather than the main
slip and piezometers may indicate standing water level in a
tube rather than actual ground water level. Dips of various
strata must be located – it may be that the slip is moving in
a different direction than the obvious and the wrong conclusions may lead to the wrong design measures being adopted.
Results from the soil survey at Alport are shown on Fig. 2.

6.   Analysis and Parameters.   The existing stability was
checked using the Method of Slices and nine possible slip
circles have been checked. The basis for setting up these
nine circles was to choose them so that they pass through
points on the ground which show tension cracks at the top and
heaving or rippling through the bottom. These circles were
taken to align with either the water bearing impermeable
layer or water table. Further circles have been chosen to
co-incide with slip indicator levels and included feedback
from moisture contents, plastic and liquid limits taken from
borehole information. The following shear strength parameters
have been used in the anlaysis:-

cohesion $C' = 0$ (failure already occurred along planes)
  $\emptyset' = 19°$ (based on test results from quick
             undrained triaxial tests, probably a
             slight underestimate from the true
             situation).
average bulk density = 1915 kg/cu.m = 18.8 kn/cu.m
                     (taken from borehole results).
Factor of Safety $F = \dfrac{c'l + Tan\ \emptyset'\ N'}{T}$ (the general
                                              equation).

From this analysis the Factor of Safety varied between 0.94
and 1.02 with one of the subsidiary slips having a Factor of
Safety of 0.74. Some of the slip circles used in this
analysis are shown on Fig. 3. Those slips affecting the road
have approximately the same Factor of Safety and, therefore,
no particular slip is critical, this means in turn that the
slip eminating from the centre of the road and ending at
river bed level is the slip that must be stabilised and, in
turn, that stabilising that slip will inevitably stabilise all
the other slips.

7.   Alternative solutions.   Many slips of this type and
with this sort of material occur in Derbyshire affecting road
stability and generally, in order of economy (cheapest first)
the following list gives the alternatives available:-

i)   Drainage solution (simple i.e. cut-off drains).
ii)   Toe loading (earth filling).
iii) Reconstituting slipped material and strengthening
      (using membranes).
iv)   Complex drainage solution (i.e. drainage of slip
      plane).
v)   Structural methods.

An important consideration in all these alternatives is
whether or not the road can be closed during construction
and, if so, for how long? This will often decide the final
solution selected.

An even more important consideration during the design
process is to check the stability of the works during construction and to properly assess the affect of the work
during service, i.e. drains at the top of cuttings to prevent
erosion and saturation of cutting material from surface water

33

are often used and they serve their purpose admirably but in some materials, slip planes have been known to start from such top drains above cuttings. An over-riding problem at this site is the situation of the river which is continually cutting in to the unstable hillside, thereby reducing the existing resisting forces towards the toe of the various slip planes. This means that stability is reduced not only due to ground water but also due to loss of toe weight. This effectively precludes any solution dependent totally on either simple or complex drainage which would also be extremely expensive to install and difficult to maintain due to the severe longitudinal and lateral cracking which had occurred already. Toe loading could be considered, however, this would have entailed a considerable volume of filling, the installation of drainage blankets and diversion of the river. The appearance of the valley would have been altered very considerably and the time taken to get the necessary approvals could have meant closing the road for a consider-able period - an unacceptable condition. At this site, reconstituting with existing materials, strengthening with membranes and installing drainage blankets would also have closed the road and due to the depth of the slip and the volume of material to be removed and replaced would have been extremely costly. Thus, the only tenable solution was to stabilise the slip by structural means.

STRUCTURAL SOLUTION - ANALYSIS

8. <u>System adopted</u>. Three basic methods of stabiliz-ation exist using structural methods, construction of wall in front of slip (can use reinforced earth); tie two parts of slip plane together using ground anchors and plates and pin the slip vertically using some form of pile or diaphragm wall. Again, all these methods have been used to stabilise slips in Derbyshire but due to site conditions and the scale of the operation, a bored pile solution was investigated and then adopted for this scheme. Such a solution is extremely good at a site such as Alport Bridge because the install-ation of the pile does not affect global stability, however, as soon as the concrete has gained strength the Factor of Safety against sliding is improved. In turn, the heavy contractors plant necessary for such an operation can work over ground already part stabilised, an important con-struction consideration. The earth pressure diagram used for analysis, together with a plan and section of the solution are shown in Fig. 4 and Fig. 5.

9. <u>Pile spacing and design</u>. Fig. 4 shows quite clearly that each individual pile is designed in accordance with classical sheet pile theory, assuming the piles to be canti-levered rather than tied. The pile was considered to carry active pressure only for the first 7 metre depth (i.e. down to the lowest slip circle with $k_a = 0.49$), together with a combination of active pressure and passive pressure below that ($k_p = 2.04$). The global shear force and bending moments

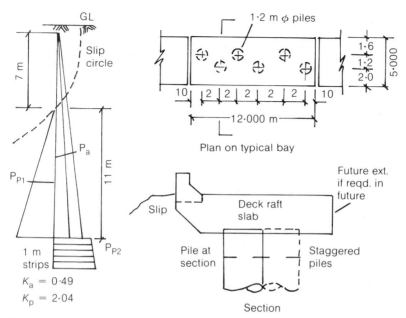

Fig 4. Earth pressure
diagram

Fig 5. Typical plan and section

approximated to zero with total embedment of 19 metres and,
therefore, the piles were specified as being 20 metres long.
A further substantial saving was made by staggering the piles
so that a raft could be constructed as a deck slab. Building
the top of the piles into the deck raft meant that the top of
the piles could be considered as encastre which reduced considerably the structural design of the piles themselves. A
further advantage of this system is that in the long term,
if frost erosion causes a problem then the deck raft (which
is designed for HA live load and checked for $37\frac{1}{2}$ units of HB
loading) can be extended to support the whole of the road in
this area without the necessity of closing the road. The
solution adopted, therefore, provides a solution to the
present problem and because it is simple to extend both long-
itudinally and laterally will provide a simple solution should
the problem extend in either direction. This is a very
important design decision when carrying out slip remedial
works. The global design of the two staggered piles and
linking raft, therefore, becomes a portal problem. The piles
were designed using SDBIAX and the top deck raft designed
using a typical STRUDL grillage. Slab width is 4.4 metres,
slab length is 12 metres and the slab was supported by six
1.2 metre diameter vertical piles staggered at 1.2 metres
centre to centre. The design and detailing of the structure
was then completed using normal, standard methods for bridge
design assuming no support from the ground under the raft.

35

CONSTRUCTION

Contract Documents and Programming

10. The completed scheme and its relation to existing carriageway is shown in Fig. 6 below.

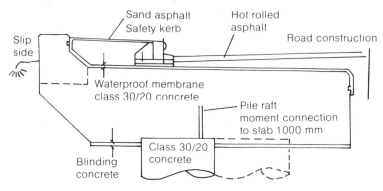

Fig 6. Cross section of solution

It can clearly be seen from Fig. 6 that during construction it was likely to prove extremely difficult to construct the large diameter bored piles and keep the road open during construction. A decision was, therefore, taken that the scheme should be priced on the basis of carrying out the piling operations with the road closed, once the piling was completed then the rest of the work would be done with single way working controlled by traffic signals. The total number of piles to be drilled, formed and cast was 36 and each pile was 1.2 metre diameter and very heavily reinforced. It was, therefore, anticipated that each pile would take a minimum of two days to form and, therefore, the closure period would be of the order of $12\frac{1}{2}$ weeks allowing for bad weather etc. The additional cost of carrying out the piling with the road open was extremely difficult to determine and so it was decided that each tendering contractor be asked to insert in his price for the job, the extra over cost of carrying out the piling and keeping half the road open. On receipt of tenders a decision would be made as to whether or not it was good value for money to pay the extra for keeping the road open throughout construction, bearing in mind the very extensive diversion routes and the status of the road as a Principal Road.

11. Method of Work. The successful tender was in the sum of £335,000 and the cost of keeping the road open was £20,000 and at this value the County Council decided that the work should be done with the road being kept open at all times. The method of piling adopted by the main contractor, Messrs Z & W Wade Limited of Whaley Bridge was to use the Lilley system and this system at this site proved extremely efficient and extremely quick. Piles were installed at the rate of one per day which was considered excellent when one considers the cramped nature of the site, the diameter and

depth (1.2m x 20 metres) of the piles, the weight of the reinforcing cages and the haul distance for the ready mixed concrete (22 miles over the Snake Pass from New Mills). An interesting feature of this contract was that all the excavation from both the bored piles and from the excavation for the raft was used to stabilise a small slip some 300 metres from the site using the toe loading system. This not only minimised the haulage distance of materials going off site but also resolved a further slip and, of course, reduced the cost of the main contract. In order to prevent erosion at the junction of the slip and the river bed a gabion wall has been installed along the whole length of the river adjacent to the slip. Gabions are a good and cheap method of providing a semi-flexible retaining and anti-scour system which can be easily and cheaply maintained in local areas of the wall.

12. <u>Environmental Considerations</u>. The road in this location is in the area of the Peak Park Planning Board and is also much used by tourists. Care is taken, therefore, in the way that remedial works are detailed. At the site, quite clearly, some form of vehicular containment is necessary due to the severe drop to the river bed. Rather than use standard steel safety fence as was installed previously, safety kerb has been adopted at this site with a verge for pedestrians behind. Pedestrians are protected by a small stone wall and this system provides an **attractive** and safe appearance to these remedial works, all of which are now not visible. The scheme was carried out over the winter period and completed in Spring 1981 and no problems occurred during construction. Photographs of the completed scheme are shown in Figs 6 and 7.

CONCLUSION

The purpose of this paper is to highlight the procedures adopted and considerations which must be made in carrying out remedial work to landslips affecting the highway and it is considered that the following points are worthy of

Fig. 6

repetition when designing remedial works caused by slope instability.

i)   make sure that the best **possible** soil information is produced.

ii)   make sure that the true water table is interpreted from piezometers.

iii)   make sure that all slip planes or slip surfaces are located.

iv)   make sure that cheap temporary works will not solve the problem before considering permanent expensive solutions.

v)   make sure that the slip direction is correctly assessed.

vi)   remember that at some time the Factor of Safety exceeded 1.

vii)   ensure that the cheaper solutions are fully investigated.

viii)   ensure that stability during construction exceeds a Factor of Safety of 1.

ix)   try to re-use slipped material or material arising from the work. The use of membranes is a powerful way of doing this. Great economies must result from re-using existing materials.

x)   in remedial design, simple text book theory will almost always be the best way to analyse the problem.

xi)   careful consideration must be given to the way the work will be done when producing designs - particulary with respect to the consequences of road closure or restriction.

Acknowledgement is made to Mr Eric Hook, CEng, MICE, MInstHT, County Surveyor for permission to write this paper, to Messrs Z & W Wade Limited who carried out the work and to the staff of the Bridge Section who designed and supervised the scheme.

Fig. 7

4. Slope failures in road cuttings through Coal Measures rocks

S. H. ROGERS, BSc(Eng), FICE, FGS, Partner, Sir Owen Williams and Partners

SYNOPSIS. The construction of a bypass to the A61 near Ripley in Derbyshire included cuttings through coal measures rocks. During excavation one cutting was affected by two failures, the second of which was very extensive and included a large area of land outside the bypass boundaries. In both cases the movements were translational and preliminary analysis showed that they could not have occurred unless the pre-existing shear strength in the failure zone was very much less than that of the intact rock. The detailed investigations led to the conclusion that the slide was seated in a zone of intraformational shears of tectonic origin.

INTRODUCTION

1. During excavation of a road cutting through easterly dipping Lower Coal Measures strata southwest of Ripley in Derbyshire the western slope was affected by two failures, the first on 15th September 1976 and the second on 3rd/4th May 1977. The cutting was part of works then under construction for the section of the A38 trunk road now bypassing the A61 between Coxbench and Alfreton. The site is near Openwoodgate at Hilltop Farm, its location and the Geology of the area are shown on the outline map, Fig.1a and the section, Fig.1b.

2. A site inspection immediately after the first failure established that it was a joint controlled block slide parallel to the bedding involving some 5000 m^3 of mudstone and sandstone. Examination of the surroundings revealed evidence of instability beyond the immediate area of the slip. A preliminary assessment of the average shear strength on the failure surface indicated that its value must have been very much less than that of the intact mudstone. A programme of detailed site investigations was put in hand to test this conclusion and establish the causes of the failure. At the same time the design of a scheme for remedial works was commenced. The preliminary assessment of the mechanics of the failure provided an adequate basis for the design of remedial works. Monitoring observations showed that some movement was occurring in the area above the slip and it was quite clear that any savings which might possibly arise from refinements resulting from the detailed investigations and analyses would

Failures in earthworks. Thomas Telford Ltd, London, 1985

39

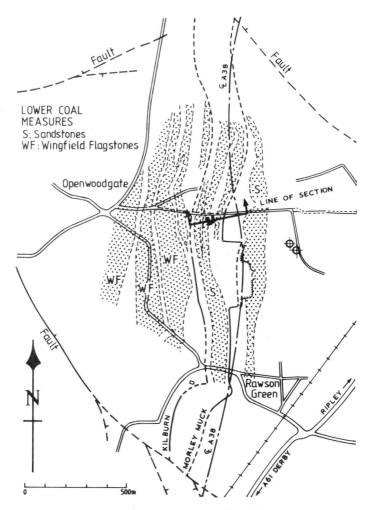

Fig.1a. Location Plan and Solid Geology

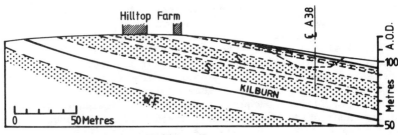

Fig.1b. Section at Hilltop Farm

be more than offset by the effects of the migration upwards of the head of the slip and by contractual disruption and delay costs if the treatment was deferred until the investigations were concluded. Accordingly reinstatement commenced in October and was completed by 20th November nine weeks after the slip occurred.

3. In addition to determining the causes of the slide the site investigation was directed also to the assessment of the stability of the area generally. A network of survey stations was established to check for ground movements. Small movements were detected over a wide area southwest of the slip and drainage works were undertaken to improve the stability of this part of the hillside. However the monitoring measurements showed that movements were continuing and by April 1977, although the March observations had shown some decrease in the rate, it was concluded that drainage works alone could not sufficiently improve the stability, that other substantial works would be required and probably that the only satisfactory solution would be general regrading to flatter slopes. Towards the end of April, before it was possible to implement any further works it was observed that the rate of ground movement was increasing. The increase continued and complete failure of a large area of the slope occurred during the night of May 3rd 1977, the volume of the slipped mass was subsequently assessed to be some 80 000 m^3. Although this slip abutted the earlier failure there was no reactivation of the reinstated area or any movements of consequence in the slopes above it. A plan of the slips of 15th September 1976 and 3rd May 1977 is shown in Fig. 2 in relation to the original layout of the roadworks, the locations of boreholes and ground movement monitoring stations are also shown.

4. The area of the second slide was some 1.5 hectares and much of it was on farmland outside the boundary of land acquired for the bypass. Various schemes were considered for the reinstatement of the area and the final proposals comprised disposal of much of the slipped material and regrading to a lower profile at a slope of about 1 in 5. These proposals including the return of the hillside to agricultural use were agreed in mid-June and the work was completed by the end of August four months after the time of the slip. Farm buildings, damaged by the movements, were demolished and the site after completion of reinstatement is shown in Fig.3.

5. A farm access bridge crosses the road near the centre of the area affected by the slips. The design of the bridge foundations had taken into account the possibility of local bedding plane slip and loss of support at the bank seat. Following the slip of September 1976 the design of the foundations and ground strut, which had then not been constructed, were modified for the possibility that the west back span might be affected by a major slope failure. Construction of the bridge was almost complete when the May 1977 slip occurred and although a substantial amount of

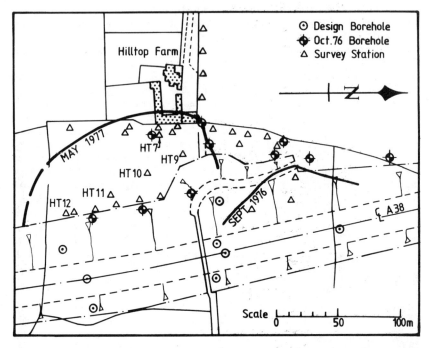

Fig.2 Original layout of roadworks showing location of slips

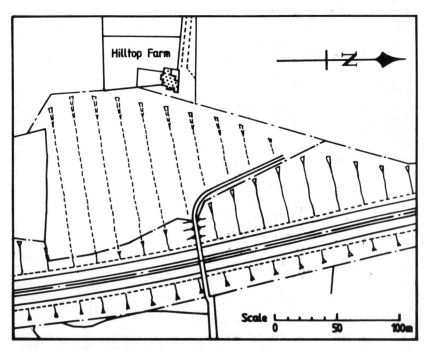

Fig.3 Plan showing works as completed

material piled up against the bank seat there was no damage other than a little superficial spalling and no displacement of the structure.

6. A general view of the cutting is given in the photograph, Fig.4, which shows in the right foreground the site of the September 1976 slip after reinstatement and beyond and above the bridge remedial works in progress to the May 1977 slip.

Fig.4 Photograph of site during reinstatement of slopes.

SITE CONDITIONS

7. The road runs in a north-south direction through a cutting on the east facing slope below Hilltop Farm. With the fall of the ground being up to 1 in 6 the western cut slopes, of maximum depth 14m as designed, are much deeper than those on the eastern side. In the southern part of the area the topography has been substantially changed by opencast mining reflected by the displaced outcrop of the Morley Muck in Fig.1a, workings for clay and other industrial activities.

8. There are almost no superficial deposits, soil cover is generally thin and the Carboniferous bedrock outcrops throughout the area. The Lower Coal Measures through which the cutting is excavated consists of a sequence of well defined mudstones or siltstones and sandstones. (See Figs. 1a, 1b). While the former have often been reduced at the surface to a soft to firm silty clay the weathering is shallow, generally no more than 3 m over the mudstones and considerably less over the sandstones. The sequence includes two coal seams of which the uppermost, the Morley Muck, outcrops at the site and was exposed in the cutting slopes. The Kilburn seam, some 30 m lower, underlies the site and outcrops well to the west beyond the crest of the hill.

9. The broad geological structure is simple, there is no appreciable folding and the general dip is easterly at some 10°-15°. While throughout the region the coal measures are faulted there are no faults in the immediate vicinity of the site which lies roughly midway between two faults which follow the main regional NNW-SSE trend and are about 1.7 km apart. (See Fig. 1a). The downthrow on each fault is to the north east, at the more southerly it is of the order of 200 m while on the northerly it is much less and only some 10-30 m.

10. The site has in the past been affected by both surface and underground workings. The Morley Muck, a seam of average thickness 0.34 m was partially worked at the site of the clay pits to the south of the site but elsewhere, as revealed by borings and on exposure during excavation of the cutting, there was no evidence of any old workings. The Kilburn seam of thickness 0.92-1.78 m underlies the site at depths less than 50 m. The seam has been extensively worked and the evidence from numerous borings including convergence of the roof and floor strata, the fragmented and collapsed condition of the mudstone immediately above the seam and the general absence of substantial cavities indicates that the workings are fully collapsed. These underground workings ceased many years ago. The Kilburn seam was being worked in an opencast site immediately south west of Hilltop Farm at the same time as the bypass was being constructed. All workings in the area are now complete.

SLOPE FAILURES

11. Construction commenced in February 1976 and by the end of the summer the earthworks were largely completed. The summer had been exceptionally hot and dry, ground water was at a low level and, other than a slip related to old opencast workings in the Morley Muck in the hillside slope above the bypass, there had been no indication of instability in the cutting slopes during excavation.

12. The first heavy rainfall to affect the cutting was on September 13th/14th. The hot summer weather had resulted in extensive shrinkage cracking of the soil and clays in the surface zone. Entry of surface water to the underlying jointed bedrock was thereby facilitated, thus creating water pressures

on the joint faces and at the same time reducing the effective pressures and therefore the shear strength on any potential surface of sliding.

13. On September 15th 1976 a slide occurred in the western cut slope immediately north of the site of the Hilltop Farm access bridge which at that time had not been constructed (see Fig.2). The movements occurred very rapidly and the sliding mass quickly came to rest. A site inspection was carried out during the afternoon of the same day and the character of the failure was readily identified as a joint controlled block slide. The length of the cutting slope affected by the failure was about 120m and its southern extent was determined approximately by the position at which the failure horizon passed below the level of excavation as existing at that time. The northern limit was roughly where the outcrop of this horizon passed from the cutting slope to the hillside above. The total area of the slide was about $1800m^2$ and its volume some $5000m^3$. Tension cracks were present above the slip and it was clear that, although the main mass had come to rest, further deterioration of the slope would occur if remedial works were delayed. As an immediate precaution, pending finalisation of the design of remedial works, the toe of the slip was loaded to minimise further movement.

14. A programme of detailed site investigations was put in hand to establish the causes of the failure, to provide further information for the re-assessment of slope stability in the cutting generally and to obtain data for the design of remedial work. Examination of the hillside above and to the south of the slide revealed indications of instability over a wider area and the investigations were extended accordingly. The programme included borings and test pits, soil sampling and rock coring, laboratory testing and observation of ground water levels. Survey stations were established in the areas of suspect stability and also in areas believed to be stable for example near to the farmhouse which was built over the outcrop of a sandstone which dipped below the base of the cutting. These stations were monitored to detect surface movements and observations were maintained on slip meters in boreholes to assess the depth of movement. These investigations and the results obtained from them are described and discussed in detail in later sections.

15. The field investigations showed that the slide was seated in a mudstone and siltstone sequence some 5 metres in thickness between two sandstone beds. The failure had occurred by shear in clay bands which at most were only a few millimetres in thickness and were often no more than partings showing as smooth slickensided bedding surfaces. From these investigations it was concluded, and subsequently confirmed during construction of the remedial works, that the stratigraphical horizon of the failure surface was approximately constant over the full area of the slip and was located near the centre of the mudstone and siltstone sequence.

16. When the geometry of the slide had been determined stability analyses were carried out for varying groundwater conditions and these led to the conclusion that on the failure surface, assuming the effective cohesion, c', to have been negligible, the average value of ϕ' was unlikely to have been more than 16° and might well have been several degrees lower. Figure 5 shows a plan of the scarp face as surveyed immediately after the slide occurred with the idealised wedge used for the analyses superimposed.

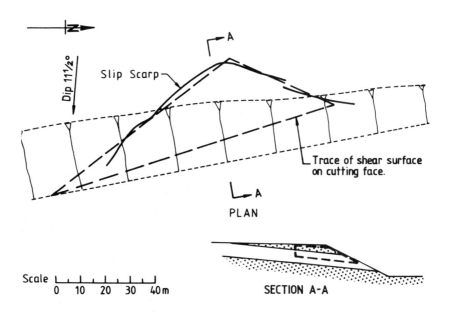

Fig.5   Idealised failure block for wedge analysis

The design of remedial works was a matter of urgency and this proceeded on the basis that the zone of low shear strength extended at the same horizon into the area immediately above the slip scarp and that, assuming c'=o, the angle of shearing resistance, ϕ', was in the range 12°-15°. The works are described in a later section.

17. The September 1976 slide was immediately to the north of the site of a three-span overbridge providing access across the bypass for farm traffic and accommodating a bridle road. At the time of design it was considered that due to the depth of the cutting it was desirable for the abutment to the west back-span to be founded at a level above that of the toe of the cutting. Although the compressive strength of the coal measures strata was more than adequate for a high level foundation the possibility existed that local bedding plane slip could endanger such an abutment. The stability of the abutment could be assured either by a very considerable

increase in the depth to founding level or by introducing a strut, below the cutting slope, between the abutment foundation and the solid sandstone below the verge. This latter solution was adopted and the bridge was programmed for construction after the cutting had been excavated. However the slip occurred before bridge construction had started and the design was reconsidered in relation to the possibility that the west back-span might be affected by a major slope failure rather than by local loss of support at the bank seat which had been the basis for the design of the strut. The loads which might be imposed on the structure if such a major slide occurred were assessed and the design of the foundation and strut were modified accordingly.

18. During the detailed examination of the hillside after the September slide several tension cracks were discovered and the barn nearest the bypass was seen to have been affected by movements. It was not clear whether these movements had been more than local although they could largely have been attributed to the poor condition of the foundations and the groundwater conditions. Furthermore the possibility of past damage due to workings in the Kilburn seam presented an additional uncertainty. The farmhouse and buildings were prime considerations in relation to any precautionary or remedial works which might be necessary to ensure safety and stability. From consideration of the geology it was concluded that the farmhouse was on stable ground but that if the shear zone which had been proved in the September slip was widespread throughout the area at more or less the same stratigraphical horizon then the barns could be at risk (see Fig.1b). Pending the outcome of the detailed investigations it was decided to provide propping and temporary support to minimise further deterioration due to local foundation movement. In addition drainage works were undertaken to cut off overland and sub-surface water flowing from the west and entering the area under consideration near the barn.

19. In order to check for ground movements survey stations were established at various times between September 1976 and January 1977, initially on a north south line passing near the barn, then on the hillside above the cutting slopes and along the road to Hilltop Farm and, after reinstatement, within the area of the September slip. The positions of the stations were monitored by precise survey techniques and the horizontal co-ordinates were determined to 1mm with a circle of error of diameter 5mm. To illustrate the measurements the results for station HT7 are given in Table 1.

20. While the result of the monitoring observations showed that movements were continuing the site investigations failed to provide conclusive evidence for the existence over the area generally of any throughgoing shear zone. The borings did however display a lithology which was favourable to the development of intraformational shears at a stratigraphic horizon similar to that found in the September slip. Although some drainage works had been carried out and others were

considered it became certain by April 1977 that drainage works alone could not ensure a sufficient increase in stability of the hill slope and that the most likely solution would be general regrading to flatter slopes. Before a conclusion had been reached regarding these further works the rate of movement increased significantly and on the night of 3-4th May 1977 complete failure occurred.

Table 1   Horizontal Movements at Station HT7

Date	Day	Movement from Origin			
		E mm	N mm	Total mm	Bearing
1 Nov 1976	0	0	0	0	–
15 Dec	44	4	2	4	63°
4 Jan 1977	64	30	18	35	59°
6 Jan	66	27	19	33	55°
10 Jan	70	28	17	33	59°
26 Jan	86	39	23	45	59°
31 Jan	91	58	31	66	62°
3 Feb	94	61	30	68	64°
7 Feb	98	71	30	77	67°
10 Feb	101	69	33	77	64°
16 Feb	107	85	34	92	68°
11 Mar	130	123	45	131	70°
16 Mar	135	121	52	132	67°
21 Mar	140	121	50	131	68°
5 Apr	155	123	47	132	69°
19 Apr	169	126	49	135	69°
21 Apr	171	148	53	157	70°
3 May	183	652	137	666	78°

NOTE: N.G. Co-ordinates on 1 Nov 1976:-
            E.437417.679; N347277.928
            Station Destroyed 3/4 May 1977

21. A site inspection and detailed survey, including measurements to the failure surface exposed at the base of the scarp face and in various trial pits, led to the conclusion that the failure was a block slide on a surface which dipped at 11-12° on a bearing of about 95° and was at a similar stratigraphical horizon to that in the September 1976 slip. The westerly extent of the failure was joint controlled and left an appreciable mass beyond in an uncertain state of equilibrium. At the farm part of the barns collapsed but the farmhouse was unaffected. Southwards the movements ceased where the mudstone dipped below the level of excavation while the northern edge abutted and slightly over-rode the

reinstated area of the September slip which remained stable.
Sandstone was at outcrop over much of the area and, together
with part of the underlying mudstone, moved in unbroken masses
many square metres in area and up to 6m in thickness. Four
monitoring stations HT9 to HT12 (see Fig.2), established on
the sandstone, remained undamaged and measurements showed
total movements at these stations of 14.14m, 14.11m, 7.69m and
6.17m respectively on bearings of 80°, 77°, 74°, 74°. A
substantial amount of slipped material piled up against the
bank seat of the farm bridge but there was no damage other
than a little superficial spalling and no displacement of the
structure. The extent of land disturbed by the slip was more
than 1.5 ha and the volume was some 80,000 m^3.

REMEDIAL WORKS

22. The remedial work to the September 1976 slip comprised
the excavation and recompaction of slipped material and
regrading of the cutting slopes generally to 1 in 2.8 instead
of the design slope of 1 in 2. Additional drainage was
installed to maintain the water table at a low level and
included a cut-off drain approximately 4m deep installed on
the hill to the west of the slip. The farm access track was
realigned and lowered to the west of the bridge, and a cut-off
drain was installed in the verge to intercept any water flow
at the slip horizon. The slipped material and the temporary
stabilising berm were removed in sections of 10m to 20m width
to a depth of 500mm into intact siltstone or to sandstone. A
blanket of well graded broken stone of 500mm thickness was
placed over the bottom of the excavation before replacing and
compacting the excavated material to a final slope of
1 in 2.8. Instructions to proceed with these works were given
on 21st October 1976 and they were substantially completed by
20th November 1976.

23. In contrast to the September 1976 slip which was
mostly within the original land take, the failure of May 1977
affected a large area of adjacent farmland. Prior to this
latter slip a scheme for regrading and reducing the cutting
slopes had been examined as a possible solution to the evident
condition of instability. The regraded land would be returned
to agriculture, albeit at a slope somewhat steeper than the
natural slopes, and the ultimate land take would be less than
required for the original scheme. Discussions were held with
the landowner who agreed to the adoption of the regrading
scheme for reinstating the hillside and to extending the
flatter slopes to include the unstable land which remained
above the slip scarp. An important factor in the decision to
adopt the regrading scheme was that suitable tipping sites
were available within a short haul for the disposal of slipped
material. The bulk earthworks and general regrading were
completed four months after the slip occurred and the works
include new cut off drains and a land drainage system in the
restored agricultural land.

GEOTECHNICAL INVESTIGATIONS

24. A programme of site investigations was put in hand immediately after the September 1976 slide. The purpose was to determine the causes of the failure and to obtain data for the design of remedial works and for the re-assessment of slope stability in the cutting generally. The investigations included ten rotary drillings to recover 100mm diameter cores and the installation of piezometers and slip-meters for the long term measurement of ground water levels and to monitor movements. Pending completion of this work, a back analysis, as described in Paragraph 16, was undertaken to determine values of the shear strength parameters to be used for the design of remedial works.

25. The ten borings were drilled in the area west and south of the slip and while the horizon of the failure zone could be established in the cores by reference to the sandstones there was no evidence of shears other than in one of the borings where a 12mm band of very soft clay was recovered. Samples of slip zone clay were obtained from a pit near the slide and from the thin layer at the base of the slide. Tests on remoulded samples of this clay using the 60mm laboratory shear box gave peak values $c=60kN/m^2$, $\emptyset=24°$ and "residual" values $c_r=0$, $\emptyset_r=15°$. Index properties were LL=43; PL=23; mc=25 and clay fraction 35%.

26. The site investigations were extended after the May 1977 slip to include five rotary drillings in the undisturbed area south of the previous investigation and the excavation of trial pits within the area of the slip for block sampling and in-situ tests. A detailed survey of the slip showed that the shear surface was at the same stratigraphical horizon as in the previous slip and the five additional drillings confirmed the uniformity of the lithological sequence throughout the area. However detailed examination of the cores from these holes did not reveal any evidence of a throughgoing shear surface or zone although a number of clay filled discontinuities were recorded. Block samples containing shear zones in the mudstone were obtained for testing in the 305mm laboratory shear box. These samples were taken by integral sampling techniques and cut so that the top of the block was parallel to the bedding. Specimens of mudstone containing natural fractures were selected for testing in the Hoek field shear box and simple sliding tests were carried out in-situ on natural blocks of mudstone. Schmidt hammer rebound values were measured on weathered joint surfaces to enable the value of ϕ_r to be assessed by the method proposed by Barton (ref.1).

27. The results obtained from the various laboratory and in-situ tests showed a wide range of values of the angle of frictional resistance on natural discontinuities in the mudstone. The "peak" values varying from 21° to 33° and the "residual" from 13° to 20° reflect not only the variability in the drainage conditions under which the tests were carried out, with only the 305mm laboratory shear box tests being fully drained, but also the variety of conditions of the

discontinuities under test. It is beyond the scope of this paper to discuss these tests and results in detail but the peak recorded on the sample in the 305mm shear box is considered representative of the behaviour of a clay covered bedding joint in the mudstone and this result together with a Hoek shear box test result indicates angles of shear resistance of 20° at initiation of failure and 15° after displacement. A test carried out on a sample of mudstone containing a thin clay seam artificially produced in the laboratory gave a 'residual' shear angle of 11.5°. From the back analyses of the failures it was concluded that the average angle of shearing resistance over the failure surface was 15°–16°.

INTRAFORMATIONAL SHEARS

28. The results of the investigations showed that the shear strengths in the critical zone must have been closer to residual values than to peak values. It followed therefore that at some time in the past the strata had been subjected to substantial shearing over a wide area. The occurrence of highly sheared zones in tectonically disturbed strata has been well known for many years and was generally attributed to bed over bed slip induced by flexure of the strata. The corollary to this view is that such intraformational shearing should not be found in flat strata such as the Coal Measures in which the cuttings have been excavated. Recent research, reference Salehy, M.R. et al 1977, (ref. 2), has demonstrated that intraformational shearing is by no means uncommon in relatively undisturbed Coal Measures and other rocks, that it is related to faulting and can extend to great distances from the faults. As with flexural shearing the development of shear zones due to faulting is influenced by the lithology and the shears most commonly occur where sandstones, limestones etc. are interbedded with argillaceous sediments. The shear zone material is generally clay in a residual or low strength condition and its thickness may be very small with the adjoining rock sound and not significantly weathered.

CONCLUSIONS

29. It has been established that the failures which affected the A38 cutting slopes below Hilltop Farm resulted from the existence of very low shear strength bedding discontinuities and that the movements were triggered by a build-up of hydrostatic pressure in major joints together with pore pressure increases in the failure zone. The low strength conditions are attributed to tectonic shearing which it is now known can result from faulting of flat lying strata in addition to the more commonly known bed over bed slip associated with folding. The investigations have shown that the identification of such intraformational shears, both in cores and exposures, even when their approximate horizon is known, may be extremely difficult. The evidence is frequently lost in conventional borings and remains undetected or its

significance not recognised in routine trial pitting. The lithology of the strata and the geological structure are therefore more than usually significant in the assessment of the site conditions and the possibility that they may have resulted in zones of low shear strength must be fully appreciated when assessing the stability of earthworks.

ACKNOWLEDGEMENTS
    30. The author wishes to thank the Department of Transport for permission to publish this paper which relates to the construction of a section of the A38 trunk road for which the author's firm, Sir Owen Williams and Partners, were Consulting Engineers. The Main Contractor for the Works was Dowsett Engineering Construction Limited and the site investigation fieldwork and laboratory testing referred to in the paper was undertaken by Soil Mechanics Limited.

REFERENCES
1. Barton, N. Discussion at the Conference on Exploration for Rock Engineering, Johannesburg Nov.1976 Proc.Vol.II pp.190–191.
2. Salehy, M.R., Money, M.S., & Dearman, W.R. 1977. The occurrence and engineering properties of intraformational shears in Carboniferous rocks. Proc.Conf. Rock Engineering – Univ. Newcastle pp.311–328

Discussion on Papers 1–4

MR F. HUGHES, Cementation Piling and Foundations,
Rickmansworth

Although you have produced idealized models for drumlins,
Professor McGown, ground investigation will still be needed.
How do you see such investigations in the future?

Are drumlins always easy to observe? There must be hidden
mounds as well as complicated arrangements as hinted in the
first paragraph. This means that the simple cavity excavation
might have to be modified. Could you take this into account
in your reply?

PROFESSOR A. McGOWN

The idealized model of a drumlin and the discontinuity
patterns within it is solely intended as a visual aid to
engineers designing their site investigations. It must not be
used as a substitute for such investigations. Also, as you
suggest, not all drumlins are exposed above existing ground
level and this must always be borne in mind in glaciated
areas. For both these reasons I therefore fully endorse your
view that boring and sampling in drumlinized tills will remain
absolutely necessary, but I believe that where drumlins are
exposed an additional most useful method for investigating
discontinuity patterns is the cavity technique. Where the
cavity technique is not practical, investigation of oriented,
continuous core tube samples may prove useful. This technique
of discontinuity investigation is presently under study at
Strathclyde University in co-operation with the Building
Research Establishment.

DR A. B. HAWKINS, University of Bristol

Professor McGown, you have given a good account of the
discontinuities present in drumlins. Your observations
support the view that in lodgement tills three main
discontinuity sets are present: a subvertical joint-like set,

a stress relief set often subparallel to the past/present ground morphology, but which may deviate to pass through particular depositional soil horizons, and a shear set related to the depositional accumulation of lodgement till.

The relevance of the shear set will depend on whether the shears are in a debris-dominated or a matrix-dominated till. If in the former they are likely to be localized and the shears are frequently not planar. In a dominantly fine material, however, or in finer horizons within a debris-dominated till, they may be widespread and more planar. As the shearing is induced during ice overriding or the lodgement of till the irregularities on the shear plane will be subparallel to the direction of ice movement. You report 1-4 mm of surface relief in the gently inclined fissures; hence the presence and orientation of the surface irregularities needs to be taken into account when appropriate effective stress parameters are being considered.

It is a pity that you did not mention that such discontinuities as those you described occur only in lodgement till. In ablation tills, because of their accumulation under atmospheric pressure, fissures are rare.

PROFESSOR A. McGOWN

I am not altogether in agreement with your classification of the various sets of discontinuities in drumlinized lodgement tills. I agree that occasionally shear discontinuities do appear in such soils but the vast majority of the discontinuities are not in my experience sheared. the near-vertical discontinuities are generally clean cracks whereas most of the low dip angle features are coated in fines. Rarely do either of these show evidence of being sheared. The near-vertical features and the low dip angle features which are clean are most probably stress relief induced and the coated low dip angle discontinuities are 'bedding' features derived from basal ice melting during the 'slip-stick' sequence at the soil-ice interface.

So far as not mentioning that such features do not appear in melt-out tills, I would stress that they do not generally occur in the same arrangement in other till types. I suggest only that in drumlinized lodgement tills do these discontinuity patterns exist.

MR R. MERCER, British Waterways Board, Leeds

With regard to the caisson drainage system at Yau Tong, could the authors of Paper 2 indicate the spacing of the caisson drains, both across and up the slope and the design basis for this selection?

MR A. J. VAIL and MR A. A. BEATTIE

The hand-dug caissons at Yau Tong were 1.5 m in internal diameter and were formed in one line, generallly at 3 m centres, in the building platform at the crest of the slope. The 3 m spacing was selected by reference to soil permeability and the extent to which the water levels had to be reduced to ensure stability of the slopes. This was confirmed by calculation as described below. Locally, the spacing of the caissons was varied slightly to avoid surface or near-surface underground obstructions. The bottom level of each caisson was determined by the level of the original ground under the fill. The lowest level to which a caisson would act as a drain was defined by either the level of the original natural ground or by the lowest level at which the outlet was formed by drilling through the retaining wall at Yau Tong Road. The minimum inclination of the raking drains was 3.8^{o} and the maximum was 20°. Caissons were interconnected by 100 mm dia. PVC pipes so that if one outfall pipe was damaged or blocked the flow could be carried to the adjacent caissons. The flow to each caisson was conservatively assumed to be the flow in a soil strip of width equal to the caisson spacing and of depth equal to the difference between the caisson invert level and the groundwater level. Flow was calculated using a soil permeability, determined by in situ testing, of 5×10^{-6} m/s.

The steady state drawdown profile of the groundwater table was assessed using a simple two-dimensional model and the classical Dupuit solution, modified to take the seepage surface within the well into account and assuming the radius of influence to be given by the expression (ref. 1)

$$R = 3000Sk^{1/2}$$

where R (m) is the radius of influence of the well, S is the drawdown in the well and k (m/s) is the permeability.

If the profile is used only in the assessment of slope stability, such an analysis is conservative. For a more precise assessment which takes account of three-dimensional variations, the groundwater profile resulting from gravity drainage by a number of independent wells can be determined by the methods given by Huisman (ref. 2). However, the variability of the materials encountered in loose fill slopes and the uncertainty of groundwater conditions, because of leaking services within development platforms, make the value of the more refined analysis questionable. Where ground conditions have been more uniform, the methods of Huisman have been used successfully by us when, for example, predicting the settlement of structures resulting from dewatering of caissons during construction.

References
1. BRAUNS, T. Drawdown capacity of groundwater wells. Proc. 10th Int. Conf. Soil Mech. Fdn. Engng, Stockholm, 1981.

2. HUISMAN, L. Groundwater recovery. Macmillan, New York, 1975.

MR A. D. LEADBEATER, Derbyshire County Council, Matlock

A general comment is to remember that we are solving remedial problems rather than designing new, very sophisticated earthworks (such as in a high embankment, a reservoir dam or reclaiming land). In these cases the extent of the analysis must by virtue encompass a very detailed analysis of not only the new construction itself, which is easy, but also a detailed analysis of the effect of the new structure on the surrounding supporting subsoils, which is very difficult.

Very rarely (indeed almost never) will remedial problems lead to an increase in weight in the active slip zone, and having recalled that as the work is remedial the structure stood up once. The purpose of the analysis here is to decide and assess what caused the failure and to attempt to resolve the problem easily, quickly and economically. Politically, it is always easier to obtain funds for major new works than to increase one's maintenance budget (as seen in current arguments on infrastructure replacement).

The purpose of the design is to increase the worst factor of safety from 1.00 to something of the order of 1.15. Analysis of such cases can often lead to the conclusion that reducing the water level by even 1-1.5 m can be sufficient, and such a process is not difficult.

Section 7 of Paper 3 shows five methods of stabilization in ascending order of cost. All have advantages and disadvantages.

(i)   The drainage solution is cheap but requires continual maintenance. If the design is poor clogging can occur (particularly mini-thrust-drilled, perforated pipes). The solution will not require land acquisition but the right to go on to land which is given in the Highways .Act. Note that drainage schemes can also help the landowner, and it is often good to sell advantages that way. During construction improper planning can result in much reduced stability.

(ii)   Toe loading is cheap (in Derbyshire), is easy to design and prove and does not require maintenance. There are problems with land acquisition - such toes can often ruin field usage - and it should be remembered that the value to the landowner or tenant is not necessarily the agricultural value of the acreage taken (mowing or seeding efficiency may be ruined). Often the fill is supported by free-draining gabions (cheap).

(iii) Reconstituting slipped material: the majority of slipped materials in Derbyshire consists of shales and clays (mining strata) and shales, mudstones and silt-clay mixes (glacial areas). It has been demonstrated that all these materials can be reconstituted using geotextiles, membranes and drainage blankets 300 mm thick. Such methods show a saving of at least £10 per metre and additionally are quicker, easier and simpler to bill and execute. Membrane use must always be looked at – it will nearly always show substantial savings.

(iv) Complex drainage solutions are almost always very expensive but in some cases can be very powerful solutions for large slips where the cost for (i)–(iii) could be prohibitive. The method consists of thrust boring or heading 1.00 m dia. pipes into the slip plane and then perforating the pipe.

These comments apply to Derbyshire remedial works carried out over the last ten years.

Method	No. of sites	Cost variation
A Drainage system	9	£4000 – £65 000
B Toe loading	4	£20 000 – £60 000
C Reconstituting slipped material	7	£8000 – £72 000
D Complex drainage and structural	2	£110 000 – £400 000
A + B	1	£95 000

DR A. R. CLARK, Rendel Palmer and Tritton, London

Mr Leadbeater, you referred in your presentation to the development of recent failures which were adjacent to but outside the limits of the stabilization programme. You also suggested that failures could occur in the future at other locations. In view of the financial constraints placed on regional and local authorities during the current period of austerity, have you considered the use of geotechnical/ engineering geomorphological field mapping techniques, as a cheap, cost-effective method of identifying landslide hazard and the relative risk of future landslide failures in different areas of the road line? This could be a useful predictive tool to identify areas where future stabilization works may be most appropriately located.

MR A. D. LEADBEATER

It is quite clear that the problems facing those making not
only 'how to resolve a problem' but also 'which problem to
solve' are facing problems which are becoming increasingly
difficult. In real terms, the amount of money available for
maintenance has reduced considerably over the past 15 years
and this has coincided with a considerable increase in lorry
axle and gross weights over the past 15 years. The vast
majority of Britain's roads are very old – like the Snake Pass
– and have been subjected to widening and 'cut and fill'
effects which alter the geological and hydrological behaviour
of the substrata. Other problems are also the presence of
many public utility mains which now need to be replaced and
the indisputable fact that politically maintenance is the very
'poor cousin' of new work in that arena.

All this means that rather than receiving sufficient funds
to carry out the work necessary the maintenance engineer is
given a cash limit and has to recommend and stand by what the
engineer believes to be the most important or 'risky' sites.
Furthermore, he has to assess the level of 'risk' that he is
willing to bear and the degree and level of work to be done.
In my case my budget is spread between land instability, the
maintenance of drystone walls and the maintenance of bridges.

I consider that a system of identifying in geotechnical and
geomorphological field mapping would be effective. However,
in the maintenance field in Derbyshire it is easy to identify
sites which 'might' slip but it is more difficult to identify
sites which will slip. Taking the Snake Pass as an example
Fig. 1 in the Paper identifies nine slip areas in a distance
of 12 km. These fit into three totally different mechanisms

(i)   deep-seated shale slips as at the study slip
(ii)   shallow-seated 'mud slides' up to 1.5 m thick
(iii) instability of disturbed sandstone layers over stable
       blocks.

I believe that such studies should take the following forms

(i)   study A (initial) – identify potential slip sites by
       walking the sites, identify waterpaths, marsh grass,
       local features and road crust performance
(ii)   study B (secondary) – large slips which have commenced
       slow movement – take boreholes, analyse, install slip
       indicators and piezometers, inspect at times of changed
       subsoil (very dry, very wet, very cold etc.)
(iii) study C (movement) – measure movements and patch repair
       on an annual basis, calculate the cost of temporary
       patching compared with final works, only do repair when
       absolutely necessary.

I reckon that on this system approximately 60–70% of slips can
be accurately forecasted, another 15% can be forecasted fairly

accurately and the remaining 15% will come 'like a bolt from the blue'.

MR I. L. WHYTE, University of Manchester Institute of Science and Technology

In section 6 of your paper, Mr Leadbeater, you give shear strength parameters of c' = 0 and ϕ' = 19$^{\delta}$ (effective strength parameters) but you report that these were obtained from quick undrained triaxial tests (total stresses). Is this correct and can you indicate how the shear strength parameters were obtained?

MR A. D. LEADBEATER

Thank you for drawing to my attention a very important mistake in this paper. Fortunately, c' and ϕ' are the correct values and it is the way that they have been obtained that is wrongly described. The description in the brackets adjacent to ϕ' should read '(based on test results from slow drained triaxial tests allowing full porewater to dissipate which confirmed the results from a back analysis based on estimated values of ϕ')'.

In other words the general basis of analysis of remedial slips in Derbyshire is to take samples, to classify them, to estimate values of c' and ϕ', to check stability and then to confirm values selected later on by the slow tests described.

MR P. S. GODWIN, West Yorkshire Metropolitan County Council, Ossett

With regard to Paper 4 it is remarkable that no faults were recorded, other than two 1.7 km apart. Many small faults can be seen within the limited confines of an opencast site. Typically, they have a 300 mm to 3 m throw and are spaced from 20 m to 200 m apart, usually parallel to the larger faults seen in the site. Such displacements are an expression of stress relief. They are difficult to locate even when investigated by the high density drilling used in opencast exploration. The shear zone was detected after failure and consequently attributed to intraformational shearing. Is this the 'chicken and egg' situation? Could the failure mechanism be explained by more common phenomena? I have seen similar thin shear planes within trenches through constructional failures in Hunsworth Cut on the M62 motorway. The intact rock was a blocky fissured mudstone with the appearance of black gravel when excavated. Each piece exhibited conchoidal fractures whose damp surfaces were often slightly clay smeared. Relatively unweathered mudstone is often softened to a clay consistency to depths up to about 2 m below the interface with an overlying fissured sandstone. This is another case of stress relief at shallow depth with water being taken in from the overlying fissures. It is noted that opencast coal workings were present uphill. If backfilled, this would have proved a ready source of water in hydraulic continuity with slip area. Conversely, if working it may have formed an effective cut-off to groundwater flow from any higher ground beyond it.

MR S. H. ROGERS

No small faults were found in the area and the clay band was traced as an essentially planar feature over the full extent of the slide. A fault shown on the 1:10 000 geological map as lying to the immediate south of the slip area could not be found during the excavation of the Works.

The ground conditions at Hunsworth Cut that you described were not the same as those at Hilltop Farm where the soft clay band lay in a fresh unweathered mudstone-siltstone sequence approximately midway between two sandstone beds.

The opencast coal workings were excavated uphill of the slip as you have stated, but these were south of the slip position. The opencast site was operational at the time of the slip and any effect on the area would have been beneficial since the excavation would have the effect of drawing down the water-table.

DR A. D. M. PENMAN, Sladeleye, Chamberlaines, Harpenden

Mr Rogers, what provision did you make to ensure the stability of the bridge? The photographs showed the bridge looking like a strut holding up the landslip area. Since the two landslips appeared to be on either side of the bridge, did it hold up the part of the slope at its abutment?

MR S. H. ROGERS

The first slip occurred before bridge construction had started. The remedial works were extended to include precautionary works to the hill slope above the bridge site. These works included excavation to below the potential failure zone, backfilling and special drainage measures. In addition the design of the ground strut, which had been included in the original bridge proposals to ensure stability of the bank seat, was modified to take into account additional load which might result from ground movements above the reinstated area. The strut was not intended to hold up the slope and the bridge did not act in this way. There was no reactivation of movements of the reinstated area when the second slip occurred although it was in part overridden by material from above, some of which, as mentioned in the paper, piled up against the bridge abutment.

DR A. B. HAWKINS, University of Bristol

The case history that you reported, Mr Rogers, is interesting but not unique. In 1973 I described a similar feature in Coal Measures rocks at Bristol and demonstrated that the speed of movement of the Blackswarth Road slip in 1957 could be correlated with rainfall. Regrettably, when that slip occurred the disturbed zones in Coal Measures strata described in the early 1970s by Walton and coworkers were not appreciated. With their description in the literature and their demonstration at a field meeting of the Engineering Group of the Geological Society in subhorizontal Liassic rocks in Dorset, the existence of the microsheared zones became better established. In the mid-1970s several such zones were observed when examining U100 samples in the gently dipping London Clay of the Hampshire Basin. They do not appear to have any unique lithological characteristics.

You refer to a band of very soft clay 12 mm thick. Having recovered this, did you examine it in detail to establish whether it was simply a softened band or whether it was a zone consisting of many small 5-20 mm shear planes? Was the very soft clay parallel to the bedding, or did it transgress the bedding? Did it have any lithological characteristic to separate it from the rest of the material which may have influenced why the zone developed at that particular level?

MR S. H. ROGERS

There were no small shear planes visible in the samples of the clay band taken in the pit and at the base of the slide. In addition to the shear box and index tests reported in the paper an X-ray diffraction test was carried out to determine the clay mineralogy. The main constituents were found to be illite and kaolinite with a small amount of chlorite. Montmorillonite was not identified but a trace of an expandable clay was found by glycolation.

The clay band was observed in a number of places over the area of the slip. In no case did it appear to transgress the bedding and it occurred within a fresh, moderately weak to weak, very thinly bedded, grey friable mudstone and lay immediately above one of many strong siltstone bands of approximately 20 mm thickness. A water seepage was associated with the clay band in some of the exposures.

5. Slope stability problems in ageing highway earthworks

A. W. PARSONS, FIHT, Principal Scientific Officer, and J. PERRY, BSc, MSc, MIMM, FGS, Higher Scientific Officer, Transport and Road Research Laboratory

SYNOPSIS   A survey of earthwork failures on selected lengths of motorway has been carried out in order to identify the basic factors affecting the stability of the side slopes of cuttings and embankments and to attempt to quantify any long-term problems. The survey mainly included areas where over-consolidated clays predominate and covered a total of about 300 km of motorway. The results reveal a significant incidence of shallow slip failures in both cuttings and embankments of the more cohesive soils. Major factors found to influence the rate of failure are geology, age of earthworks and geometry of slope.

INTRODUCTION

1.   An inevitable feature of earthworks is that some deterioration of the slopes of cuttings and embankments will occur with age. In general problems with instability of side slopes are expected to be associated with cohesive soils and the critical condition of a slope will arise at an age that will depend on the degree of over-consolidation, the rate of pore pressure equilibration (refs. 1 and 2) and design factors such as the height and gradient of the slope and methods of drainage.

2.   At the present time there are over 2600 km of motorway in the United Kingdom (ref. 3), in addition to extensive lengths of major road involving considerable earthworks. To aid the planning of maintenance strategies, whether involving repairs where failures have occurred or preventative measures in areas identified as being at risk, information on factors affecting the long-term performance of the earthworks is needed. In addition, such information could provide data for use in the design of earthworks in new construction so that future maintenance requirements are reduced.

3.   The construction of the motorway system in the United Kingdom began over twenty five years ago and, over considerable lengths of the system, a sufficient time has therefore elapsed to allow a meaningful study to be made of the performance of the earthwork side slopes. This Paper describes the

results of a survey of the earthworks of selected lengths of
motorway. The objectives of the survey were to determine the
scale of deterioration of earthworks, to determine the factors
that affect the performance of the side slopes of cuttings and
embankments and to attempt to quantify any long-term problems.

SURVEY PROCEDURES

Preparatory work

4.   The areas of the survey were selected to include a high
proportion of earthworks in over-consolidated clays.
Selected lengths of the M1, M4 and M11 Motorways were studied,
each area being delineated by the maintenance boundaries of
the County Council concerned;   earthworks for side road
diversions built at the same time as the motorway were also
included.   The total length of motorway surveyed exceeded
300 km, and the ages of the earthworks at the time of the
survey (ie the elapsed times since opening to traffic)
varied from 3 to 22 years.   The survey was carried out in
the period 1980-83.

5.   Plans of the lengths of motorway to be surveyed were
obtained, the most commonly used scale being 1 : 2500, and the
geology of the area superimposed from information provided by
geological survey maps and site investigation reports.   Borehole logs were particularly useful in establishing the depth
of drift deposits in cutting slopes.   The materials in
embankments were established as accurately as possible from
'mass-haul' data prepared during construction, from records
made by supervisory staff of day-to-day earthmoving operations, or from 'as-constructed' motorway plans.   Most of the
information regarding materials in embankments was obtained
with the generous assistance of the Consulting Engineers or
Local Authorities who supervised the motorway construction.

The survey

6.   Those areas of earthworks delineated on the plans as
less than 2.5 m high were observed from a slowly moving
vehicle operating on the hard shoulder of the motorway;   all
earthworks in excess of 2.5 m high were inspected on foot.
The inspection was concerned with locating any problems
associated with or that might affect the stability of the
slopes.   These problems comprised:-

   a.   slips that had been repaired;
   b.   unrepaired slips;
   c.   incipient slips, where tension cracks had
        developed, usually near the top of the slope;
   d.   settlement emanating, either within the fill,
        or in the subsoil;
   e.   seepage of water from within the slope;
   f.   erosion of material at the base of the slope
        (toe erosion).

Of the above problems, the most difficult to identify were the incipient slips and those repaired slips in areas where top soiling and seeding of the repairs was normal practice. Any omissions due to these difficulties are considered to be very small in number, and will have resulted in an under-estimate of the number of problems.

7.   Whenever a problem was encountered full details of the characteristics of the slope and of the problem were noted, including the results of measurements of the geometry using an Abney level and a 30 m tape. If no problems were encountered details of characteristics and geometry were only noted for those areas of earthworks with heights in excess of 5 m. Major characteristics noted in addition to full descriptions of any problems were orientation of slope, drainage, vegetation and a visual description of the soil type. All details were noted on a specially designed survey form.

Processing of the data

8.   As a first stage of analysis of the information obtained the locations of all slope failures, both repaired and unrepaired, were collated and passed for information to the appropriate County Council responsible for maintenance operations. An assessment of proportionate lengths of the slopes with heights in excess of 2.5 m that had failed was also made and early results were published in a TRRL Leaflet (ref. 4).

9.   To aid detailed analysis it was considered necessary to use a computer, and all information has been extracted from the plans and the special survey forms and stored in the TRRL CYBER 720/815 computer system. With the aid of an appro-priate program the length of slope with any given combination of characteristics can be extracted, and the results given in this Paper were obtained by this means.

SLOPE FAILURES

10.   The survey has revealed a significant incidence of slope failures in side slopes of both cuttings and embank-ments. In the 300 km of motorway surveyed, accumulated lengths of over 3 km of cutting slope and over 11.5 km of embankment slope have failed. The type of slope failure found varied from distinct slab-type to shallow circular-type, but with most slips having a combination of transla-tional and circular movement. The depth of failure plane below the surface of the slope rarely exceeded 2 m. In most cases the area of failure extended from top to bottom of the slope.

11.   The method of reinstatement of failures has been similar for all areas included in the survey. The failed material has been removed to below the failure plane and the

resulting excavation backfilled with a granular free-draining
material such as gravel, brick rubble or crushed rock. Topsoil has been added in some regions, this having the effect
of obscuring the remedial treatment and providing a more
attractive appearance than that of the uncovered free-draining
material.

ANALYSES OF DATA

Factors studied
12.   Conclusions reached during the survey were that the
principal factors having an influence on the observed extent
of failures were the geology, the age of the earthworks, and
their geometry. The effects of these factors have, therefore, been studied in detail. Additional factors which were
considered to be worthy of study were the type of drainage
employed (at the base of embankment slopes and at the top of
cutting slopes) and the orientation (compass bearing) of the
slope. Because of space limitations the results of analyses
of these various factors are given only briefly in the
following paragraphs. It is intended to publish the results
of the analyses in greater detail at a later date.

Geology
13.   The overall results for each of the principal geologies
that were encountered are given in Table 1. Only those
materials that occurred in lengths of slope in excess of
2.5 km have been included and they are listed in order of
decreasing severity of failure rate. Failure rate is defined
as the length of failed slope (parallel to the centre line of
the road) expressed as a percentage of the total length of
slope involved. Only repaired and unrepaired slips were
considered as failures in the analyses; incipient slips,
where only tension cracking had occurred, have been treated
as contributing to the length of stable slope. In fact, had
incipient slips been included, the failure rates would have
been increased on average by factors of about 1.4 in cuttings
and about 1.2 in embankments. Observations have revealed
that some areas of slope classified in the survey as
'incipient slip' have subsequently failed.

14.   It must be remembered that variations in geometry have
not been taken into account in the overall failure rates given
in Table 1. Overall, the Table shows that the failure rates
for embankments were generally higher than for cuttings, the
one exception being Gault Clay, where the failure rates were
fairly similar at over 9 per cent. Failure rates in embankments in the range of 4 to 8 per cent were exhibited by
Reading Beds, Kimmeridge Clay, Oxford Clay and London Clay.
Although not included in the Table, combinations of London
Clay with either Plateau Gravel, Reading Beds or Boulder Clay,
and combinations of Reading Beds and Chalk also had failure
rates in embankments in the same range.

Table 1. Major materials encountered, total lengths of
side-slopes, proportions failed and the predominant slopes
(Data are for earthworks comprising only one geology with
total lengths in excess of 2.5 km)

Geology	Age of earthworks when sur- veyed (years)	Total length (km)	Failure rate (%)	Predominant slope (v : h)
Cuttings				
Gault Clay	10, 22	6.4	9.7	1 : 2.5
Oxford Clay	10, 22	14.6	3.2	1 : 2
Reading Beds	10	20.2	2.7	1 : 3
Lower Coal Measures	14	2.8	1.4	1 : 2
Plateau Gravel	6, 10	2.6	1.1	1 : 3
Boulder Clay	3,6,7,14,22	44.5	1.0	1 : 3
Middle Coal Measures	14, 15	48.9	0.7	1 : 2
Valley Gravel	10, 22	3.1	0.3	1 : 2
London Clay	3,5,6,7,10	20.2	0.3	1 : 3.5
Clay-with-flints	10, 22	29.4	0.2	1 : 3
Glacial Gravel	3, 22	8.3	0	1 : 3.5
Chalk	3,10,14,22	48.2	0	1 : 2
Lower Greensand	10, 22	2.7	0	1 : 1.75
Kimmeridge Clay	10	5.1	0	1 : 2.5
Kellaways Clay	10	5.2	0	1 : 3.5
Cornbrash	10	3.8	0	1 : 1.5
Great Oolite Clay	10	9.0	0	–
Upper Coal Measures	15	6.6	0	1 : 2
Embankments				
Gault Clay	22	4.8	9.1	1 : 2.5
Reading Beds	3, 10	40.8	7.8	1 : 2
Kimmeridge Clay	10	16.7	6.1	1 : 2
Oxford Clay	10, 22	33.8	5.7	1 : 2
London Clay	5,6,10	60.7	4.4	1 : 2
Middle Coal Measures	14, 15	51.1	0.8	1 : 2
Clay-with-flints	10, 22	12.7	0.7	1 : 2
Boulder Clay	3,6,7,22	33.9	0.4	1 : 2
Chalk	3,10,19,22	28.4	0.1	1 : 2
Lower Greensand	22	4.1	0.1	1 : 2.5
Glacial Gravel	22	3.5	0	1 : 2
Coral Rag	10	3.6	0	1 : 2
Great Oolite Clay	10	7.5	0	1 : 1.75
Lower Coal Measures	14	3.0	0	1 : 2

15.   Except for the Gault Clay, mentioned in 14 above,
cutting slopes in other geologies do not exhibit failure
rates in excess of 4 per cent. In the range of 1 to 4 per
cent, the materisls concerned are Oxford Clay, Reading Beds,
Lower Coal Measures, Plateau Gravel and Boulder Clay.

16.   Many other materials encountered in the survey over
significant lengths of cutting and embankment slopes
exhibited failure rates less than 1 per cent or had no
failures of any kind. These also have been included in
Table 1.

Age of earthworks
17.   The ages of the earthworks encountered in the survey
have been included in Table 1. (Age is assumed to accumulate
from the date of opening of the particular length of motorway.)
For those geologies exhibiting the more severe failure rates
over differing time scales analyses have been made for given
ranges of height and slope and the results are given in
Table 2. The necessity to compare similar geometries at
different ages has limited the comparisons to two geologies
each for cuttings and embankments. The height bands used
(0 - 2.5 m; 2.5 - 5 m;; more than 5 m) were selected for
convenience; the slopes are the predominant slopes as given
in Table 1.

18.   For cuttings the general trend is for failure rate to
increase with age, as would be expected. Boulder Clay
cuttings exhibit serious failure rates only with earthworks
of 22 years of age, but with Gault Clay a failure rate in
excess of 3 per cent has occurred even at an age of 10 years
in 1 to 2.5 side slopes, 2.5 to 5 m high.

19.   For embankments the trend with age appears completely
different to that of cuttings; an apparent illogicality is
demonstrated with the failure rates decreasing with age in
five out of six combinations of geology and geometry.
Clearly, if the same earthworks had been surveyed at the
appropriate time to relate the rate of failure to age the
trend could only be for the accumulated failure rate to
increase with age or remain constant. However, as earthworks
constructed at different periods are being compared to
assess the effect of age, other factors must be predominant
to produce the effects registered. To explain this, further
studies will be required of soil properties, specifications
and construction practices, in particular with regard to any
changes that have been introduced over the years since motorway construction began.

20.   To determine the effect of age on the performance of
given areas of earthworks over a long time scale, dates of
occurrence of slope failures are required, but such information is not readily available from the Authorities

Table 2.   Comparisons of failure rates in earthworks of
different ages for given combinations of geology and
geometry.

Geology	Slope (v : h)	Height (m)	Age (years)	Total length (m)	Failure rate (%)
Cuttings					
Gault Clay	1 : 2.5	0 - 2.5	10	348	0
			22	202	0
		2.5 - 5	10	353	3.8
			22	299	4.3
Boulder Clay	1 : 3	0 - 2.5	6,7	2003	0
			22	812	0.5
		2.5 - 5	6,7	1825	0
			22	363	9.9
		More than 5	6,7	2645	1.1
			22	804	0
Embankments					
Oxford Clay	1 : 2	0 - 2.5	10	3567	2.0
			22	1417	1.8
		2.5 - 5	10	1191	23.0
			22	1500	7.0
		More than 5	10	512	41.4
			22	1263	36.3
London Clay	1 : 2	0 - 2.5	5,6	551	0
			10	7426	0
		2.5 - 5	5,6	625	14.6
			10	9463	6.2
		More than 5	5,6	122	30.9
			10	7452	21.8

responsible for the maintenance of the particular lengths of
motorway.   An attempt is to be made, therefore, to determine the effect of age on specific lengths of motorway by
studying aerial photographs taken in the lifetime of the
earthworks. Locations will be selected where high failure
rates have been observed. It is hoped that photographs may
have been taken sufficiently frequently to allow the pattern
of failures on a time base to be established.

Geometry of slope
21.   In general, the slope angles of the earthworks tend to
be designed to be fairly uniform for any given age and
geology, especially with embankments, although sufficient
variability occurs in practice to provide some indication
of the effect of slope angle on the occurrence of failures.

Results are given in Figs. 1 and 2 for those combinations of geology and age that showed the higher rates of failure; failure rates are shown plotted against slope angle (equivalent to the slopes of 1 to 3.5; 1 to 3; 1 to 2.5; 1 to 2; 1 to 1.75) for the same ranges of height given in 17. Only results for total samples in excess of 100 m of slope are included.

22.   Figs. 1 and 2 illustrate the very high rates of failure that have occurred with certain combinations of slope angle, height, geology and age. Thus, more than half the cutting slopes have failed in particular geometries of Gault Clay (22 years old) and Oxford Clay (22 years old). In embankments high failure rates have occurred in particular geometries of Gault Clay, Oxford Clay, Kimmeridge Clay and the combination of Plateau Gravel and London Clay.

23.   For both cuttings and embankments there is clear evidence that the height of slope has a logical effect on failure rate, ie slopes more than 5 m high have the highest failure rate at a given slope angle in the majority of cases; see Figs. 1 and 2.

24.   The effect of slope angle is not so clear, however, and in some instances flatter or intermediate slope angles have yielded the higher rates of failure. Where single geologies are concerned this is most pronounced with cuttings of Gault Clay (22 years old), Boulder Clay (22 years old), and Reading Beds (10 years old) and with embankments of Gault Clay (22 years old), Reading Beds (10 years old) and Kimmeridge Clay (10 years old). In some instances earthworks of the same geology but of different ages appear to exhibit different effects of slope angle, eg Gault Clay cuttings in Fig. 1, and Oxford Clay embankments in Fig. 2. Note that in Fig. 1 ranges of slope angle and vertical scales are different.

25.   A possible explanation for the association in some instances of higher failure rates and flatter slope angles involves the rate of pore-pressure equilibration at different slope angles. In over-consolidated clay soils high strengths would initially prevail following construction because of the negative pore pressures resulting from stress relief. This would apply to the material remaining in the side slope of a cutting or placed and compacted in the side slope of an embankment. In the course of time the pore pressures will equilibrate as the clay takes up water, the source being either ground water or rainfall in the case of cuttings and rainfall on the surface of the slope in the case of embankments; this assumes the integrity of the pavement drainage system. Subjective assessments made during the survey indicate that in most instances the material 'softens' from the outside with the lowest strengths

occurring initially on the surface of the slope. The reductions in strength may eventually reach a critical depth when,
in association with positive pore pressures set up in periods
of heavy rainfall, failure on a relatively shallow slip
plane occurs. Access to water may be further assisted by
the presence of shrinkage cracks initiated in hot, dry spells.

26. It is assumed that all slope angles above a maximum
'safe' angle could be susceptible to failure when the soil
has reached a fully softened condition, then those slopes
with easier access to water could well be the first to fail.
Thus it is conceivable that of the slope gradients in the
range liable to fail in the long term, the less steep
angles, with less ability to shed water, will be in a

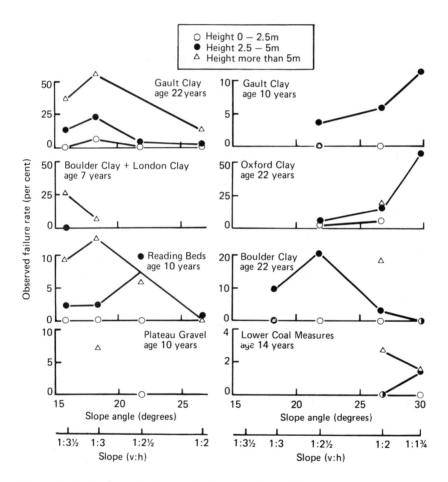

Fig. 1. Relations between failure rate of cutting slopes and
geometry of slope

saturated state for longer periods, and therefore progress more rapidly to the fully softened condition at the critical depth for failure. This hypothesis could explain the data obtained with cuttings of Gault Clay and Boulder Clay in Fig. 1 and many of the materials in embankments in Fig. 2. It also implies that in the longer term the steeper slope angles will reach at least the same rates of failure as those angles currently exhibiting the highest rates of failure.

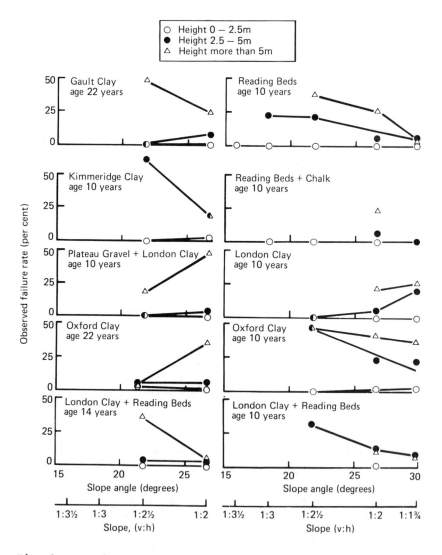

Fig. 2. Relations between failure rate of embankment slopes and geometry of slope

27.   An alternative explanation that, on first consideration, is plausible is that the earthworks with the reduced
slope angles may have been designed in areas where the soil
exhibited the lower strength characteristics; in the event
the reduced slope angle may not have compensated sufficiently
for the lower strength. Additionally, in the case of cuttings, the ground water regime may have been more severe
where the flatter slope angles were used. However, apart
from Reading Beds materials, where soil types varied from
sand to clay, the randomness with which the variations in
slope angle occurred throughout the earthworks surveyed does
not support this explanation.

Type of drainage
28.   Different types of drainage have been observed at the
top of cutting slopes and at the base of embankment slopes.
The principal types of drainage in both locations have been
categorised as 'none', 'open ditch', and 'French drain'. A
limited study of the effects has been made.

29.   With cutting slopes no combination of geology and
geometry incorporates examples of all three types of drainage.
Additionally, no one type of drainage is predominant for cutting slopes in general. Where the types of drainage at the
top of the slope are either 'none' or 'open ditch' (six
examples), four indicate higher rates of failure with 'open
ditch' drainage and only one (Reading Beds, 2.5 – 5 m high,
1 to 2.5) shows a significant reversal of this trend. With
the three examples where the types of drainage are 'none' and
'French drain' there is no clear trend for any one type to
have the more serious failure rates. The presence of the
open ditch or French drain indicates that the need for a
cut-off drain was recognised during design, probably due to
general ground-water movements towards the slope of the
cutting. Any instances where higher failure rates were
associated with cut-off drains may be caused, therefore,
mainly by the ground water regime.

30.   In the case of embankments the highest proportion of
observed lengths of slope had open ditch as the type of
drain at the foot of the slope. A study of the failure
rates for slopes with different types of drain shows that in
the majority of cases (6 out of 10) distinctly higher
failure rates occurred with the use of open ditches. In the
ten examples studied ranges of failure rates are as follows:-

        Open ditch   21 to 62 per cent
        French drain   0 to 45 per cent
        None   0 to 34 per cent

In embankment design the presence of the drain at the foot
of the slope is associated with the need to protect the subsoil from ground water and is not connected with the main-

tenance of **the stability of** the slope. However, in the light
of the above results, consideration must be given to the
possibility, in potential failure situations, that the open
ditch configuration may remove toe restraint, or that the
maintenance operations using mechanical plant to clean out
open ditches may give rise to disturbance of the toe of the
slope.

Orientation of slope

31. In a limited study slopes of given combinations of
geology and geometry were classified according to their
orientation in 90° quadrants averaging north, east, south
and west facing directions. For cuttings and embankments
variations occurred between the failure rates of slopes of
varying orientations, but there was no consistent pattern of
behaviour over the range of geologies studied. However, with
embankments of Reading Beds material, north facing slopes
(bearing 315° to 360° to 045°) showed significantly higher
failure rates than other orientations, as shown in Table 3.

32. It could be expected that ground water movements,
which would be independant of orientation, may have greater
influence on cutting slopes and that the effect of orientation would, therefore, be random, as the data indicate.
With embankments, however, the climatic influence of
orientation, either the slower drying rates of north facing
slopes, or the greater likelihood of shrinkage cracking on
south facing slopes, could both contribute to higher rates
of failure. Apart from the data for Reading Beds given in
Table 3, no evidence has been obtained to indicate whether
or not these factors have an influence, or which of them
predominate.

Table 3. Effect of orientation on the failure rates of
Reading Beds embankment slopes (10 years of age)

Height (m)	Slope (v : h)	Orientation	Failure rate (%)
More than 5	1 : 2.5	North	71
		East, South, West	15
More than 5	1 : 2	North	49
		East, South, West	17
2.5 - 5	1 : 2.5	North	43
		East, South, West	6

Table 4.   Maximum slopes required to restrict failure rates to below 1 per cent within 10 to 22 years of construction as indicated by the results of the survey.

Geology	Maximum slope (v : h)		
	Height :– 0 – 2.5m	2.5 – 5m	More than 5m
Cuttings			
Gault Clay	1 : 3.5	1 : 4	1 : 4
Oxford Clay	1 : 2.5	1 : 3	1 : 3.5
Reading Beds (cohesive)	1 : 4	1 : 4	1 : 4
Reading Beds (non-cohesive)	1 : 2.5	1 : 2.5	1 : 3
Lower Coal Measures	1 : 1.75	1 : 2	1 : 2.5
Plateau Gravel	1 : 2.5	1 : 2.5	1 : 3.5
Boulder Clay	1 : 4	1 : 4	1 : 4
Middle Coal Measures	1 : 2	1 : 2.5	1 : 3
Valley Gravel	1 : 2.5	1 : 2.5	1 : 2.5
London Clay	1 : 3.5	1 : 3.5	1 : 3.5
Clay-with-flints	1 : 2	1 : 2	1 : 2
Glacial Gravel	1 : 2	1 : 2	1 : 2
Chalk	1 : 1.5	1 : 2	1 : 2
Lower Greensand	1 : 1.75	1 : 1.75	1 : 1.75
Kimmeridge Clay	1 : 2.5	1 : 2.5	–
Kellaways Clay	1 : 2	1 : 3	1 : 3.5
Cornbrash	1 : 1.5	1 : 1.5	1 : 1.5
Upper Coal Measures	1 : 1.75	1 : 1.75	1 : 2
Embankments			
Gault Clay	1 : 2.5	1 : 3	1 : 3.5
Reading Beds (cohesive)	1 : 3	1 : 4	1 : 4
Reading Beds (non-cohesive)	1 : 1.75	1 : 1.75	1 : 1.75
Kimmeridge Clay	1 : 2.5	1 : 3.5	1 : 3.5
Oxford Clay	1 : 3	1 : 3.5	1 : 3.5
London Clay	1 : 2	1 : 3	1 : 3
Middle Coal Measures	1 : 2	1 : 3	1 : 3
Clay-with-flints	1 : 2	1 : 3	1 : 3.5
Boulder Clay	1 : 3	1 : 3	1 : 3
Chalk	1 : 2	1 : 2	1 : 2
Lower Greensand	1 : 2	1 : 2	–
Glacial Gravel	1 : 2	1 : 2	1 : 2
Coral Rag	1 : 2	–	–
Great Oolite Clay	1 : 1.75	1 : 1.75	1 : 1.75
Lower Coal Measures	1 : 2	1 : 2	1 : 2

DESIGN OF SIDE SLOPES IN NEW CONSTRUCTION

33.   The survey has revealed a significant incidence of
shallow slip failures in both cuttings and embankments in
cohesive materials, mostly of an over-consolidated nature.
From the data conclusions can be reached as to the maximum
slopes to be employed for given ranges of height in order to
minimise the risk of failures within an approximate 20 year
life span. (Data from the survey only apply to a maximum of
22 years of age, with many results only for an age of 10
years.)

34.   The results of an analysis of the maximum slopes
allowable to minimise failures are given in Table 4. It has
been assumed that a total failure rate of 1 per cent in a
20 year period is acceptable. The geologies are given in the
same order as in Table 1, ie in order of decreasing severity
of slip problems. A comparison can be made between the maximum advisable slope given for each range of height (0 - 2.5 m;
2.5 - 5 m; more than 5 m) in Table 4 with the observed predominant slope given in Table 1. Gault Clay cuttings, for
example, exist predominantly at 1 to 2.5 slopes in the areas
surveyed (Table 1), but the results of the survey indicate
that slopes should be 1 to 3.5 for heights less than 2.5 m
and 1 to 4 for heights in excess of 2.5 m.   Note that this
corresponds very closely to the quoted upper bound residual
angle of internal friction for Gault Clay defined in ref. 5.

35.   The values given in Table 4 should be used only as a
general guide. If the results of laboratory tests and design
analyses suggest steeper slope angles than those given in the
Table, it might be advisable to ensure that the design information is sufficiently accurate, and especially that due
account has been taken of possible low effective stress conditions at shallow depths in the slope.

CONCLUSIONS

36.   The results of a survey of the condition of earthwork
slopes on selected lengths of motorway have been analysed and
the following conclusions have been reached:-

1.   A significant incidence of shallow slip failures have
      occurred since construction. In the 300 km of motorway
      surveyed, failures in cuttings have a total length in
      excess of 3 km and in embankments over 11.5 km of
      slope have failed.

2.   The slips have occurred mainly in plastic over-consoli-
      dated materials.

3.   In cuttings, the principal geologies associated with
      failures are Gault Clay, Oxford Clay, Reading Beds,
      Lower Coal Measures, Plateau Gravel and Boulder Clay.

4. With embankments, the principal geologies associated with failures are Gault Clay, Reading Beds, Kimmeridge Clay, Oxford Clay and London Clay.

5. With cuttings, some failures were observed in earthworks of 10 years of age, but the majority were shown to have occurred in the 22 year old earthworks.

6. More recently constructed embankments of Oxford Clay and London Clay had higher failure rates than earlier constructions.

7. Within the range of ages of the earthworks studied it is not necessarily the steepest slope angle that contributes the highest failure rate. Variability of soil properties or in rates of pore pressure equilibration could be the reason for this.

8. There is some evidence that the method of drainage at the foot of embankment slopes has an effect on failure rate, with open ditches causing the slopes to be more susceptible to failure.

9. There is no evidence that the orientation of the slope generally affects the failure rate, although north facing slopes of Reading Beds embankments have significantly greater rates of failure than the remainder.

10. General guidance on the design of earthwork side slopes can be gained from the results of the survey.

11. The failure of cutting and embankment slopes is a continuing process. The identification of areas at risk of failure in the longer term will be aided by reference to the maximum advisable slope angles deduced from the results of the survey.

RECOMMENDATIONS

37. In the light of the findings from the survey, and the need to understand further the mechanism by which the observed slope failures occur, it is recommended that:-

1. A better understanding is required of the pore pressure regime at shallow depths in embankment and cutting slopes of over-consolidated clays. The effect of slope angle and age of earthworks on the pore pressure regime should be taken into account in the study.

2. In soils prone to failures a better understanding is required of the shear strength characteristics for the potentially low effective stress conditions on the failure planes. Attempts should be made to develop procedures that allow the incorporation of that understanding into the design of the geometry of earthwork slopes.

3. The reasons for higher failure rates in more recently constructed embankments should be determined. The

study needs to take into account any changes in design standards, specifications and construction practices that have been introduced since motorway construction commenced.

4. The use of aerial photographs taken in the lifetime of the earthworks to determine the effect of age on the accumulated failure rates should be explored; useful information might thus be obtained on the possible future pattern of failures.

5. A further extension of the survey should be made to gather data of a similar kind on other geologies that occur widely in the United Kingdom.

ACKNOWLEDGEMENTS
38. The work described in this Paper forms part of the programme of the Transport and Road Research Laboratory and the Paper is published by permission of the Director. The co-operation of the Local Authorities concerned with the maintenance, and the Consulting Engineers concerned with the original construction, of the lengths of motorway studied, is gratefully acknowledged.

REFERENCES
1. VAUGHAN P R and H J WALBANCKE. Pore pressure changes and the delayed failure of cutting slopes in over-consolidated clay. Geotechnique, 1973, 23, (4), 531–539.
2. CHANDLER R J and A W SKEMPTON. The design of permanent cutting slopes in stiff fissured clays. Geotechnique, 1974, 24, (4), 457–466.
3. DEPARTMENT OF TRANSPORT. Transport statistics, Great Britain, 1972 – 1982. HM Stationery Office, London, 1983.
4. ANON. Studies of slope stability problems in highway earthworks. TRRL Leaflet 943, Transport and Road Research Laboratory, Crowthorne, 1983.
5. HUTCHINSON J N, E N BROMHEAD and J G LUPINI. Additional observations on the Folkestone Warren landslides. Q.J. Eng. Geol., 1980, 13, 1 – 31.

Any views expressed in this paper are not necessarily those of the Department of Transport.

6. Shallow slips in highway embankments constructed of overconsolidated clay

J. R. GREENWOOD, Regional Geotechnical Engineer,
D. A. HOLT, Geotechnical Engineer, and G. W. HERRICK,
Geotechnical Engineer, Department of Transport

SYNOPSIS Highway embankments constructed of overconsolidated clay at conventional side slopes have been prone to shallow slope failure a few years after construction. The extent of the problem and areas currently at risk are identified. Factors likely to increase the risk of a slip and probable mechanisns of failure are described. Effective stress parameters are suggested for use in analysis and problems in determining these parameters are discussed. Back analysis and parametric sensitivity studies are described. The investigation of a slipped area and field trials of various repair techniques are briefly reported. The techniques, some including the use of geogrids, are considered as alternatives to the conventional 'excavation and granular replacement' repair method. Comment is made on the future design of embankments in overconsolidated clay.

INTRODUCTION

1.   The problems of slope failures in embankments and cuttings has long been recognised. It is well documented that cutting failures in overconsolidated clays generally occur many years after construction. The time dependent nature of these failures has been well researched (ref. 1) and is taken into account in highway cutting design.

2.   Embankment failures have generally occurred during or very soon after construction due to the presence of weak foundation soils and the generation of excessive pore water pressures. Consequently the idea that 'embankment design using undrained analysis is sufficient for stability and that embankments become stronger with time' was fostered amongst road design engineers.

3.   Although this approach is often valid for the foundation condition, experience has shown that embankments constructed of overconsolidated clay are at risk from long term shallow failure due to softening with time in much the same way as occurs in cuttings.

4.   The trunk road network in the South and East England has developed rapidly over the past 20 years and in order to satisfy the stringent environmental and design criteria a large number of high embankments and deep cuttings have been constructed.

5.   The geological map of Eastern England shows a mantle of glacial deposits covering virtually the whole area.   This glacial mantle is often relatively thin and the deeper road cuttings and borrow pits have penetrated the underlying solid deposits.   Fig. 1 shows the solid geology of the area with the major trunk roads superimposed.   It can be seen that with the exception of the wide tract of chalk running NE-SW across the area the majority of the geology consists of overconsolidated clay,   ie Lias,   Oxford,   Ampthill   and   Kimmeridge   Clays (Jurassic) Gault Clay (Cretaceous) and Woolwich and Reading and London Clay (Eocene).

6.   Embankments constructed of these soils have been built typically at 1 in 2 side slopes up to heights of about 10m based on undrained ($\phi = 0$) design criteria.   Many of these embankments are now experiencing shallow slope failures.   The frequency of slips has increased over the past few years. During the winter of 1982-83 following periods of wet weather a large number of embankment slips occurred on the M11, A604 (Bar Hill), A10 (Hoddesdon and Ware), M25(A1-A111), and A45 (Cambridge).   Slips   have   also   occurred   on   the   M1 (Bedfordshire) and A12 (Colchester).   Approximately 40 slips occurred in Cambridgeshire alone during a 12 month period.

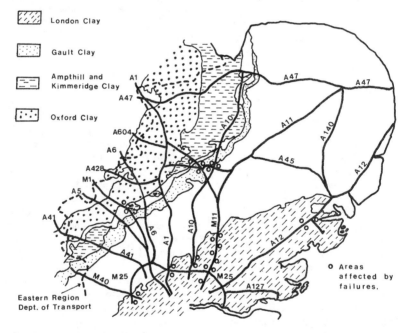

Fig. 1.   Distribution   of   overconsolidated   clays   in   Eastern England   and   road   sections   affected   by   shallow embankment failures

7.   The location of the embankment slips has been confined to a number of distinct areas shown on Fig. 1.   All the slips reported   occurred   in   embankments   constructed   of

overconsolidated clays. Parsons (ref. 2) has studied the occurrence of both cutting and embankment failures on the M1, M4 and M11 motorways and has shown that in certain sections up to 20% of embankments in overconsolidated clays have failed.

8. With growing concern over the number of failures and the cost of repairs, a section of failed embankment on the A45 at Cambridge was selected for a programme of investigations, monitoring and trial constructions of various repair techniques.

9. The following sections of this paper consider the mechanism of failure and the factors affecting the occurence of shallow embankment slips. The problems of determining appropriate design parameters are discussed and a simple theoretical analysis is applied. The A45 Cambridge trials are briefly described and the merits of alternative repair and preventative measures discussed.

MECHANISM OF FAILURE

10. A typical profile of a failed slope is shown in Fig. 2. Isolated slips sometimes occur over a width of 10-15m but often extend to give a continuous zone of failure perhaps 50-60m in width.

11. The development of the slips is rarely observed because, unlike cutting slopes, embankment batters cannot be seen from the highway. Slips are usually reported by the maintenance authority after prolonged wet periods, although the report cannot generally pinpoint the exact date of failure because of the infrequency of maintenance inspections.

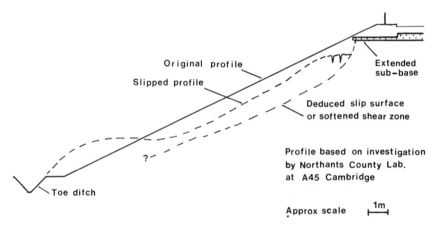

Fig. 2. Cross section of typical shallow slope failure

12. A study of the slipped material soon after failure reveals a very wet slipped mass with the material at the base of the back scarp being extremely soft. The back scarp frequently intersects the extended granular sub-base or 'capping' layer. Water may be seen seeping from this layer (as it was intended to do) and the clay material beneath is often very soft to the touch.

13. The question to be considered is how an embankment constructed typically of firm to very stiff overconsolidated clay can wet up, soften and ultimately fail by shallow slippage within a few years of construction. The mechanism of failure is believed to be as follows: The overconsolidated clay is excavated from cutting or borrow pit from an increasing depth below the original ground surface. The deeper material tends to be placed on the top of embankments. Clay from depth will exhibit considerable suction when taken from the ground. Reworking during placement and compaction is unlikely to relieve this suction and compaction induced shearing may further increase suction forces. When water is available it is drawn into the clay by the suction pressures with consequent softening. With the negative (suction) pressures relieved a continued supply of water (eg from carriageway drainage channels, direct rainfall, tree planting holes and tension cracks filling with water or a steady seepage from the road sub-base) will give rise to positive pressures locally within the clay slope and a subsequent decrease in the factor of safety until failure occurs.

14. Farrar (ref. 3) measured the negative pore pressures existing within embankment fill. Subsequent results (ref. 4) show only a gradual reduction of those negative pressures after 10 years. At the same time positive values were recorded at shallow depth in the embankment slope. Anderson and Kneale (ref. 5) showed the transient nature of the positive pore pressures at shallow depth following rainfall and also identified a marked decrease of the permeability of the clay fill below about 1m at which depth shallow failure occurred.

FACTORS AFFECTING THE RISK OF EMBANKMENT FAILURE

15. Shallow embankment failures are brought about by a combination of factors; properties of the fill, climatic conditions, design features, and construction techniques. These factors are considered in turn.

16. Plasticity index of clay fill. The higher the PI of the clay the more susceptible it will be to seasonal moisture changes and the formation of shrinkage/tension cracks during dry periods and swelling during wet periods. Any tendencies for downslope creep will be more pronounced.

17. Stress history and structure of clay fill. As noted, in para 13, the overconsolidated clay will exhibit suction when taken from the ground. The stress history of a particular sample will govern the suction pressures developed in the fill. The degree of fissuring and brecciation can affect the rate of softening when water is present.

18. Moisture content and compaction of clay fill. The upper moisture content permitted for clay fills has been typically 1.3 x Plastic Limit (PL). If the clay is close to this upper limit it will be relatively easy to compact to a low air voids ratio giving a material of low permeability and a moisture content close to equilibrium at shallow depth. However most overconsolidated clays are in a dryer condition

(ie < 1.1 x PL) when excavated and consequently are more difficult to compact in fill. Smooth wheeled rollers tend to ride over hard clay lumps leaving large air voids. Compaction may be particularly poor at the edges of embankments due to lack of lateral restraint and loose material being pushed over the side. Inadequate compaction will allow the ingress of water and permit swelling due to lack of confinement, and hence more rapid softening. On recent contracts a tamping type roller has been specified for clays having a Liquid Limit > 50, and clays with a moisture content < 0.9 x PL have been excluded from embankment fill.

19. _Topsoil depth_. With the trend towards topsoil thickness of up to 300mm to assist planting there is a greater chance of more moisture retention on the slope. However the topsoil and root system may protect the underlying soil to some extent against excessive shrinkage and swelling.

20. _Tree planting_. Many of the failed slopes have been subject to tree planting schemes. Planting often leaves depressions around each sapling where water is encouraged into the slope; holes are sometimes dug and left unfilled. A line of planting holes may encourage the development of tension cracks. The development of a mature root system should give some resistance to shallow slope failure but the weight of larger trees with shallow root systems could be detrimental to stability.

21. _Slope aspect_. It might be expected that north and west facing slopes would be more at risk but to date the records of slips in Eastern England do not support this.

22. _Slope angle_. Clearly the flatter the slope the more stable it should be. However it is possible that water ponding on a shallow slope could cause softening at a faster rate and with resulting higher local pore pressures.

23. _Base Drainage Blanket_. Suction pressures at the toe may, to some extent, draw moisture from a high water table in the foundation soils. An efficient base drainage blanket should help to control water pressures in the toe region. However evidence indicates that the prime source of water is rainfall and surface seepage and that, overall, the presence of a base drainage layer would have little effect on the occurrence of shallow slips. An inefficient drainage blanket may increase the risk of failure by providing a source of water under pressure (ref 6).

24. _Carriageway drainage_. Road gully discharge pipes are sometimes present within the failed mass. The road drainage design frequently requires discharge pipes to run down the embankment to the toe drain every 20-30 m and therefore their presence in embankment failures may be due to probability. However they are sometimes observed near the centre of slipped masses, the granular surround is often wet and in some instances the clay pipes are broken or the rubber sealing rings are missing.

25. _Pavement drainage_. The granular sub-base is designed to accept and discharge water which finds its way in between the pavement and sub-grade. The permeability of road

pavements may be as high as 5×10^{-6} m/s (ref. 7) and the sub-base (and capping layer) may be recharged at times of heavy rainfall from inefficient central reserve drains. A sub-base extension (Fig. 2) is designed to permit water to drain to the edge and flow down the slope of the embankment. However these layers are usually covered up by topsoil and high pore pressures in the upper edge regions are therefore likely. Many of the embankment slips intersect this extended sub-base and the soil surrounding it is observed to be wet.

26. <u>Chemical Weathering.</u> It is possible that waste products from petrol combustion, salts used during winter maintenance and 'acid rain' effects might produce contaminated run off waters which could affect the electro-chemical bonding of the clay minerals with subsequent weakening.

27. It is hoped that as more data on failures is collected (ref. 8) the significance of the above factors might become clearer.

FIELD DATA AND PARAMETERS FOR DESIGN

28. Few detailed studies of shallow slipped areas have been reported. Studies inevitably run into problems of variable embankment materials and conditions which were not necessarily those at the time of failure.

29. Trial excavations taken through slipped masses on the A45 (Fig. 3) revealed moisture contents close to the assumed placement values of 20-30% below 1.5m. Above 1.5m moisture contents had increased on average by 8% and densities and vane shear strengths showed a marked decrease.

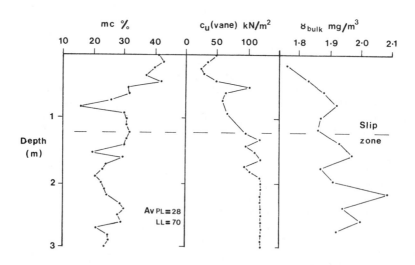

Fig. 3. Variation of moisture content, vane strength, and bulk density with depth at mid height on a failed Gault Clay slope at A45 Cambridge.

30. It would be convenient if the softening process could be modelled in the laboratory. Measurement of appropriate strength parameters in the field or laboratory is not easy because the clay at construction does not reflect the properties it will exhibit at the time of failure.

31. The stiff overconsolidated clay is observed to retain much of its structure during placement and compaction in embankments. Tests on laboratory samples remoulded to a higher moisture content to simulate softening may not reflect the remnant 'fragmented' clay structure which can be seen in the field.

32. To simulate field conditions it is necessary to test samples under very low stress conditions of say 10 to 30 kN/m^2. Conventional triaxial apparatus is not suitable for these low stresses unless it can be modified for increased accuracy.

33. The simple shear box might also be used. It has the advantage that low stresses can be more readily controlled (subject to measurement of apparatus friction) but has the disadvantage that the state of the sample cannot be easily monitored. Creep tests under constant load in the shear box may be worth further consideration.

34. Extrapolation of effective stress parameters from tests carried out under higher confining stresses is unreliable because of the probable 'curvilinear' profile of the failure envelope (ref. 10).

35. Holt (ref. 11) is investigating, inter alia, the possible relationship of undrained shear strength to the equilibrium moisture content predicted from the swelling line of e/log p' curves for remoulded clay samples.

36. Until a reliable laboratory or field method can be established to measure softened strength parameters we must rely, as for natural cutting slopes in these soils, on data obtained from the back analysis of slips.

ANALYSIS

37. A typical slip section is presented in Fig. 4. Slip surfaces are considered running parallel to the batter slope and the factor of safety for each surface is calculated based on assumed parameters using the simple method described by Greenwood (ref. 12).

38. The variation of factor of safety with depth is shown on Fig. 4 for the cases a) where c' is assumed at 1.5 kN/m^2 and b) where c' is assumed at 5 kN/m^2. Pore water pressure ratios (ru) of 0.2 and 0.4 are considered for each case and horizontal earth pressure coefficients (K) of 0 and 0.4 are considered. These plots demonstrate that a minimum factor of safety exists at 1-2m below ground if c' = 1.5 kN/m^2 or 2.5-3.5m below ground if c' = 5 kN/m^2. It may tentatively be deduced that c' = 1.5 kN/m^2 is more appropriate as slips tend to occur at about 1.5m depth (although it is quite possible for a c' of 5 kN/m^2 to operate with a localised higher pore water pressure to produce the same factor of safety at 1.5m depth).

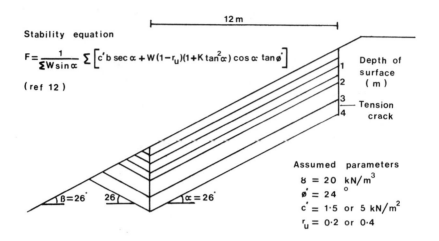

Stability equation

$$F = \frac{1}{\Sigma W \sin \alpha} \Sigma \left[c'b \sec \alpha + W(1-r_u)(1+K \tan^2 \alpha) \cos \alpha \cdot \tan \phi' \right]$$

(ref 12)

Depth of surface (m)

Tension crack

Assumed parameters

$\gamma = 20$ kN/m^3
$\phi' = 24°$
$c' = 1.5$ or 5 kN/m^2
$r_u = 0.2$ or 0.4

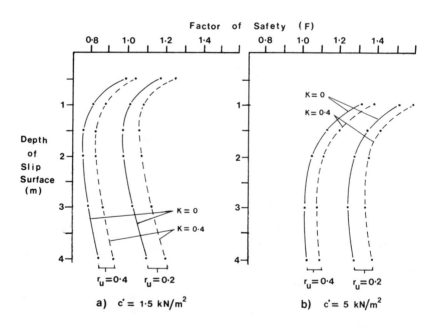

Fig.4.   Simple Analysis of shallow slip surfaces - Variations in factor of safety with depth.

39.   By combining the possible parameter options in Fig. 5, a curve of likely variation of F of S with depth is derived. It is apparent that the likelihood of deeper slips occurring is small because three factors are combining to contribute to a higher F of S at depth:-

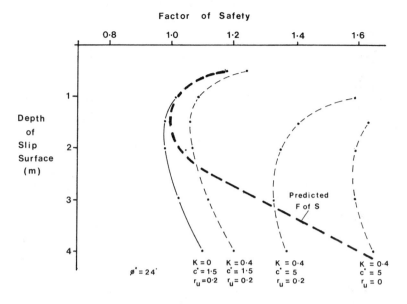

Fig.5. Predicted variation of factor of safety with depth at time of failure.

(a) The cohesion is less likely to have reduced to the softened value as increased overburden restricts softening.

(b) The pore water pressures are still negative and are less likely, because of the low permeability, ever to reach peak positive values as high as those closer to the surface.

(c) The contribution to shear resistance from the horizontal stress is more reliable at depth (ie K may be assumed = 0.4 rather than 0).

40. The slip surface at 1.5m depth is considered in more detail in Fig 6 where the effect on the F of S of varying individual parameters is shown. The analysis is particularly sensitive to the values of c' and ru applied, both of which are difficult to establish in the field with any certainty along the entire length of the slip surface. Flattening the slope angle β from 26° (1 in 2) to 19° (1 in 3) pushes the factor of safety up from 0.98 to 1.36. Oversteepening of a slope, as sometimes happens when land take has not been sufficient, may reduce the F of S considerably.

41. It is concluded that with the inherent variability of the embankment fill materials and their varied stress histories prior to and during placement it is unlikely that precise analysis can be carried out. However parameters appropriate to remedial works may be derived by back analysis of failures and by parametric studies to determine the sensitivity of the design to the parameters selected.

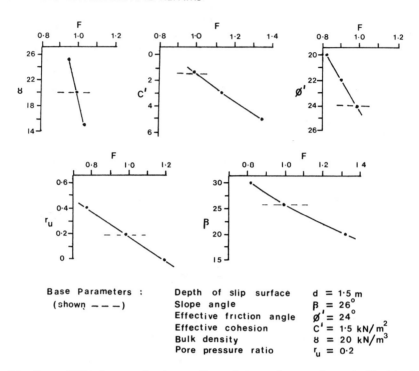

Base Parameters :
(shown ─ ─ ─)

Depth of slip surface	$d = 1.5$ m
Slope angle	$\beta = 26°$
Effective friction angle	$\phi' = 24°$
Effective cohesion	$c' = 1.5$ kN/m^2
Bulk density	$\delta = 20$ kN/m^3
Pore pressure ratio	$r_u = 0.2$

Fig.6. Effect on factor of safety of varying individual
parameters with others held at constant base values.

REMEDIAL MEASURES

42. The shallow embankment failures do not usually cause
immediate danger to the highway but if left unrepaired
secondary failure can occur with subsequent carriageway
cracking. The slip debris tends to block toe ditches and
damage fence lines and can be very unsightly.

43. The normal remedial technique to date has been to
excavate the slipped material, remove it to an offsite tip and
replace it with a granular fill. The use of course brick
rubble or limestone cobbles has led to problems with top-
soiling and replanting. Repairs have sometimes, for economy,
been left without topsoil leaving an obvious scar on the
landscape. Occasionally there has been a recurrence of a
shallow slip because insufficient softened material was
removed.

44. The cost of the conventional repair with granular
replacement has been approximately £25 per cu m with typical
repair costs amounting to £10-15,000. Recently a number of
alternative repair techniques have been considered. To
evaluate the practical benefits and problems associated with
alternative solutions a series of trial repairs were initiated
during the Winter of 1983/84 by the Department of Transport's
Eastern Regional Office in association with the Transport and
Road Research Laboratory.

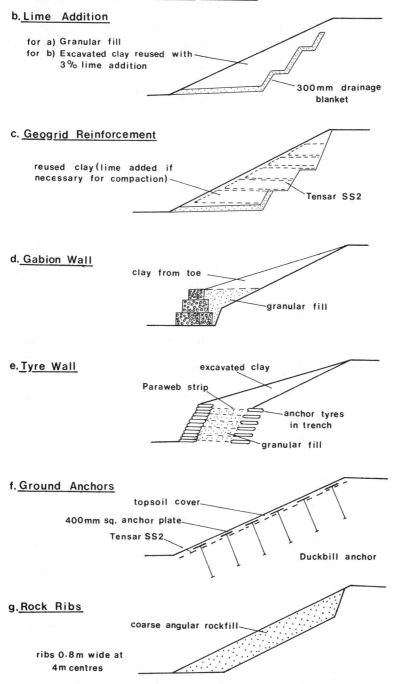

a. Excavation and Granular Replacement

b. Lime Addition

for a) Granular fill
for b) Excavated clay reused with
3% lime addition

300mm drainage
blanket

c. Geogrid Reinforcement

reused clay(lime added if
necessary for compaction)

Tensar SS2

d. Gabion Wall

clay from toe

granular fill

e. Tyre Wall

excavated clay

Paraweb strip

anchor tyres
in trench

granular fill

f. Ground Anchors

topsoil cover
400mm sq. anchor plate
Tensar SS2

Duckbill anchor

g. Rock Ribs

coarse angular rockfill

ribs 0·8m wide at
4m centres

Fig.7. Alternative remedial and preventative Techniques
(A45 Cambridge trials).

45.   The trials were carried out on a 150m length of Gault clay embankment on the A45 Cambridge Northern Bypass.   The 8m high embankment was constructed in 1976 at 1 in 2 side slopes and had experienced recent shallow slippage.   The field repair work was carried out by Cambridgeshire County Council assisted by Northants County Laboratory for the investigation and installation of instrumentation.   The work was monitored by TRRL and details will be reported in due course.   (ref 9)

46.   In addition to the conventional excavate and replace technique, four alternative repair techniques and two preventative methods were tried.   These are summarised in Fig. 7.   All the techniques were regarded as alternatives to slackening the slope which can be costly and often time consuming in obtaining the additional land.

47.   All the methods were successfully constructed over the winter period using conventional plant, ie dragline, backhoe and dozer.   Limited trials with the recently introduced Priestman V.C.15 showed it to be particularly suitable for embankment batter repairs.

48.   Of the remedial methods the Geogrid reinforced repair was readily constructed and possibly offered most economy of materials, plant and labour.   It has been suggested (ref. 13) that the geogrid return at the face of the slope might, for ease of construction, be omitted and replaced by a short intermediate layer of geogrid.   The tyre wall produced an aesthetically pleasing face but even though old tyres are available at low cost the additional fill and manpower required might reduce the competitiveness of this solution unless a steeper batter is required.   The Gabions produced a similar end result to the tyre wall but were more costly and time consuming to erect.

49.   The addition of lime to make the on site clays re-usable is attractive particularly when used in association with the geogrid to prevent local bogging of construction plant.   There may be a question as to the long term benefit of the 'lime only' treatment because of the difficulties of adequately mixing the lime with the clay on a small scale and the uncertainty of whether or not the resulting mixture, having a lower density and higher permeability than the untreated clay, is more prone to continued softening.

50.   The trials demonstrated that all the methods were feasible.   In considering the method to use at a particular site factors such as the availability of plant and materials and restrictions on access to the slope are likley to determine the most economical solution.

51.   Whichever remedial method is employed it is important that the site is properly prepared by sufficient excavation below the softened debris and benching into the remaining material.   Drainage measures require careful consideration for each location.

52.   The Department of Transport's advice note (ref. 8), dealing with the detection, reporting and repair of shallow slips on trunk roads and motorways, recommends the inclusion of a free draining layer at the base of the backfill to

prevent the build up of pore water pressures. This drainage layer is essential if water is allowed to penetrate the embankment fill but it is preferable that water seeping from granular pavement layers or within the embankment fill is directed away from the slope to avoid recurrence of the failure by further softening. Consideration might be given to reinstating the extended sub-base/capping layer on top of a polythene sheet and not covering it with topsoil to encourage seepage water to evaporate at the slope surface rather than seeping into the clay beneath the topsoil. For persistent seepage flows a cut off drain may be desirable but access for construction of such a drain is often restricted by crash barriers, service ducts and other roadside furniture.

53. For stabilization of areas at risk prior to failure the rock ribs may give a buttressing effect but their comparitively shallow depth may encourage access of water and continued softening of the surrounding clay. The use of ground anchors such as the Duckbill has interesting potential. At a unit material cost of around £10 a maximum tensile load of 1 to 2 tons is available from each anchor. Installation is achieved by driving the Duckbill on a rigid mandril which is then withdrawn and tension introduced by jacking against a galvanised steel bearing plate.

54. In the trials the Duckbills were driven to only 2.2m depth because of difficulty in withdrawing the driving mandril. From Fig. 5 a depth of 2.5 to 3m would seem most appropriate for optimum benefit. The Tensar SS2 grid placed below the topsoil, whilst not essential for the anchors operation, should restrict surface shrinkage movements, provide a bond for the topsoil and help load spread on to adjacent anchors should one yield excessively. The trial anchors were spaced on a staggered 2.5m grid.

CONCLUSIONS

55. Many of the motorway and trunk road embankments built over the past 20 years in South East England are at risk from shallow slope failures because they are constructed at relatively steep slope angles of stiff overconsolidated clays which are susceptible to softening when given access to water.

56. Numerous construction and environmental factors contribute to the risk of slip but the key factor is the availability of water. Prolonged seepages from the extended pavement sub-base and capping layers (and possibly from gulley drainage pipes) are thought to be responsible for many of the slips.

57. Laboratory simulation of the softening process is difficult as it involves testing under very low stress conditions. Parameters derived from back analysis are considered the most reliable for application to remedial works. Back analysis of the failures and parametric studies have demonstrated appropriate parameters consistent with the typical slip depth of 1 to 1.5m and have indicated that the likelihood of deeper slips occurring is small.

58. Trials of various repair methods have demonstrated feasible alternatives to the usual excavation and replacement with imported granular material. The use of geogrid reinforcement layers appears to offer savings by enabling the softened slipped material to be reused.

59. Novel techiques such as the addition of simple low cost ground anchors may help to stabilise slopes at risk but further research and development work is required.

60. For the future construction of embankments the use of overconsolidated clays should be avoided where possible in slopes steeper than 1 in 3. Where they are used at steeper slopes the maintenance requirements must be recognised or measures included, such as geogrids, to maintain stability as softening occurs. Attention to details such as the selection of more stable soils for the embankment batters, the adequate compaction of the embankment shoulders and control of seepage waters will help to prevent the onset of instability.

ACKNOWLEDGEMENTS

61. The work described was carried out in the Eastern Region of the Roads and Local Transport Group of the Department of Transport and the Paper is published by permission of the Deputy Secretary, Roads and Local Transport. The Authors are grateful to all those who assisted with this work. The views expressed are those of the Authors and should not be attributed to the Department of Transport.

REFERENCES
1. CHANDLER R.J. and SKEMPTON A.W. The design of cutting slopes in stiff fissured clays. Geotechnique 1974,4, 457-466.
2. TRANSPORT AND ROAD RESEARCH LABORATORY. Studies of slope stability problems in highway engineering. TRRL Leaflet LF934.
3. FARRAR D.M. Settlement and pore-water pressure dissipation within an embankment built of London Clay. Clay fills, Institution of Civil Engineers, London, 1978, 101-106.
4. FARRAR D.M. Long-term changes in pore-water pressure within an embankment built of London Clay. Transport and Road Research Laboratory (in preparation).
5. ANDERSON M.G. and KNEALE P.E. Pore water pressure changes in a road embankment. Journal of the Institution of Highway Engineers, May 1980, 11-17.
6. FINLAYSON D.M. GREENWOOD J.R. COOPER C.G. and SIMONS N.E. Lessons to be learnt from an embankment failure. Proc Inst Civ Engrs, Part I 1984, 76, 207-220.
7. INGOLD T.S. Geotechnical aspects of pavement drainage. Journal of the Institution of Highway Engineers, November 1981 9-15.
8. DEPARTMENT OF TRANSPORT. Maintenance of highway earthworks. Departmental Advice Note HA26/83.
9. JOHNSON P.E. Construction costs of seven methods of embankment control. TRRL (in preparation).
10. CHARLES J.A. An appraisal of the influence of a curved failure envelope on slope stability. Geotechnique 1982, 4, 389-392.
11. HOLT D.A. MSc Thesis, University of Surrey (in preparation).
12. GREENWOOD J.R. A simple approach to slope stability. Ground Engineering 1983, 16, No.4 45-48.
13. DEVATA M. Geogrid reinforced earth embankments with steep side slopes Symposiumn on Polymer Grid Reinforcement, Institution of Civil Engineers March 1984.

7. Performance of embankments and cuttings in Gault Clay in Kent

C. GARRETT, BSc, MSc, DIC, FIHT, FICE, County Soils
and Materials Engineer, and J. H. WALE, BSc, MSc, MICE, FGS,
Principal Soils Engineer, Kent County Council

SYNOPSIS. The M26 Motorway, an adjacent section of the M25
Motorway and sections of the M20 Motorway in Kent run along
the outcrop of the Gault Clay which is notorious for its
swelling/shrinkage behaviour. In addition the near-surface
layers of the clay have been severely affected by periglacial
action forming relatively weak soliflucted and cryoturbated
zones. The paper describes various trials carried out for
the embankments and cuttings on these schemes, the resulting
earthworks designs, and their subsequent performance
particularly where failure occurred.

INTRODUCTION

1. In the South-East of England the Gault Clay outcrops
around the perimeter of the Weald as a gently sloping vale
between outcrops of more resistant strata, the Chalk of the
North and South Downs and the Lower Greensand Escarpment
(Fig 1). In Kent the outcrop of the Gault Clay adjacent
to the North Downs is typically 1 to 2 kilometres wide and
most of the outcrop is utilised as pastureland or woodland.

2. There are relatively few buildings and no villages built
directly onto the Gault Clay in Kent probably because of its
severe seasonal swelling/shrinkage behaviour and associated
changes in strength. Similarly the principal traffic routes
in Kent through the ages have avoided wherever possible the
Gault Clay outcrop and have been located on strata such as
the Lower Greensand and Chalk outcrops which provide better
all-weather surfaces. Subsequently centres of habitation
developed at suitable locations along these routes often
extending outwards to the start of the Gault Clay outcrop.

3. The Gault Clay outcrop has also been associated with a
number of major landslides in Kent such as that at Folkestone
Warren (ref. 1). In addition other areas of the outcrop,
particularly those adjacent to the Chalk escarpment, had
previously been identified as lobate areas of periglacial
origin with only marginal stability similar to those adjacent
to the Lower Greensand escarpment encountered along the route
of the A.21 Sevenoaks By-Pass (ref. 2).

4.   Selecting routes for the motorways to replace these earlier principal routes posed many problems because of the presence of such built-up areas on the more competent strata as well as topographical, geological and environmental features such as areas of outstanding natural beauty. As a result of these somewhat conflicting constraints the Gault Clay outcrop was selected as the corridor for several sections of the finally adopted motorway network in Kent which essentially consists of the M.25 Orbital route around London and the M.2 and M.20 motorways from London and the M.25 motorway to the English Channel ports via the Medway Towns and Canterbury and via Maidstone and Ashford respectively. The location of these motorways is shown in Fig 1.

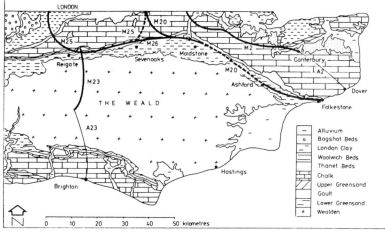

Fig. 1   Geology and Motorway Location plan

GEOLOGY

5.   The Gault Clay was deposited under marine conditions during the Lower Cretaceous period before being consolidated by the effects of several hundred metres of Chalk and later deposits.   Later the Wealden Anticline was eroded away, re-exposing the Gault Clay at the ground surface.   Subsequently in Kent frost action under periglacial conditions during the Pleistocene Devensian Glaciation (13,000 to 80,000 years before present) resulted in extensive disturbances of the upper layers of the Gault Clay by cryoturbation and solifluxion.   According to the Institute of Geological Sciences (ref. 3) the thickness of the Gault Clay stratum increases westwards from approximately 40 metres at Folkestone in East Kent to approximately 70 metres in West Kent between Maidstone and Sevenoaks and to over 80 metres at Reigate in Surrey.

6. In its undisturbed state Gault Clay is essentially a dark blue or greyish homogeneous heavy clay becoming very sandy and dark green as a result of the presence of glauconite at its base where phosphatic nodules, often large, are also present in layers. Detailed work on the lithology of the Gault Clay has established the existence of thirteen zones related to fossil ammonites, the Jukes-Browne classification (ref. 4), shown on Fig 2. Subsequently other classification systems for the Gault Clay have been developed based on microfossils. However, in general the near surface periglacial and other effects present are of much greater significance for engineering purposes and therefore the simplified sub-divisions given in Table 1 were adopted which are related to the physical structure and chemical weathering.

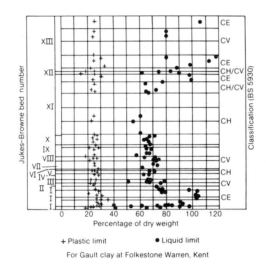

+ Plastic limit   ● Liquid limit

For Gault clay at Folkestone Warren, Kent

Fig. 2   Gault Clay - Lithology and Classification

7. The natural ground water regime in the Gault Clay is generally hydrostatic consistent with a water table between 1 and 2 metres below ground level depending upon the time of year. The groundwater table has generally been found to fluctuate rapidly as a result of the fissured structure. There are typically two principal sources of water affecting the water table, namely rainfall and springs at the base of the more permeable Chalk escarpment. At certain times and locations two or more water tables are present such as a perched water table in the near-surface soliflucted material with a different water table in the lower zones.

95

Table 1 - Sub-Divisions of Gault Clay

SOLIFLUXION DEPOSITS - Usually consist of remoulded Gault Clay with Chalk pellets, tufa nodules and coarse flint fragments, typically up to 150 mm in size. Generally 1m - 3m thickness with shear surfaces or zones, particularly at its base, approximately parallel with the ground surface.
CRYOTURBATED GAULT CLAY - This material contains a network of shear planes which are more randomly orientated and relatively steeply inclined. Fissuring is present at a spacing usually less than 25mm with generally irregular orientation. This layer is of variable thickness, being absent in some places, but seldom extending below about 6 metres from the ground surface.
WEATHERED GAULT CLAY - As a result of chemical weathering, superficial alteration has occurred resulting in the predominantly brownish grey colour. It is normally less stiff than Unweathered Gault Clay with irregular fissuring, typical spacing being 50-100mm. Typically extends down to 9 metres or so beneath ground level.
UNWEATHERED GAULT CLAY -Unweathered Gault Clay is essentially a stiff, fissured dark-grey plastic clay, the fissure spacing being typically 100mm. Some fissures display a high degree of polish and slickensiding but the majority are planar with a matt surface texture.

GEOTECHNICAL PROPERTIES

8. State of stress in-situ. As a result of its geological history the Gault Clay exists at present as a heavily over-consolidated clay whose near-surface layers have been significantly affected by periglacial action. During 1977 as part of the investigation for the proposed retaining walls at Dunton Green, near Sevenoaks (ref. 5) the lateral earth pressures were measured by a Cambridge Self-Boring Pressuremeter to a depth of approximately 15 metres below original ground level. The results of these tests indicated an existing coefficient of lateral earth pressure (ratio of effective horizontal pressure to the effective vertical pressure), K, increasing from approximately 2 at 15 metres depth to a maximum value of approximately 3.5 at about 3 metres depth (Fig 3)

9. Plasticity Indices and Moisture Content. As shown on Fig 2, the Plasticity Indices for the Gault Clay at Folkestone, and particularly the Liquid Limit and the Plasticity Index vary significantly throughout the stratum, with mean values of 75% and 47% respectively, whereas its Plastic Limit varies less markedly, being on average approximately 28%. These values correspond to a clay of very high plasticity (CV) whereas towards the top and bottom of the stratum in particular the Gault Clay is of extremely high plasticity (CE) with Liquid Limit and Plasticity Index typically in the range 100-120% and 70-90% respectively.

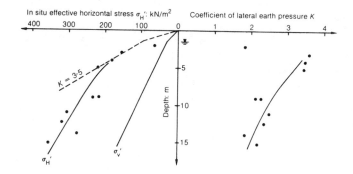

Fig. 3 Original state of stress in ground

10. The typical variation in Plasticity Indices with depth at
the Dunton Green site is shown on Figure 4 and these indicate
that the Gault Clay there exhibits higher than average Liquid
Limit and Plasticity Index. The associated moisture content
– depth profile shows the general decrease in moisture
content with depth to values close to the plastic limit in
the unweathered Gault Clay. In contrast, the moisture
content in the near-surface layers varies markedly with
rainfall and temperature throughout the year, resulting in
the swelling and shrinking behaviour for which Gault Clay is
notorious.

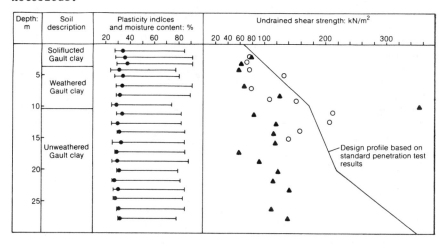

Fig. 4 Typical geotechnical profile

11. Undrained Shear Strength. The determination of the undrained shear strength of such a stiff fissured clay using conventional laboratory tests on 38 mm and 100 mm diameter specimens obtained by undisturbed sampling equipment produced a wide scatter of results and a poor correlation between strength and depth. In contrast a relatively good correlation was obtained between Standard Penetration Test (S.P.T.) resistance, N, and depth from site investigations for the M.25 and M.26 motorways. The undrained shear strength was obtained from these values using the correlation proposed by Stroud (ref. 6) which, for typical Plasticity Indices of the Gault Clay, becomes:-

Undrained shear strength, $Cu = 4.4 \times N \text{ kN/m}^2$

This resulted in the design undrained shear strength – depth profile shown on Fig 4 along with the values obtained for that site from laboratory tests on undisturbed samples and from in-situ tests by the Cambridge Self-Boring Pressuremeter.

12. Effective Shear Strength. The effective shear strength parameters were found to vary according to the "weathering" classification of the Gault Clay and from the results obtained from tests carried out on samples from sites along the recent motorways in Kent, the values given in Table 2 were deduced which were those normally used in design.

Table 2 – Design Effective Shear Strength Parameters

Sub-division of Gault Clay	Effective Cohesion, C' kN/m^2	Effective Angle of Shearing Resistance, ϕ'
Soliflucted	0	14°
Cryoturbated	0	14°
Weathered	13	24.5°
Unweathered	13	24.5°
Remoulded	10	23°

13. These parameters reflect the effect of the structure and particularly any discontinuities present in the clay, the parameters for the soliflucted and cryoturbated Gault Clay being residual shear strength values, whereas for the less affected weathered and unweathered zones, significantly higher values were adopted. Typical stress – strain relationships obtained from drained reversal shear box tests on Gault Clay from approximately 1.4 metres depth are shown on Fig 5 along with the resulting peak and residual effective shear strength parameters. These results demonstrate the large drop in shear strength from peak to residual, a brittleness index of approximately 70%.

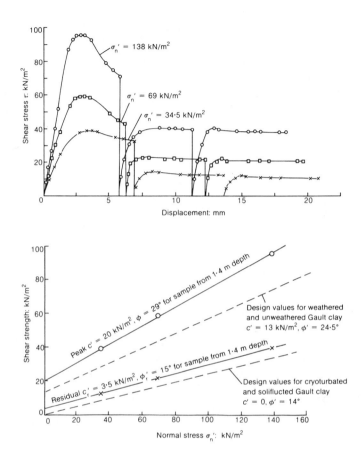

Fig. 5   Reversal shear box test results

M.20 MAIDSTONE BY-PASS

14. This dual two-lane highway was constructed during the late 1950's/ early 1960's. The section of the Maidstone By-Pass east of A.249 crosses an undulating section of the Gault Clay outcrop on a series of embankments and cuttings (Fig 6). Cuttings up to approximately 15 metres in depth were excavated generally at an inclination of 1 (vertical) in 3 (horizontal). In the deeper cuttings a 3 metre wide berm was provided at approximately 6 metres above the finished road level. Open ditches with unpaved inverts were excavated along the up-slope side of all cuttings. For this contract it was decided that all the Gault Clay obtained from the cuttings was unsuitable for re-use. Folkestone Sand from the adjacent section of the By-Pass to the west and other materials were imported for embankment construction.

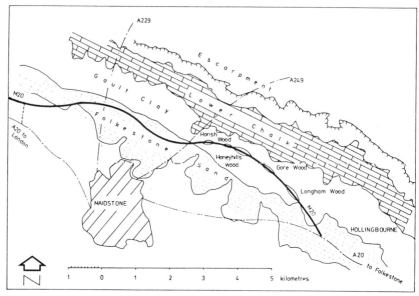

Fig.6   Plan of M20 Maidstone By Pass Showing Geology

15. The earthworks were completed in 1960 without any major
incident during construction. Three or four minor slips took
place before seeding of the slopes had started but no further
movements occurred after the cuttings were reinstated to the
original profile. In the winter of 1965, five years after
the end of construction, varying degrees of failure occurred
in nearly all the cuttings and particularly on the northern,
upslope, side of the motorway. In the subsequent years
movements both of a progressive and retrogressive nature have
continually taken place.

In all cases failure was observed to start near the toe of
the cutting and progress towards the rear of the section. The
history of failures and details of the remedial works and
trials carried out at two of the cuttings which typify the
problem and alternative solutions to it are summarised below.

16. <u>Longham Wood Cutting.</u> Longham Wood forms a very marked
spur extending across virtually the whole of the Gault Clay
outcrop. The axis of the spur is approximately perpendicular
to the line of the motorway and the natural ground slope was
between 1 and 2 degrees in this direction. The By-Pass cuts
through this spur in a cutting with a maximum depth on the
north side of 16.2 metres.

17. Failure was first observed at the toe of the northern
cutting face during the winter of 1966 and successive slips
subsequently occurred further up the slope giving rise to
step-like features before there was a widespread failure in
December 1968 involving the whole of the lower section of the

cutting and parts of the upper section above the berm.
Separate failures initially occurred in each section of
cutting above and below the berm in which an interceptor
drainage system had been installed. During the following 5
years no major movements occurred in the cutting although
minor movements were taking place continually in the slipped
mass and above it resulting in the open drain at the top of
the cutting slope being severed by 1969. The profile of the
slope became more regular and smoother with time, with a
corresponding lengthening of the zone around the toe where
the slopes of the slipped mass, extending on to the hard
shoulder of motorway, were significantly flatter than the
average slope of the cutting. The slipped material from the
upper part of the cutting slope by this time had over-run the
berm and was extending onto the lower part of the cutting.

18. The next major movement occurred during the 1973/74
winter after a period of above-average rainfall and the whole
cutting slope was affected over a length of approximately 300
metres. In this failure, the berm and its drainage system
were severed and the toe of the slipped mass advanced further
to the edge of the running lanes of the carriageway entirely
covering the hard shoulder (Fig 7).

19. As a result of the detailed investigation into the
failure of the cutting during 1968 and the subsequent
monitoring of the ground movement and watertables by
instrumentation remedial works were proposed during 1972, to
regrade the north cutting slope to 1 in 6 instead of the

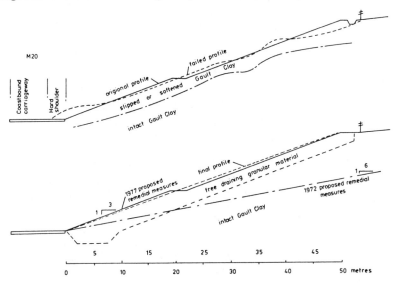

Figs. 7   Longham Wood Original and Failed Profiles
      8   Longham Wood — 1972 and 1977 Proposed Remedial Works

original slope of approximately 1 in 3 (see Fig.8). These remedial works were not carried out partly because of the greatly increased landtake which would have been required but particularly because such a slackening of the side slope would have removed trees in the wood at the top of the cutting which provided visual screening of the motorway.

20. After the 1973/74 failure of the slope additional investigations were undertaken with a view to stabilising the cutting slope within the original landtake, the back scar of the slip having by then receded to the boundary fence line. As a result in 1977 the solution shown on Fig.8 was proposed which essentially consisted of digging out the slipped and softened Gault Clay, excavating a toe dig-out zone extending approximately 3 metres below motorway level and restoring the slope to its original profile using free-draining granular material. These works were subsequently carried out during 1978 as proposed except that a continuous slope without a berm was adopted. These remedial works have continued to perform totally satisfactorily to date and during this period trials have been undertaken in order to develop appropriate vegetation cover for the slope without recourse to topsoil on the free draining material which is essentially 40mm beach shingle.

21. Horish Wood Cutting. The eastern section of the Maidstone By-Pass with the Gault Clay subgrade is now well over 20 years old and on several occasions during this period the carriageway has been improved or rehabilitated but several features of its design such as the hard shoulders are still substandard and also many slipped areas are present within the cuttings. In addition the road pavements have little residual life and therefore complete rehabilitation of the road pavements and cutting slopes is proposed for completion within the next few years. In order to accommodate carriageways to present standards, the width of the motorway at finished road level will have to be increased but because the landtake cannot be increased, the cutting slopes will have to be steepened further.

22. In order to determine the cost and effectiveness of different solutions to stabilising existing slipped areas and improving the stability of at present stable areas whilst, in both cases steepening the slopes, the cutting slope on the north side of Horish Wood cutting was stabilised during September and October 1983 using lime, dig-out/recompaction and counterfort drainage techniques. All of the works were carried out with the adjacent carriageway closed and with contra-flow traffic arrangements in operation on the other carriageway. During the relatively short time, approximately 1 year, since these trials were completed all sections of the cutting slopes have remained in a stable condition, the performance of the cutting slopes being monitored by

inclinometers and piezometers. For the conditions pertaining during the trial the construction costs of lime stabilisation method, the recompaction approach and the installation of counterfort drains were respectively approximately 0.50, 0.33 and 0.17 times the cost of the previously-utilised method of replacement with granular materials.

M.25/M.26 MOTORWAYS

23. Proposals for an orbital route of London south of the Thames have existed since at least 1937 when the Highway Development Survey of Greater London, known as the Bressey Plan, was published including an orbital route as a development required during the next 30 years. In 1966 the intention of establishing a south orbital road as an East-West motorway, M.25, from Staines via Sevenoaks to Wrotham in Kent was announced. It was from this date that the M.25 developed into its present form as a complete ring route around London with the spur from Sevenoaks to Wrotham being re-designated as the M.26 motorway.

24. During the conventional site investigations in the late 1960's the results obtained reinforced the previously expressed fears and uncertainties concerning the suitability of Gault Clay for use and re-use in the roadworks and particularly how it would behave under normal site handling conditions. For example, the Gault Clay was only marginally suitable for use as a fill material according to the then current Ministry of Transport Specification (ref. 7), certain facies of the Gault Clay with apparent silt size particles in reality showed the characteristics of a clay and two basic forms of slip surface existed in the Gault Clay, namely principal slip surfaces and non-continuous slip surfaces. A trial embankment and a trial cutting were therefore constructed during 1970 at a site near Otford north of Sevenoaks close to the junction of the M.26 Wrotham Spur with the M.25 London Orbital motorway.

25. <u>Otford Trial Embankment.</u> The site of the trial embankment was adjacent to that of the trial cutting, the material excavated from the latter being used for the former. In order to achieve the objectives of the trial it was necessary to design an embankment which would fail during construction whilst being similar to the proposed embankments along the M25/M26 route whose maximum height was 10.5 metres. The sideslopes of the embankment were determined from stability analyses for this height of embankment using the following assumptions:-

(i) principal slip surface present at approximately 4.5 metres below original ground surface on which residual effective shear strength parameters pertained, $c' = 0$, $\emptyset' = 14°$.

103

(ii)   pore pressure parameter B = 0.5 for construction
       pore water pressures.

(iii) for the section of the failure surface within the
       fill material peak effective shear strength
       parameters were operative, c' = 14 kN/m2, $\emptyset$' =
       23°.

For these predicted performance conditions, side slopes of 1
(vertical) in 2 (horizontal) were adopted resulting in a
minimum calculated factor of safety of 0.7 which was
considered suitable to ensure failure during construction.

26. Due to the limited amount of Gault Clay fill available it
was decided to use granular Head material from the excavation
site for the ramps at either end of the trial embankment,
restricting the use of the Gault Clay to the main 50 metre
length of the embankment which was considered long enough to
result in essentially two-dimensional stress and strain
conditions. Complete failure occurred along the whole length
of the embankment when the maximum height at the centre-line
had reached 10.65 metres, the 34 day construction period
including a 7½ day standstill during which no fill was placed
because of wet weather (Fig.9).

27. The first signs of instability were noticed five days
before complete failure when a bulge developed in the lower
section of the embankment and this became more marked during
the succeeding days. On the day before failure cracks were
observed along the lower section of the embankment (see
Fig.10), the uppermost of which increased in width from a
thin hair-line crack at 8 am to 60 mm wide 10 hours later. At
this time the area affected by ground heave extended for a
maximum distance of 9 metres beyond the toe of the embankment
with a maximum heave of 0.3 metres. By early the next

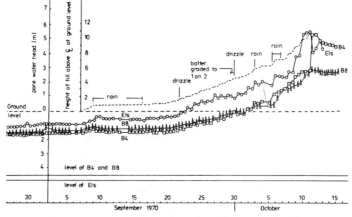

Fig.9   Pore pressures and Rate of Construction

morning, the day of the failure, the ground heave had
increased to 1.3 metres above original ground level but there
was no further extension of the affected area. Deterioration
of the embankment continued during the day and towards
evening the lower section of the slope consisted of a number
of displaced blocks bounded by cracks in the Gault Clay fill,
although at this time there were no signs of failure in the
top section of the embankment slope. Complete failure
occurred at 10 pm, the lower section of the slope having been
displaced as a series of wedges, the top section of the slope
then slumped behind the displaced wedges leaving a polished
and striated rear scarp.

28. Details of the behaviour of the embankment during its
construction and failure were recorded by conventional ground
survey methods and by high air entry piezometers installed in
the foundation strata. After the failure this information
was supplemented by two additional boreholes through the
slipped mass and a trial pit excavated at the toe of the
slipped mass. Particularly important findings of the failure
investigation were:

(a) the relatively high pore pressures below the lower
    section of the embankment which was consistent with the
    mode of failure commencing in the vicinity of the toe.
(b) the reduced strength of the Gault Clay from the slip
    zone, an undrained shear strength of only 17 kN/m2
    compared to typical values at that depth of 60 - 80
    kN/m2.

(c) the dissipation of excess pore water pressure that
    occurred overnight during construction particularly
    during the early stages.

(d) the rapid increase in pore water pressure which occurred
    near failure (Fig.9).

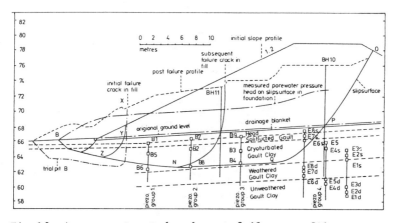

Fig.10 As - constructed and post-failure profile

105

Further details of the Otford Trial Embankment investigation
are given in Simons (ref. 8) and particularly Chinsman (ref.
9).

29. <u>M.25/M.26 Embankment Design.</u> From these results and
observations from the Otford Trial Embankment the concept of
"toe dig-out" was developed to increase the shear strength in
the vicinity of the toe of the embankment by destroying the
pre-existing slip surfaces in the zone in which the failure
path was likely to be located. Initially two alternatives
were considered, backfilling the toe dig-out zone with either
imported granular material or recompacted Gault Clay. The
re-use of Gault Clay was subsequently adopted as granular
material, whilst acting additionally as a means of reducing
the excess pore water pressures in the vicinity of the toe,
provided no higher factor of safety and was considerably more
costly. Further improvements in stability were effected by
flattening the side-slopes to 1 (vertical) in 4 (horizontal)
and the use of a drainage blanket beneath the shoulders of
the embankment in order to assist and control vertical
dissipation of excess pore water pressure. Details of the
embankment design and particularly toe dig-out zone adopted
for most of the embankments on the M.25 and M.26 motorways as
well as similar sections of the M.20 motorway are shown on
Fig.11.

30. Similar dig-out treatment was used at structures not only
for the final arrangement but also for intermediate stages
during the construction, the extent of the treatment
depending upon the height of the embankment, the inclination
of its side slopes and the depth to the interface between the
cryoturbated and weathered Gault Clay. At overbridge sites
significantly more extensive dig-out zones were required in
order to achieve the required Factors of Safety with the
embankment slopes steepened to 1 (vertical) to 1½

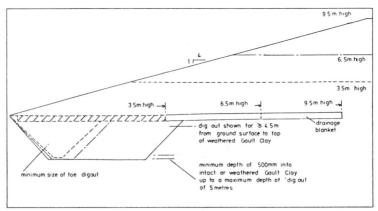

Fig. 11   Embankment Construction with toe dig-out

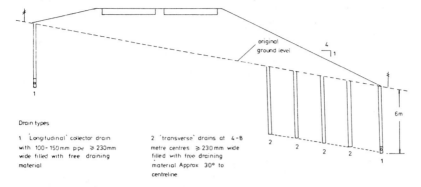

Drain types

1 'Longitudinal' collector drain
with 100-150mm pipe ≥ 230mm
wide filled with free draining
material

2 'transverse' drains at 4-8
metre centres ≥ 230 mm wide
filled with free draining
material Approx 30° to
centreline

Fig.12   Embankment Construction with Drainage Trenches

(horizontal) for reasons of bridge economy and/or aesthetics
in the granular backfill to such structures.

31. At two locations on the M.20 motorway the underlying
Gault Clay was landslipped and only marginally stable. The
dig-out approach was therefore considered to be inappropriate
and likely to reactivate the potential instability of these
areas. Instead, as shown on Fig.12, a system of trench
drains were used beneath the embankment and around its
perimeter linked to the drainage blanket.

32. All of the motorways using these designs of embankment
were completed successfully without any instability during
the period between 1976 and 1981 . To date on these
motorways with a total length of 9km on Gault Clay there has
only been one recorded embankment slope failure which was
relatively shallow and confined to the side-slope, resulting
from a concentrated water discharge within the top of the
slope.

33. Otford Trial Cutting. The layout of the site, shown on
Fig.13, essentially consisted of borrow area, up to
approximately 8 metres depth, constructed along the line and
approximately to the shape and level of the future motorway
cutting at this location, including:-

   (i)   a north cutting slope where the inclination varied
        continuously from 1 (vertical) in 2½ (horizontal) to
        approximately 2 (vertical) on 1 (horizontal).

  (ii)   a south cutting slope with an inclination of 1
        (vertical) in 5 (horizontal) which included trials
        of counterfort drains for the lower half of the
        cutting, the full depth of the cutting as well as a
        control section without counterfort drains.

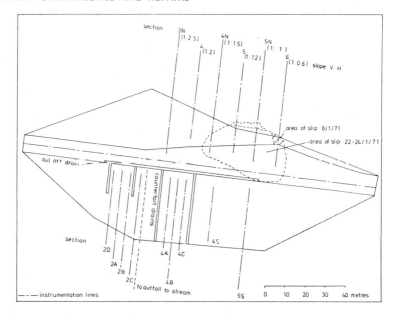

Fig.13 Layout of Otford Trial Cutting

In both cutting slopes piezometers and slope indicators were
installed in advance of the excavation and further
piezometers were installed after the excavation. The
objectives of the trial were not only to measure the
performance of the cuttings, with and without drainage both
in the short-term and longer term but also to obtain further
information concerning the in-situ permeability and strength
of the Gault Clay.

34. The trial cutting site at Otford was located towards the
centre of a moderate spur which runs in a NE–SW direction
with a natural ground slope of approximately 3.5° towards the
flood plain of a tributary of the River Darent. The original
groundwater table was between 2 and 3 metres below ground
level over the site, the groundwater flow being essentially
along the spur southwards towards the floodplain. The strata
present at the cutting site were approximately 2 metres of
topsoil and granular head material overlying 0.5 metres of
soliflucted Gault Clay, 1 metre of cryoturbated Gault Clay,
and 1 metre of weathered Gault Clay overlying the unweathered
Gault Clay at a depth of approximately 4.5 metres below
original ground level.

35. At the end of construction none of the cutting slopes
showed any signs of instability. The first slip, a minor
surface failure, occurred just to the east of Section 6, 25
days after the end of construction and after a period of
moderate rainfall. During another wet spell, 87 days after

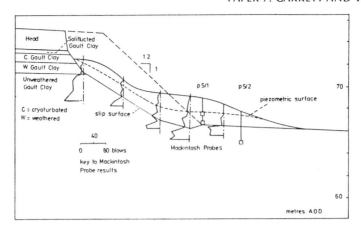

Fig.14 Original and post-failure profile - Section 5

the end of construction, a new larger slip occurred which was
centred on Section 6 where the initial slope was
approximately 1.7 (vertical) on 1 (horizontal). This was
followed 14 days later by a major slip which continued for
two days and which extended from Section 6 through to Section
4N where the original slope was 1 (vertical) in 1.5
(horizontal). The original and post failure profile at
Section 5 with a 1 (vertical) in 1.2 (horizontal) original
slope is shown on Fig.14. During the subsequent life of the
trial cutting until the construction of the motorway in this
area some years later no further major slips were recorded
although there was continued recession of the rear scarp and
minor movements in the slipped mass. Further details of this
work are given by Chinsman (ref. 9).

36. M.25/M.26 Cutting Slope Design. From analyses utilising
the results obtained from this trial cutting and from the
cuttings on the M.20 Maidstone By-Pass two alternative
cutting slope designs were developed namely:-

   (i)   1 (vertical) in 6 (horizontal) without counterfort
         drainage - utilised where there were no landtake
         constraints permitting possible return of the land
         to agriculture outside motorway boundaries.

   (ii) 1 (vertical) in 4 (horizontal) with 4 or 5 metre
         deep counterfort drains at 8 metre centres where the
         cutting depth greater than 3.5 metres. Counterforts
         were installed over the full cutting slope where the
         depth of the cutting exceeded 5 metres but only over
         the lower half of the cutting where its depth was
         between 3.5 metres and 5 metres.

37. All of the motorways using these designs for cutting slopes were completed successfully without any instability during the period between 1976 and 1981. To date on these motorways with a total length of 14 km on Gault Clay there have been no recorded cutting slope failures.

CONCLUSIONS

38. The results of the investigations in Kent during recent years into the properties and behaviour of Gault Clay were consistent with the previously held qualitive views concerning the difficulty and/or undesirability of using or re-using Gault Clay in highway construction. However by utilising appropriate investigations, full-scale trials and methods of analysis, technically and economically sound designs have been developed which have been validated in practice both during construction and subsequently in service.

ACKNOWLEDGEMENTS.

The authors would like to thank Mr. M.N.T. Cottell, the present County Surveyor of Kent, Mr. A.D.W. Smith, the previous County Surveyor, and Mr. P.M. Lee, the Director of the South Eastern Regional Office (S.E.R.O.) of the Department of Transport for their support in this work and allowing the paper to be published. The authors would also like to express their thanks to the numerous members of staff in the S.E.R.O. and in K.C.C., particularly those at the K.C.C. Highways Laboratory, who have contributed over the years to the work described.

The work described was carried out in the South East Region of the Roads and Local Transport Group (RLT) of the Department of Transport and is published by permission of the Deputy Secretary, RLT. The views expressed in this paper are those of the authors and do not necessarily represent those of the Department of Transport or of the Kent County Council.

REFERENCES

1. HUTCHINSON, J.N. (1969)
A reconsideration of coastal landslides at Folkestone Warren, Kent, in terms of effective stress, Geotechnique 19 No. 1.

2. SKEMPTON, A.W. and WEEKS A.G. (1976)
The Quaternary History of the Lower Greensand Escarpment and Weald Clay Vale near Sevenoaks, Kent. Phil.Tran. R. Soc. Series A. 283

3. INSTITUTE OF GEOLOGICAL SCIENCES (1965)
The Wealden District, H.M.S.O., London

4. JUKES-BROWNE, A.J. (1901) The Cretaceous rock of Great Britain - Gault and Upper Greensand. Memoir of the Geological Survey.

5. GARRETT, C. AND BARNES, S.J. (1984) The Design and Performance of the Dunton Green Retaining Wall, Geotechnique 34 No. 4.

6. STROUD, M.A. (1974) Standard Penetration Tests in Insensitive Clays and Soft Rocks, European Symposium on Penetration Testing, Swedish Geotechnical Society, Stockholm.

7. MINISTRY OF TRANSPORT (1969) Specification for Road and Bridge Works, 4th ed., London, H.M.S.O.

8. SIMONS, N.E. (1976) Field Studies of the Stability of Embankments on Clay Foundations in Laurits Bjerrum Memorial Volume - Contributions to Soil Mechanics, Norwegian Geotechnical Institute, Oslo.

9. CHINSMAN, B.W.E. (1972) The Influence of Geological Factors on the Engineering Properties of the Gault Clay in the South East of England, Ph.D. Thesis, University of Surrey.

Discussion on Papers 5–7

MR A. W. PARSONS and MR J. PERRY, Transport and Road Research
Laboratory, Crowthorne

In reviewing the geologies in which the greater number of
slope failures occurred (Table 1 of Paper 5) we should like to
add two further materials from later surveys which, in
embankments, showed significant rates of failure: these are
Weald Clay and Lower Lias (Table 1A).

Table 1A

Geology	Age of earthworks when surveyed (years)	Total length (km)	Failure rate (%)	Predominant slope (v:h)
Embankments				
Weald Clay	9	12.0	1.6	1:2.5
Lower Lias	13	28.0	1.1	1:2

Further details of surveys of earthwork side slopes in these
materials can be found in the Technical Note entitled
'Incidence of highway slope stability problems in Lower Lias
and Weald Clay' by J. Perry.

MR P. S. GODWIN, West Yorkshire Metropolitan County Council,
Ossett

With regard to Paper 6 the proposal for a minimum slope of 1:3
is viewed with concern in terms of landtake. Simple
calculations indicate that about 100–300 hectares of
additional land would have been required for the construction
of the M1 motorway between Sheffield and Leeds. This

Failures in earthworks. Thomas Telford Ltd, London, 1985   113

represents one or two farms. This aspect must also be considered by the responsible design engineer. Each case should be judged on its merits and the material design modified appropriately. The remedial and preventative techniques illustrated in Fig. 7 of the paper are equally applicable to cuttings. For embankments, there are two other methods which could be used to overcome some of the problems described in the paper. These are composite construction and horizontal drainage layers. Are the published cost comparisons a fair basis for assessing the best form of remedial works? Often slippage must be quickly repaired by direct labour maintenance gangs. The chosen materials must be readily available locally, and the method of working is often influenced by access. Granular replacement and deep counterforts can usually be installed easily and quickly. Shallow-seated embankment failures in Yorkshire have usually fallen into one of three categories: sub-base seepage, tree planting or a pocket of particularly water-susceptible fill.

MR J. R. GREENWOOD, MR D. A. HOLT and MR G. W. HERRICK

You are rightly concerned at the additional landtake necessary for 1:3 embankment slopes. However, flatter side slopes are only one of the various options suggested in paragraph 60 for reducing the risk of failure in overconsolidated clay embankments. The design engineer must determine the most suitable design for each particular location taking due regard for the risks involved and the consequence of failure.

The remedial and preventative techniques discussed are in general also applicable to cutting failures although in cuttings more attention is often required to the interception of groundwater behind the slipped mass.

Reducing the instability risk by composite embankment construction (i.e. more stable soil on batters) was mentioned in paragraph 60. Both composite construction and intermediate horizontal drainage layers have been used on recent schemes in the Eastern Region. It is debatable whether intermediate drainage layers will help and they may contribute to the problem by concentrating water on the slope as the extended sub-base does, particularly if their permeability is not sufficiently high. The inclusion of geogrid layers would seem a cheaper, more effective way of maintaining 1:2 side slopes.

MR M. P. O'REILLY, Transport and Road Research Laboratory, Crowthorne

In Table 1 I present cost information on the treatments described in paragraphs 44-54 of Paper 6, and which is to be published by the Transport and Road Research Laboratory (ref. 1). These costs are specific to the site on the A45 near Cambridge.

Table 1. Embankment treatment costs

Technique	Time taken (days)	Total cost (£)
Reinstatement methods		
Gabion wall	18	8360
Granular replacement	5	5020
Anchored tyre wall	8	4760
Lime stabilization	7	4730
Geogrid containment	6	3430
Preventative methods		
Geogrid plus anchors	7	3430
Rock ribs	6	2160

In the period since the completion of the trial sections, none of the reinstatement or strengthening methods have shown any signs of movement. However, it is still early and the success of the seven methods depends on their long-term performance. Long-term monitoring will continue by periodic visual inspection and by measurements between reference pegs driven into the trial sections and datum points beyond the toe of the slope.

The success of granular replacement does show that zoning of embankment fill with the better materials in the outer zones on the slopes of the embankment and the less stable material in the core would produce stable embankments since any 'failures' of the latter materials would be contained. Whether such zoning would be practical is another question: for example, would it really be possible to use chalk fill on the outer slopes of embankments with a Gault Clay core?

The cheaper methods of slope treatment require greater engineering input: a similar point was made by Professor McGown in the opening paper to this symposium.

The papers we are discussing highlight the extreme importance of symposia such as this to ensure that information on what is being done and more importantly on what has been done successfully is readily available to our profession. The Otford Trial Embankment (paragraphs 25-28 of Paper 7) carried out in 1970 is a particular example of this where the results have received scant attention until now.

Finally could we have an indication of the cost of repair to the 40 slips in Cambridgeshire during the period of 12 months mentioned in paragraph 6 of Paper 6?

Reference
1. JOHNSON, P. E. Construction costs of seven methods of embankment control. Transport and Road Research Laboratory.

MR J. R. GREENWOOD, MR D. A. HOLT and MR G. W. HERRICK

The cost of the A45 trials, as you have noted, are specific to
that particular location and the methods of working. For
example much of the cost of the 'geogrid plus anchors' method
was in stripping topsoil and importing new topsoil. We agree
with Mr Godwin that the local conditions and factors
appropriate to each site must be carefully assessed in
determining the most economical repair technique. Any
standard repair method should be reviewed periodically so that
new technology can be assessed and possibly incorporated.

The exact costs of slip repairs are not readily available
but it is likely that the 40 slips in Cambridgeshire during
the 12 month period would have cost about £300 000 to repair.

MR C. GARRETT and MR J. H. WALE

Mr O'Reilly you comment that the results from the Otford Trial
Embankment which was constructed in 1970 have received scant
attention until now and you have stressed the importance of
symposia in ensuring the dissemination of such information to
the profession. While we agree with the general point that
such valuable information should be published early and
widely, as indicated in our paper some of the information in
that particular case had been published previously, both in
general terms and in some detail, albeit in publications of
somewhat limited accessibility and circulation.

The main question is how is such work to be organized and
funded? Typically at present no funds are available from the
Client, the Engineer or the Contractor. In addition the staff
associated with the work are often moved immediately on to
other new funded work resulting in loss of continuity with the
data or even the effective loss of the data themselves.
Conversely at present there are a wide range of separately
funded/organized research initiatives in search of projects or
information. It is suggested that we should reinvest part of
the overall cost of construction projects in achieving this
objective of obtaining rapid and reliable feedback, the aim
being to advance our design procedures including the provision
of value for money by whole-life costing. The formation of a
British Geotechnical Institute would be appropriate to manage
this work, including co-ordinating the various types of
academic and commercial bodies involved in it as well as
arranging appropriate monetary funding and staff input into
suitable projects.

MR C. GARRETT, Kent County Council, Aylesford

The subject of this symposium is 'Failures in earthworks' but
the underlying object must be to utilize failures to the
greatest benefit whenever they occur. In this connection it

is useful to consider what failures are and the reasons why they occur.

The term failure is normally associated with large-scale structural collapses which make the news headlines but there are other forms of failure. The three principal types are

STRUCTURAL   FUNCTIONAL   ECONOMIC

Essentially the structural failure results in partial or complete collapse of the construction, damaging it to prevent its use. A functional failure occurs when a construction cannot carry out the function for which it was built and this may or may not be accompanied by structural damage. When the cost for using a construction partially or totally restricts its use an economic failure has occurred and this may or may not be associated with structural and/or functional failures.

Similarly failures can arise from a number of causes which may operate singly or in association. Some of the principal types are

CONCEPTUAL   DESIGN   COMMUNICATION

A conceptual failure occurs when the basic concept of what is constructed is at fault as opposed to faults in the actual design of what would otherwise have been a successful project. Either of these can also involve communication failures in which essential vital knowledge obtained in the past or available elsewhere was not taken into account.

The term 'failure' is generally equated with adverse effects such as disrupted construction, additional costs and adverse publicity. However, even such an unintentional failure can have a beneficial effect if a thorough post-mortem investigation is carried out and lessons are learned from it both in terms of procedures and design methods. Furthermore failures can be used intentionally from the outset either in the form of a trial to optimize the design of the project and/or in the form of a percentage of the works which are knowingly designed to fail to optimize the cost of the project, total exclusion of failures not necessarily equating with optimum short-term and long-term costs of the project. From this active use of failures, positive benefit can be gained both by the project concerned and by feedback into future designs and design methods. The full process of utilizing failures is summarized in Fig. 1.

The work reported in the paper concerning failures in earthworks in Kent involving Gault Clay illustrates the positive use of information obtained from both unintentional and intentional failures to assure quality and to optimize the design of the earthworks for recent motorways in Kent constructed on the Gault Clay outcrop.

In this connection, on a more general note, the geotechnical engineer appears to be viewed by clients and by our colleagues in other areas of civil engineering as some form of rescue

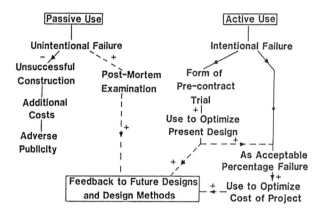

Fig. 1

service which is only called in when failure is imminent or
has already occurred. Previously I have defined successful
geotechnical engineering as 'the art of dealing in a safe,
scientific and economic manner with situations where the
factor of ignorance is greater than the factor of safety'. As
a 'factor of safety' is only a euphemism for a 'risk of
failure' the non-use or abuse of geotechnical engineering is
likely to lead to increased numbers of the more obvious
structural and functional failures and the less obvious
economic failures: total safety is not synonymous with optimum
cost. Additionally as Peck pointed out 'the Act of God of
today is the criminal negligence of tomorrow'. A question
that I think we should ask individually and corporately is
'are we guilty of a form of failure - a failure to communicate
adequately these and other benefits of geotechnical
engineering?'

MR D. H. BARKER, Geostructures Consulting, Edenbridge

Direct shear tests of soils with well-established deep-rooted
vegetation, such as certain grasses, shrubs and trees, have
shown that the root matrix can increase significantly the
strength of saturated soils. There are two further positive
effects due to plant transpiration: increased soil suction and
a reduction in moisture content within the root zone which can
extend to a depth of 1-1.5 m - the depth of the majority of
the shallow translational failures discussed in Papers 5 and
6. It is possible through the use of models established by
various workers, such as Wu et al., for forested slopes (which
can be modified for other types of vegetation such as grasses)
to quantify enhanced factors of safety of a slope against

shallow translational failures, due to root reinforcement etc., of the order of 10-15% at least.

Since such small increases in soil strength and resulting factors of safety can often be sufficient to prevent the onset of downslope movement, I suggest that the beneficial effect of vegetation on slope stability merits consideration by engineers. Would the authors of Paper 6 in particular care to comment?

Finally, I have another definition of a geotechnical engineer to add to that given by Mr Garrett: all too often he is someone brought in at the last moment to share the blame!

MR J. R. GREENWOOD, MR D. A. HOLT and MR G. W. HERRICK

If suitable plants are available which can develop the necessary root system while remaining manageable and environmentally acceptable, trials should be carried out without delay. It is an area where close liaison between the geotechnical engineer and the landscape/horticultural experts is necessary to explore the possibilities that nature has to offer.

PROFESSOR C. P. WROTH, University of Oxford

For problems of long-term stability, the 'drained' strength of the soil is represented by the parmeters c' and ϕ' interpreted from the classical Mohr-Coulomb failure envelope. It is well known that in a simple limit analysis of the long-term stability of a slope or cutting

(i)   the value of c' plays a crucial role in the calculation of the factor of safety
(ii) it is difficult to select an appropriate value of c'.

What should a designer do?

To meet this challenge, it is suggested that a more sophisticated interpretation of standard tests can provide a rational method for selection of c' and ϕ'; it is illustrated by a case history of an embankment for a major road constructed of London Clay.

It has been argued by Schofield and Wroth (ref. 1) that the assumption of a single Mohr-Coulomb failure envelope for a clay is erroneous, and that the proper way to take account of the effect of water content and stress history on drained shear strength is to use the interpretation introduced by Hvorslev, as early as 1937, in his doctoral thesis. The essential feature is that the cohesion intercept c' is not a constant but an exponential function of the water content.

The principles are illustrated in Fig. 1 where the Hvorslev failure criterion is expressed in terms of the triaxial variables (p_f', q_f', w_f) and forms a ruled surface consisting

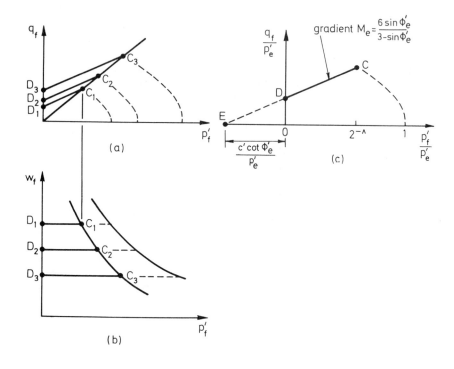

Fig. 1. Representation of Hvorslev failure surface in effective stress and consolidation spaces

of parallel lines C_1D_1, C_2D_2 etc., with the points C_1, C_2 ... tracing out the critical state line. By dividing by the equivalent pressure, the set of lines in Fig. 1(a) reduces to the single line CD, having the gradient and intercept OE on the abscissa as shown in Fig. 1(c).

For the embankment in question, a number of samples of the London Clay were taken. Each sample of 100 mm diameter was trimmed to form three triaxial specimens 38 mm in diameter. For each trio (having an identical geological history) the three specimens were reconsolidated under different prescribed cell pressures and then subjected to conventional drained triaxial compression tests. For each trio a failure envelope was obtained which closely defined a pair of drained strength parameters (c', ϕ'). However, there was considerable variation in the 14 sets of data, ϕ' varying between 16.2° and 25.0° with an average of 21.2°, and c' varying dramatically from 6.9 kN/m^2 to 56.2 kN/m^2 with an average of 30.8 kN/m^2. This large range of c' presents a major problem to the designer. It must be emphasized that this standard interpretation only takes account of the failure stresses and neglects all data of water contents at failure.

For each specimen, the value of w_f was known, so that given the position of the isotropic normal consolidation line for

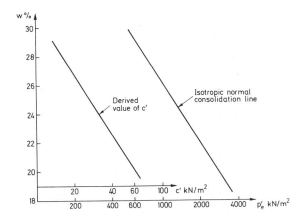

Fig. 2. Cohesion values and isotropic normal consolidation of London Clay

the clay this could be converted to the equivalent pressure $p_e{}'$. The best fit to the data available, making use of the comprehensive series of tests on London Clay reported by Henkel (ref. 2) gives the isotropic normal consolidation as that drawn in Fig. 2.

Adopting this line, the relevant values of $p_e{}'$ have been estimated and used to obtain the results plotted in the normalized stress space of Hvorslev in Fig. 3. It is notable that the data lie close to the line CDE, calculated to be the best-fit straight line by the standard method of least-squares regression. This line has a gradient of 0.815 and intercept OE of −0.0573, which gives $\phi' = 21°$ and $c' = 0.022p_e{}'$. This latter relationship can be converted to give c' as a function of water content by calculation, or by plotting in Fig. 2 (note that the scale for c' is different from that chosen for $p_e{}'$).

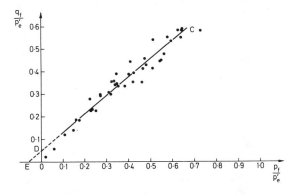

Fig. 3. Normalized Hvorslev failure envelope for London Clay

In Fig. 3, if the Hvorslev relationship had not been used, the data suggest that the failure envelope would be slightly curved to pass through the origin and would not have an intercept on either axis.

The interpretation of the data by Hvorslev's method provides a rational explanation for the large range of c' obtained from the 14 trios of Mohr's circles. Moreover it leads to a relationship between c' and w, so that the designer has available a unique prediction of drained shear strength in terms of w and p'. The relationship can also be used directly to provide a rational basis for controlling the quality of fill material by monitoring water content.

References
1. SCHOFIELD, A. N. and WROTH, C. P. Critical state soil mechanics. McGraw-Hill, London, 1968.
2. HENKEL, D. J. The effect of overconsolidation on the behaviour of clays during shear. Géotechnique, 1956, vol. 6, no. 4, 139–150.

MR J. R. GREENWOOD, MR D. A. HOLT and MR G. W. HERRICK

Professor Wroth's development of the Hvorslev method presents a valuable design aid by relating drained shear strength to the moisture content and effective stress conditions. This approach has been used by consulting engineers in the Eastern Region to assess the stability during construction of large embankments on London Clay foundations.

MR C. GARRETT and MR J. H. WALE

Professor Wroth's contribution concerning the interpretation of shear strength properties in terms of c' and ϕ' is particularly relevant and important. As he rightly states, the value of c' plays a crucial role in the calculation of the factor of safety of the long-term stability of a slope or cutting and that the typical range of the value of c' obtained from laboratory testing presents a major problem to the designer. For example, he cites values ranging from approximately 7 kN/m^2 to 56 kN/m^2 obtained by the standard interpretation for a particular series of tests on a sample of London Clay. In other areas of geotechnical engineering the c'-factor of safety relationship is even more extreme than for slope stability and hence accurate estimation of appropriate values of c' are even more important. For example, the factor of safety of a cantilever bored pile wall with 5 m exposed height and 10 m penetration apparently changes by a factor of 2, from 0.75 to 1.50 with a change in c' from zero to 10 kN/m^2 although in reality there has been no change in the risk of failure of the wall, just a different interpretation of shear strength parameters.

MR P. C. HORNER, Consulting Engineering Geologist, Luton

On the M25 Godstone to Westerham section, major shallow
(typically about 2 m deep) slips occurred in two cuttings,
beneath a temporary diversion and beneath one of the topsoil
stockpiles. They were believed to have taken place along
shear planes within the soliflucted or cryoturbated Gault Clay
(Figs 1 and 2).

Fig. 1

Fig. 2

The information in the tender document provided a clear
indication to a geotechnical specialist or those with previous
experience of solifluction-induced failures that stability
problems were likely, but may not have been so obvious to
others. The potential instability was taken into account in
the temporary works: for example 1:2 battens were adopted in
bridge excavations compared with the more usual 1:1. The
possibility of slips in the permanent work was identified and
a similar view was held by the designers.
The remedial measures for the first area of slips consisted
largely of replacing the slipped material and the slip planes
with suitable Gault Clay and predictably failures reoccurred.

Subsequent remedial works included the replacement of the
slipped material with suitable clay placed as a basal drainage
blanket, together with the emplacement of an intercepter drain
3 m deep at the top of the cutting.

At the time many of the estimators, planners and site staff
had not heard of solifluction, cryoturbation etc. and
therefore a significant part of the in-house geotechnical
engineer's role was essentially educational. The importance
of the communication of geotechnical information and
developments between geotechnical specialists and non-
specialists cannot be underestimated.

MR J. R. GREENWOOD, MR D. A. HOLT and MR G. W. HERRICK

The theme of good communication has been brought out both by
Mr O'Reilly and by you. Design and construction failures
inevitably result from communications failures. It is the
responsibility of individuals, encouraged by the profession,
to communicate their experiences, both good and bad, to
colleagues. This symposium has played a major role in
reporting and developing discussion on 'failures' and we are
grateful for the opportunity to participate. It is essential
for the profession that free exchange of technical information
and experience can continue without unnecessary restriction.

MR C. GARRETT and MR J. H. WALE

Mr Godwin, Mr Barker, Mr O'Reilly and Mr Horner all give
useful examples of alternative methods of ensuring the
stability of slopes of embankments and cuttings during their
original construction or during remedial works to failures.
This is an area in which generalizations are difficult and
potentially dangerous as each case should be considered on its
own merits: a successful solution related to a certain problem
in a particular engineering/geotechnical/geological and
geographical environment can be a technical or an economic
failure when used in other circumstances.

Concerning the initial design of embankments Mr Godwin and
Mr O'Reilly both mentioned the use of composite or zoned
construction as methods of reducing slope stability problems.
In Kent typical examples used in recent years have been

(i)   zoned Gault Clay embankments with high liquid
      limit/plasticity index clay in the core surrounded by
      lower liquid limit/plasticity index clay in the
      shoulders and beneath formation level
(ii) composite embankments with a clay core surrounded by
      granular material in the shoulders and beneath formation
      level.

Suffice it here to say that such solutions have not been
without their own particular technical and financial problems.

124

The adoption of such solutions is very dependent on the
availability of suitable material for this purpose, the
logistics of using it, possibly the disposal of additional
less suitable material and thus the overall relative cost-
benefit of the proposal. In this context Mr Godwin also
mentions the use of horizontal drainage layers. These or
similar zones of more permeable material are also not a
universal solution and can have little or even an adverse
effect. For example, their use adjacent to moisture-hungry
overconsolidated clays typically results in detrimental
effects in the short term due to expediting the softening and
swelling of the adjacent clay with associated loss of
undrained shear strength/rutting problems etc.

Similarly it must be stressed that a full investigation of
all failures should be carried out before the execution of
remedial works to ensure that the cause is dealt with and not
just the final outward manifestations of it. For example, as
Mr Godwin indicates, the cause of earthworks failures is often
the unsatisfactory provision made for groundwater drainage
(e.g. sub-base seepage) or for surface water (e.g. over-the-
edge drainage) and hence the remedial works must adequately
deal with that aspect, not just the earthworks, to be
successful. In connection with the M25 Godstone to Westerham
section in Surrey, Mr Horner mentioned the unsuccessful
stabilization even in the short term of slipped cutting slopes
using suitable Gault Clay, an experience which apparently
conflicts with the successful use of this technique in Kent in
similar circumstances albeit combined with some attention to
drainage details. The relative initial costs of these various
forms of remedial works in this trial in Kent and its
performance to date indicate that in certain circumstances
this could be the most appropriate option:

Ballast replacement	£19/m^3
Lime stabilization	£8.50/m^3
Excavate out and recompact	£6.50/m^3
Install counterfort drains	£2.20/m^3

It must be reiterated that no assessment can yet be made of
the long-term relative costs of these forms of remedial works
for cuttings because all the sections have remained similarly
stable to date during the first 18 months' life of the trial.

Similarly Mr Barker proposes the use of vegetation to effect
improvements in the stability of slopes whereas Mr Godwin
cites the apparently contradictory experience of tree planting
being one of the causes of shallow-seated embankment failures.
It appears that the vegetation solution is more appropriate to
the initial design stage/long-term solution than to the short-
term condition and the repair of failed slopes. Of particular
value in this context is vegetation with tap-roots which in
addition to effecting groundwater level reduction in a similar
manner to well pointing also provides a natural form of ground
reinforcement in a form similar to ground anchors. This is

again an area where particular attention to detail in the design and the execution of the works is necessary if the correct result rather than the opposite adverse result is to be obtained.

8. Cracking of a PFA embankment over soft alluvium

N. R. ARBER, BEng, PhD, MICE, Senior Geotechnical
Engineer, Travers Morgan & Partners

SYNOPSIS.   The construction of a road embankment over soft
alluvial clay is described.   After the completion of the
first of the two construction lifts cracks were observed on
the surface of the pulverised fuel ash fill.   The cracking
was attributed to differential movements caused by a large
variation in the depth of the alluvial clay across the
width of the embankment.   A crushed limestone rockfill
was used to complete the embankment.

INTRODUCTION
    1. The 11km long Stage 1 of the new A55 North Wales
Coast Road runs from Llanddulas east of Colwyn Bay to Glan
Conwy near the River Conwy estuary (Fig. 1).   Construction
began in August 1981 and is due for completion early in 1985.
The alignment of the new A55 over the final 2km takes it on
an embankment along the floor of the Afon Ganol valley.
This valley at one time formed part of the River Conwy estuary.
    2. The embankment had to be constructed over alluvium
consisting of very soft to soft silty clay, varying in
depth up to 15m, which contained 0.5m to 4.0m layers of
peat.   For most of its length the embankment was supported
by a piled raft where the peat layers were found within the
alluvium.   At Glan Conwy, however, the alluvium was at its
deepest and there were no clearly defined layers of peat.
Here, the embankment incorporated two slip roads to a low-
level rotary interchange and was constructed directly onto
the alluvium through which band drains had been installed
to promote primary consolidation.
    3. The alluvium at Glan Conwy was instrumented with
hydraulic piezometers, inclinometers and hydrostatic profile
gauges.   This instrumentation was installed to provide a
check on the performance of the band drain system and to
monitor the effect of embankment construction on the alluvium.
In this respect the embankment was considered a form of trial
from which information would be used in the design of a
similar embankment to be constructed later as part of the
A55 Stage 2.

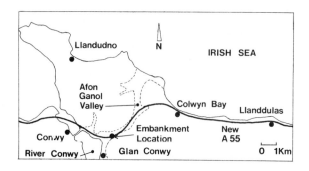

Fig.1   Site Location.

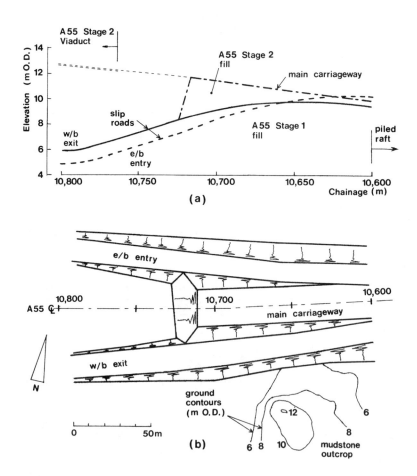

Fig.2   Embankment (a) profiles and (b) plan.

4. Shortly after the start of the consolidation period
for the first of the two embankment construction lifts some
fine cracks were observed at the surface of the pulverised
fuel ash (PFA) fill along one side of the embankment.
The cracking was attributed to differential movements
caused by a large variation in the depth of alluvium across
the width of the embankment.

5. A brief description of the ground conditions along
the Afon Ganol valley and of the response of the instrumentation at Glan Conwy to the construction of the embankment
first lift has previously been given (ref.1). This paper
describes the properties of the alluvium at Glan Conwy
and the response of the instrumentation in more detail.
It also describes and discusses the cracking of the embankment and the subsequent remedial measures made to complete
the embankment.

GROUND CONDITIONS

6. The alluvium overlies sand and gravel and boulder
clay which rests on mudstone bedrock. Ground investigation
boreholes usually only penetrated to the boulder clay
which was intercepted at depths varying up to 30m. The
mudstone outcropped at ch.10,625 (Fig. 2b) just beyond the
southern toe of the embankment.

7. The ground level was generally 3.8m OD with the
groundwater table 0.5m to 1.0m below this. The alluvium
was found to increase in depth below ground level from 0m
to 12m across the embankment width with the depth increasing
with distance from the mudstone outcrop.

8. Consecutive undisturbed sampling of the alluvium in
four boreholes at ch.10,710 indicated three layers. The
upper layer, which included a desiccated crust, was a soft
grey silty clay and contained vertical plant stems up to 2mm
diameter and flecks of peat. The decomposition of the plant
stems varied greatly. The middle layer was a soft grey
silty clay with fine sand partings and occasional plant stems.
It was not clear whether the sand partings were continuous.
The lower layer was very similar to the upper layer, but had
a higher peat content.

9. Atterberg limit and bulk density profiles from one of
the four boreholes clearly showed the presence of the layers
within the alluvium (ref. 1) and these are reproduced in
this paper as Fig.3a and Fig.3b. Due to the variability
of the alluvium, it was not possible to ascertain whether
these layers were continuous over the area of the embankment.
Composite Atterberg limit and bulk density profiles for
all the boreholes in the area indicated the same general
trends as those shown in Fig.3a and Fig.3b but there was a
wide scatter of results.

10. The variation of preconsolidation pressure (p_c') with
depth is shown in Fig.3c along with the in-situ effective
overburden pressure (p_o'). Below 4m depth the over-
consolidation ratio (OCR), p_c'/p_o', lay in the range 1.2 to

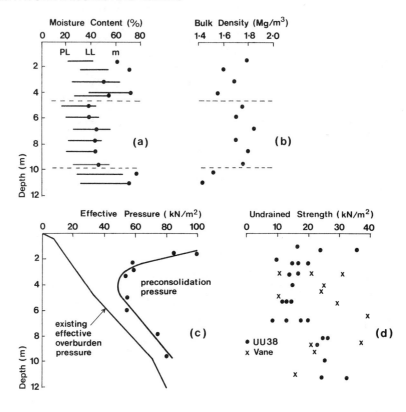

Fig.3   Alluvium (a)Atterberg limits, (b)bulk densities,
         (c)effective overburden and preconsolidation
         pressures, (d)undrained strengths.

1.4 and decreased with depth.   If the depth to the minimum
undrained shear strength was taken as the lowest previous
groundwater level (ref.2) then an OCR of approximately 1.5
was indicated.   This was higher than the measured OCR and
it may be that the oedometer had underestimated the field
p_c' or, alternatively, the over-consolidation below 4m depth
was produced primarily by secondary compression.

   11.   The undrained shear strength (c_u) of the alluvium was
measured by in-situ vane testing and unconsolidated undrained
triaxial tests on 38mm diameter samples.   There was a wide
scatter of results (Fig.3d).   The general trend of the
laboratory results was for c_u to decrease from an average
of 20kN/m^2 just below the desiccated crust to 15kN/m^2 at
3.5m depth.   It then increased uniformly to 28kN/m^2 at 10m
depth.

   12.   Constant head in-situ permeability tests were carried
out at various locations and depths within the alluvium.

The permeabilities from these tests, when combined with
laboratory measured values of the coefficient of volume
compressibility (m_v), indicated a wide variation in the
horizontal coefficient of consolidation (c_{vh}) which generally
decreased with increasing effective stress.   The range of
results varied from $1.5m^2$/yr to $30.0m^2$/yr. Triaxial dissipation tests were also carried out and these indicated a
lower bound c_{vh} of $1.2m^2$/yr and this value was used to
determine the band drain spacing.

EMBANKMENT CONSTRUCTION
Fill material
     13.   There was not sufficient suitable fill available ex-
site and it was decided at an early stage in the design
of the road to import PFA.   This could readily be achieved
by railway from storage lagoons at Fiddlers Ferry Power station
near Warrington.   It also reduced the environmental impact
of importing large quantities of fill onto site by road.
Advantage could also be taken in embankment design of the
low bulk density of PFA compared to normal fill.
Construction sequence
     14.   It was originally intended to construct the embank-
ment to slip road level (Fig.2a) with the remainder of the
fill to the main carriageway placed at a later date during
the A55 Stage 2 construction when a viaduct or embankment to
carry the road over the low-level rotary interchange was to be
constructed.   The embankment was to be placed in two construction lifts.   The first lift was to 8m OD at the main
carriageway centre line with 1:40 crossfalls to the embankment edge.   This lift included a 1m thick drainage blanket
placed directly on the original ground surface.   A twelve month
minimum consolidation period was then to follow to allow
90% primary consolidation.   The second lift was then to
follow with the PFA fill placed to an overfilled profile
to allow for settlement resulting in an embankment height
at the edge of 5.6m above original ground level.   A $10kN/m^2$
surcharge was then to be applied to the overfilled slip
roads and the embankment left for a further twelve months
minimum, after which the surcharge was to be removed and
the pavement laid to the slip roads.   The second lift,
however, was modified as will be described later.
     15.   Fill placement.   Placement of the first lift PFA
began during early October 1982 and was substantially completed by late November 1982.   A small amount of PFA was
placed in mid-December 1982 to complete the lift at ch.10,600.
The PFA was delivered to site in two train loads per day
Monday to Friday, each train load bringing in approximately
900 tonnes.   It was then transported from a specially built
discharge structure to the embankment by dump trucks, spread
and compacted by vibrating roller in accordance with clause
608 of the Department of Transport 'Specification for Road
and Bridge Works' (ref.3).   The measured compaction moisture
contents and densities are summarised in Table 1.   An average

relative compaction of 95% was achieved with the range of
results varying from 93% to 96%

Table 1.   PFA compaction moisture contents and densities

	Field Moisture Content (%)	Field Dry Density (kg/m³)	Field Bulk Density (kg/m³)	BS 1377 Compaction Test 12 (2.5 kg)	
				OMC (%)	MDD (kg/m³)
Range	24–29	1165–1280	1450–1640	21–26	1160–1380
Average	26.3	1240	1555	23.8	1297

Instrumentation response

16.   The instrumentation installed at ch.10,650 is shown
in Fig.4a.   Similar instrumentation was installed at ch.10,600
and ch.10,710.   This section describes the response of the
instrumentation at ch.10,650 to placement of the first
lift PFA and briefly compares this with that at the other
two embankment cross-sections.

17.   Settlement.   During the installation of the band
drains and instrumentation approximately half the thickness
of the drainage blanket was used as a working platform
to protect the surface crust of the alluvium.   These
were installed during July and early August 1982 with the
drainage blanket generally completed to full height soon
after.   The instrumentation was commissioned during early
September 1982.   Ground levels taken in trial pits excavated
through the drainage blanket indicated that 100mm to 200mm
settlement had occurred between start of placement of the
drainage blanket and start of placement of PFA.   It appeared
from the initial piezometer and settlement readings that
the alluvium had consolidated very rapidly under the loading
imposed by the drainage blanket once the band drains had
been installed.

18.   From the start of PFA placement settlement of a
large area of the southern half of the embankment could
not be detected.   The settlement profile measured in hydrostatic profile gauge HPG33 on 16th December 1982 is shown
in Fig.4b.   This corresponded to the time the cracks in
the surface of the PFA were first noticed.   The embankment
appeared to have settled almost symmetrically about a line
mid-way between the northern embankment toe and a point
10m to 12m south of the main carriageway centre line directly
beneath the zone of cracking.   The maximum settlement was
400mm.   For comparison, the settlement profiles on 16th
December 1982 of the embankment cross-sections at ch.10,600
and ch.10,710 are shown in Fig.5a and Fig.5b respectively.
Fine cracks were observed at the southern edge of the

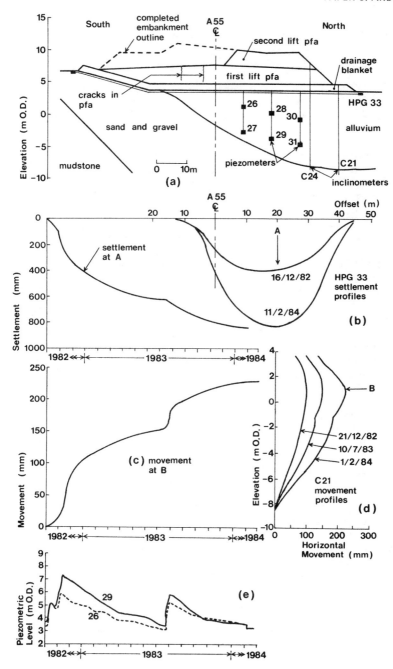

Fig.4   Ch.10,650 (a)instrumentation,(b)settlement,
        (c)inclinometer movement,(d)inclinometer
        profiles,(e)piezometric levels.

embankment at ch.10,600 and, as at ch.10,650, the hydrostatic profile gauge indicated little or no settlement
between the southern end station and the zone of cracking.
No cracks could be seen on the surface of the PFA at ch.10,700.
As at ch.10,650, the settlement profiles were very nearly
symmetrical despite the varying depth of alluvium across
the embankment width.

19. Inclinometer movements. The maximum horizontal
movement and the movement profiles normal to the embankment
for inclinometer C21 are shown in Fig.4c and Fig.4d
respectively. The maximum horizontal movement on 21st
December 1982 was 95mm and Fig.4c shows that during December
1982 the rate of inclinometer movement was decreasing
rapidly from that occurring during PFA placement. The
ratio of the increment of maximum inclinometer movement (ΔH)
to the increment of maximum settlement (ΔV) had decreased
to 0.2. A similar ratio was calculated for the movements
at ch.10,600 and ch.10,710 (Fig.5c).

20. Piezometric levels. The piezometric levels for
piezometers 26 and 29 are shown in Fig.4e. The two increases
in piezometric level, one during October 1982 and the other
during November 1982, are accounted for by the fact that the
PFA fill was placed in two increments to complete the first
construction lift. A plot of excess pore pressure against
embankment pressure due to the weight of PFA placed over
a piezometer tip location (Fig.5d) showed an initial $\bar{B}$ of
0.4 which then increased to 1.0. This indicated an average
$p_c' - p_o'$ of 12kN/m^2 which increased to 24kN/m^2 when a correction
was made to allow for rapid consolidation of the alluvium due
to the drainage blanket loading. The corresponding measured
value of $p_c' - p_o'$ varied from 15kN/m^2 to 30kN/m^2 over the
depth range of the piezometers.

Actions subsequent to discovery of the cracks
21. The response of the instrumentation at the three
instrumented embankment cross-sections indicated that the
embankment cracking had not noticeably affected the response
of the alluvium and that there was no threat to embankment
stability. As it was early in the consolidation period
it was decided not to investigate the cracks until later
when more settlement had occurred. Immediate action
would have meant opening up the embankment during the
winter. The crack locations were surveyed·and these are
shown in Fig.6. The survey confirmed the cracks at ch.
10,650 and ch.10,600 were almost directly above the point
of zero settlement which was not the point of zero depth
of alluvium.

22. The piezometric levels measured by the piezometers
continued to decrease during early 1983 and an analysis of
the dissipation curves (ref.4) indicated that the apparent
c_{vh} had decreased from above 10m^2/yr to within the range
1.75m^2/yr to 2.25m^2/yr. This rate of consolidation
presented the opportunity of reducing the twelve month

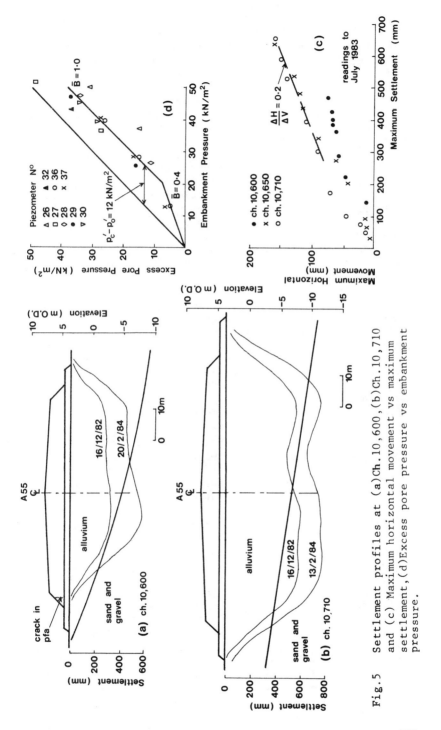

Fig. 5 Settlement profiles at (a) Ch. 10,600, (b) Ch. 10,710 and (c) Maximum horizontal movement vs maximum settlement, (d) Excess pore pressure vs embankment pressure.

135

consolidation period. Before proceeding with early placement of the second lift PFA it was decided to construct a
30m x 30m area of second lift PFA centred around ch.10,650
over the northern half of the embankment and to investigate
the cracks further.

23. Second lift PFA. The small area of second lift PFA,
which was 2.25m to 2.5m high, was placed during late July
1983 in three equal increments with one full day separating
each increment. The aim was to compare the response of the
alluvium with that during the placement of the first lift
PFA. The piezometers responded to fill placement with an
average $\bar{B}$ of 0.75. This low $\bar{B}$ probably resulted from
assuming the PFA load acted directly on the alluvium,
whereas it was acting through the first lift PFA. The
$\Delta H/\Delta V$ ratio was 0.85 during fill placement, but this later
decreased to 0.2 after completion of filling. The initial
ratio agreed reasonably well with a ratio of 1.0 for
undrained loading (ref.5), but the second ratio was nearly
twice that anticipated. The post-construction $\Delta H/\Delta V$ ratio,
however, agreed very well with that prior to placement of
the second lift PFA.

24. The second lift PFA was left in place until February
1984, and from Fig.4e it can be seen that the piezometric
levels had almost returned to their July 1983 levels.
This was further confirmation of an apparent c_{vh} value
in the range 1.75m^2/yr to 2.25m^2/yr. The settlement had
increased to 840mm and the settlement profile of HPG33
for 11th February 1984 is shown in Fig.4b. The inclinometer
C21 profile for 1st February 1984 is shown in Fig.4d and
the maximum horizontal movement was 225mm. For comparison,
the settlement profiles at ch.10,600 and ch.10,710 for
February 1984 which were for first lift PFA loading only
are shown in Fig.5a and Fig.5b. The settlement profiles
were again nearly symmetrical with the settlement under the
southern half of the embankment generally slightly larger.

25. Investigation of cracks. At the time the small area
of second lift PFA was placed at ch.10,650 the cracks in
the surface of the first lift PFA had not visibly increased
in width since they were first discovered. In some places
it was still not possible to trace the line of the cracks
continuously along their length. At two locations (Fig.6),
where there appeared to be localised low spots in the
surface of the PFA, the PFA at the edges of the crack was
eroded at ground level during the winter of 1982/83. It
was decided to excavate trial pits at these locations.
These were excavated during mid-August 1983 and they revealed
at approximately 0.5m depth from the surface of the PFA, a
10mm to 20mm wide crack which ran across the floor of both
pits. A visual inspection of the PFA in this upper 0.5m
indicated that it appeared to have a loose and lumpy structure.
This may have been caused by frost heave. This breakdown
of the surface compacted PFA probably explained why the
cracks had not shown themselves to their full width at the

surface of the PFA. A wide shallow trench was subsequently
excavated along the line of all the visible surface cracks
and again at depths of up to 0.5m the fine surface cracks
developed into 10mm to 20mm wide cracks (Fig.7). Probing
indicated the cracks extended to the full depth of the PFA
and this was subsequently confirmed when a short trench was
excavated to the full PFA depth as part of a trial of remedial
measures. It was not clear whether the drainage blanket
had been affected.

26. The settlement profiles indicated that a convex
settlement profile had caused zones of tension within the
embankment. The resulting major crack was generally radial
and followed a similar path to the alluvium isopachytes.
There were other cracks running off this radial crack which
appeared to run across the embankment. These were probably
caused by differential settlements between adjacent embankment
sections. These secondary cracks appeared to run normal
to the alluvium isopachytes.

REMEDIAL MEASURES

27. Two methods of treating the cracks were investigated.
The first involved pouring various consistencies of a
PFA/water slurry into a short section of one crack. This
proved not to be a very practicable method and was quickly
abandoned. The second method involved excavating a narrow
trench along a short length of the crack and compacting a
stiff slurry of PFA by a vibrating poker. This produced
satisfactory compaction, but it became apparent that the
method was not suited to the scale of the problem.

28. Another concern was that placement of the second
lift PFA would cause additional settlement. This would
probably have resulted in the widening of the existing
cracks or, if they were filled, the development of new
cracks along the line of the previous cracks. It was also
likely that the second lift PFA would crack. In fact,
this proved to be the case as the corner of the small area
of second lift PFA placed at ch.10,650 covered one of the
cracks. Within a few days of completion of its placement
a crack was visible in the surface of the second lift PFA
along the line of the crack in the first lift PFA.

29. The possibility of removing the first lift PFA was
considered, but was discounted on the grounds of cost.
Circular slip stability calculations indicated there was no
serious reduction in the factor of safety of the completed
embankment when allowance was made for a longitudinal
crack in the first lift PFA. The major risk in leaving
the first lift PFA in place was considered to be the possibility of erosion and subsequent collapse of the cracks
resulting in localised depressions which would affect the
slip road and main carriageway pavements. The decision
was taken, therefore, to use a fill in the second lift which
would not crack and which would bridge any local collapse
of the cracks in the first lift PFA. A crushed limestone

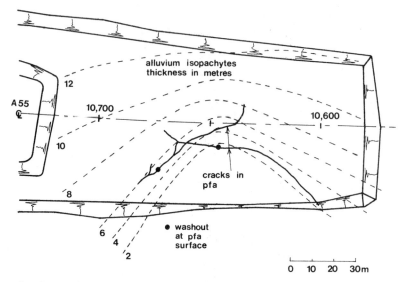

Fig.6   Embankment plan showing crack locations.

Fig.7   Crack in PFA at depth.

with a maximum nominal size of 150mm was selected as this
could be readily produced in sufficient quantities by a
local quarry.

30. The bases of the trenches excavated along the line
of the cracks were covered with polythene sheeting and the
trenches were then backfilled with compacted PFA. This
was to minimise water infiltration into the cracks. The
embankment was then left until February 1984 when the small
area of second lift PFA was removed and the whole surface
area of the first lift PFA was recompacted and regraded to
a 1:40 crossfall prior to placement of the crushed limestone
rockfill.

31. It was decided not to try and seal the surface of
the PFA because of the difficulty of regrading the embankment
to the correct crossfalls to allow for the differential
settlements across and along the embankment. Any impermeable
membrane would have needed protection against puncturing
by the rockfill and there was also concern that preferential
flow paths could develop which might channel water into
the cracks. It was not envisaged that large quantities of
rainfall would reach the PFA surface because of the impermeable
road formation and the associated drainage. It was considered
that any that did reach the PFA would permeate into the PFA
or run-off as it had done during the winters of 1982/83
and 1983/84.

32. At the start of construction it had not been decided
whether the future main carriageway would be carried over
the low level rotary on a viaduct or an embankment and the
A55 Stage 2 fill indicated over the A55 Stage 1 embankment
(Fig.2a) had not been included in the present embankment.
During construction the viaduct option was selected and as
a result it was decided to place the A55 Stage 2 fill during
the A55 Stage 1 construction to minimise differential
settlements between the embankment and the viaduct.

CONCLUDING REMARKS

33. It was considered that the convex settlement profile
of the embankment described caused tension zones in the PFA.
The cohesive properties of the PFA resulted in tension
cracks forming and its strength enabled them to be sustained.
In this respect, any fill material having cohesive properties
would probably also have cracked.

34. The settlement of the alluvium at depth was not
measured. The varying depth of alluvium across and along the
embankment would have resulted in a non-uniform stress and
strain state in the alluvium. One of the most interesting
aspects of the response of the alluvium, however, was the
near symmetry of the embankment settlement profiles.

35. Placement of the second embankment lift incorporating
the A55 Stage 2 fill was commenced during late February 1984.
In the interest of embankment stability the rate of fill
placement was restricted to 0.25m per week to allow some
consolidation of the alluvium to occur during fill placement.
The embankment was completed during late June 1984.

ACKNOWLEDGEMENTS
The Author wishes to thank the Director of Engineering,
Transport and Highways, Welsh Office for permission to
publish this paper.

REFERENCES
1. ARBER, N.R. The design and construction of the Afon
   Ganol valley section of the North Wales Coast Road (A55).
   Quart. Jour. Enq. Geol. Vol.17, 1984, (to be published).
2. PARRY,R.H.G. Overconsolidation in soft clay deposits.
   Geotechnique, Volume XX, No.4, December 1970, 442-446.
3. DEPARTMENT OF TRANSPORT. Specification for road and bridge
   works. HMSO, London.
4. NICHOLSON,D.P. and JARDINE,R.J. Performance of vertical
   drains at Queenborough bypass. Geotechnique, Volume XXXI,
   No.1, March 1981, 67-90.
5. TAVENAS,F. and LEROUEIL,S. The behaviour of embankments
   on clay foundations. Canadian Geotechnical Journal, 17,
   No.2, 236-260.

9. Failure of a large highway cutting and remedial works

G. PILOT, Head, Soil Mechanics Division 1, Laboratoire Central des Ponts et Chaussées, Paris, J. P. MENEROUD, Head, and B. MANGAN, Engineer, Engineering Geology Group, Laboratoire Régional des Ponts et Chaussées, Nice, and B. BESCOND, Head, Soil Mechanics Group, Laboratoire Régional des Ponts et Chaussées Aix-en-Provence

SYNOPSIS. The building of the A 08 motorway in southeastern France, near the Italian border, involved the execution of a large cutting, the slopes of which are as much as 40.m high. These slopes were the site of several slides while the work was being done and just after the motorway was opened to traffic.

This article describes the slides, the investigations carried out, and the remedial work done, which now ensures the stability of these large cutting.

Emphasis is placed on the difficulties of the investigation, in a site having a very complex geological structure, with remolded, weathered, or fissured soils, and on the advantages, in such a case, of a practical approach using the back analysis of a characteristic slide in order to plan remedial works.

INTRODUCTION

1. Near Nice, the motorway along the Riviera between France and Italy passes through mountainous terrain requiring large bridges tunnels and large earthworks. The latter include, in succession, the large hillside embankments of "Ardisson" and of "La Borne Romaine", followed by the "Chez Tredez" cutting, the subject of this article, all of which involve the same geological formations. There were slides of the slopes of this cutting on a number of occasions between 1976 and 1981, while the earthmoving work was being done and then when the motorway was in use. These failures greatly aggravated the difficulties of executing the work, then disrupted the operation of this major artery of communication.

2. Very thorough geological and geotechnical investigations, described below, were necessary to discover the cause of these movements, which turns out to be a very peculiar local geological structure combining a number of factors detrimental to stability : the positions of the soil layers, the weathering and remolding of the materials, and finally the level of the water table.

GEOLOGICAL AND HYDROGEOLOGICAL CONTEXT

Geology

3. The motorway alignment crosses the subalpine chains of the Nice area, structured by foldings having a general East-West orientation, overthrusting to the South. It encounters a series of Jurassic limestone anticlinal secondary chains interspersed with cretaceous synclines made up of marls and marly limestones. Figure 1 is an extract of the local geological map, showing the position of the cutting, which goes through one of these marly formation resting on the back of the anticline, continuous with the underlying limestone bedrock. This geological series is cut by a set of transverse faults and has a pronounced northward dip.

4. When intact, this cenomanian marl is compact and very hard (and the earth-moving work must be facilitated by blasting). Otherwise, it is highly sensitive to the action of water, in contact with which it readily decomposes, yielding a plastic material in which the slip surfaces encountered during the work develop. Figure 2 shows a series of transverse cross-sections in which the formations described above can be seen, together with the special nature of the site : the presence of a "gutter" opened up in the cenomanian marl in the course of a phase of quaternary erosionn, subsequently filled with debris (pebbles) in a clayey matrix. The thickness of this fill ranges from 15 to 30 m in the centreline of the "gutter". It can be seen clearly that the earthworks uncover a geometrical configuration of weathered marls that lends itself readily to deep landslides.

5. This overdeepening of the marls has a general west-north-west inclination, and so is skewed with respect to the motorway. Along its centreline, its geometry is highly complex, being a succession of sills and troughs.

Hydrogeology

6. The hydrogeological situation is also very complicated. There is, first of all, a water table, located above the wearhered marl, fed basically by rainwater infiltration. To this are added deep outflows from the jurassic limestone, in which there is a strong but very irregular water table ; these flows have two effects :

- one is to feed water to the fill of clayey debris via the marly limestones, which act in this case as a hydraulic temporary reservoir ;
- the other is to create a head, under the relatively impermeable weathered marl, in a deep water table in the fissured intact marls.

The localized combination of these actions results in the generation of very high pore pressure values.

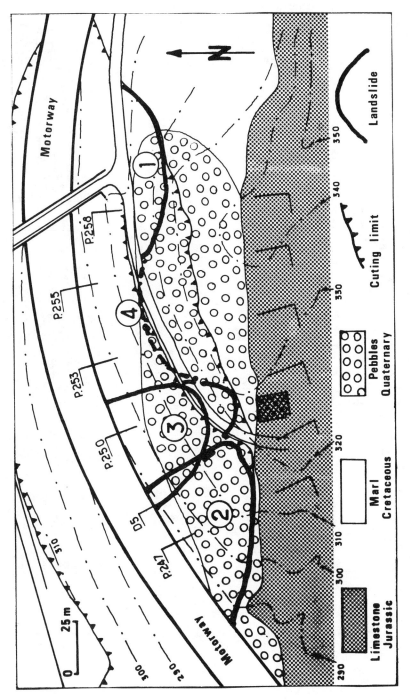

Figure 1. local geological map

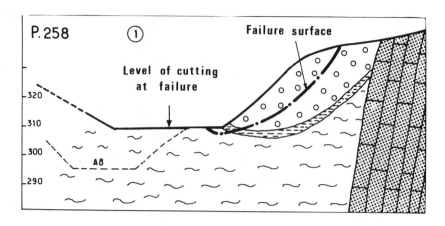

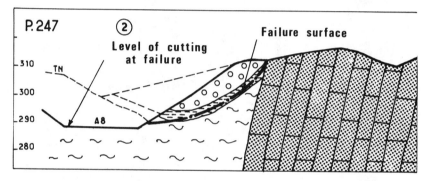

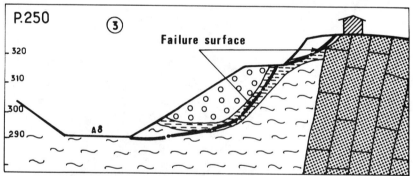

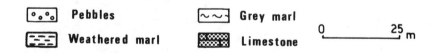

Figure 2. P 258 Slide (1) in February – March 1976
P 247 Slide (2) in June 1976
P 250 Slide (3) in March – April 1977

DESCRIPTION OF THE LANDSLIDES

7. Practically as soon as the earth-mowing work was started (February-March 1976), there was a first slide in the eastern part of the cutting, which was still 16 m above its final level (slide (1) of figure 1 and profile 258 of figure 2). This movement involved the debris formation, the base of which was intersected by the excavation.

8. In June 1976, there was a second slide, at the western end of the cutting, when the cutting had practically reached the final motorway level (slide (2) of figure 1 and profil P247 of figure 2). At this location, the work was in contact with debris and weathered marls. The slide was stabilized by the construction of a coarse limestone rockfill replacement. A first series of eight subhorizontal drains was also built. Seventeen more drains were built at the end of 1976 and beginning of 1977.

9. Following the rainfall of the winter of 1976, this second slide was reactivated in March-April 1977 and spread, constituting a third movement shifted eastward (slide (3) of figure 1 and profile P250 of figure 2). Local treatments, primarily substitutions of clayey materials, temporarily stabilized this zone.

10. From 1977 to 1981, various disorders of lesser importance appeared, in particular between slides (1) and (3). They were troublesome in so far as they foreshadowed a generalized movement of the embankment. Figure 3 is a view of the site and of the south slope of the cutting, after the opening of the cutting, while drainage works were being put in.

11. Following a long dry spell, there were heavy rains from the end of September to the end of December 1981 and the disorders spread to form a new slide (slide (4) of figure 1) :
- formation of an upward scarp reaching an amplitude of 2 to 3 metres and affecting more or less all the zones that has previously slipped ;
- aggravated bulging of the slope in the zone of the third slide ;
- swelling of the pavement as the toe bulge moved north. Figure 4 shows the toe of the slide, which appears in the form of bulging of the right-hand lane of the motorway.

DETAILED GEOLOGICAL AND GEOTECHNICAL INVESTIGATIONS

12. Following the spreading of the troubles, and because of the gravity of those of 1981 (generalized reactivation, spread of the disorders to the pavements), detailed investigations were conducted in 1981 and 1982. They were based on data collected in earlier years and on a new campaign of exploration, testing, and instrumentation.

Figure 3. General view of the cutting after slides (1), (2), (3). Drainage works are in progress

The purpose of these investigations was a better knowledge of the positions of the geological formations in the site, of the movements, of the changes in ground water conditions, and of the properties of the materials. All that in order to be able to assess by calculation the gain in stability contributed by alternative remedial measures.

Figure 4. Slide (4) in 1981 - Bulging of the pavement

Additional geological information

13. From the geological viewpoint, the major interest of this study was the quaternary cover (clayey debris with pebbles) and the shape of the debris-marl contact. In this respect, the general shape of the transverse profile of the "gutter" was investigated by boreholes and by geophysical means. It is more or less uniform when going in an east-west direction. At an altitude, it is uneven and has, in particular, a sill (figure 5) separating a trench, to the east, from a rather steeply sloping surface, to the west. This geological peculiarity in fact determines the instability of site, both because the slip surfaces are guided by the debris marl interface and because the ground water regime is directly affected by the longitudinal flows that start from the sill.

14. The respectives shapes (in particular the depth of failure) and amplitudes of the four movements were monitored by the instrumentation implanted from 1980 to 1982 :
- 29 benchmarks and two inclinometers on slide 2 ;

147

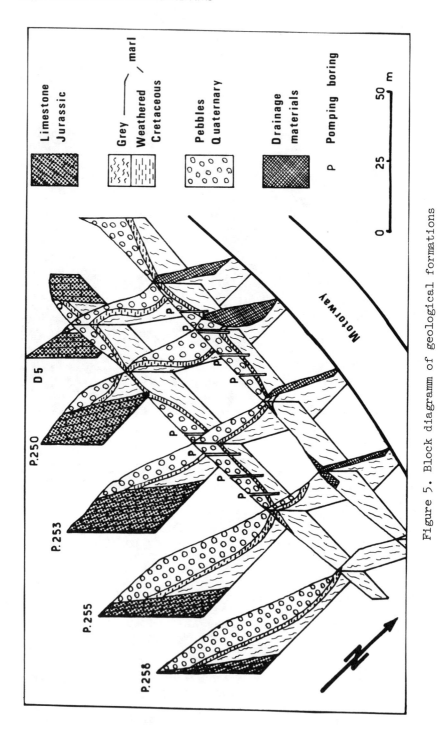

Figure 5. Block diagramm of geological formations

- 20 benchmarks on slides 3 and 4 ;
- 6 micrometer-screw tiltmeters on slide 3.

Site hydrogeological conditions

15. The piezometric measurements made prior to the 1972-73 work revealed that in winter the water table in the surface formations is close to the level of the natural ground. It was also found that it is the "gutter" that drains this water table, from east to west.

16. The digging of the motorway cutting resulted in a significant lowering of the water table, accentuated by the drainage works put in rockfill and drainage ditches). The zones of debris-marl contact are places where the flow of water tends to concentrate : the most abundant seepages from the slope occurred where these zones of contact outcropped. Furthermore, the subhorizontal drains that delivered substantial flows all intersected this level.

Geotechnical properties of the materials

17. The materials of the site lend themselves poorly to the taking of intact and representative samples for shear tests. The rockfill used as drainage facing materials was arbitrarily assigned the characteristics $c'=0$ and $\phi'=45°$.

The debris in clayey matrices cannot be sampled and arbitrary values were assigned to them too ; $c' = 0$ and $\phi' = 30°$ were chosen. The weathered marls, dense ($\gamma = 22$ kN/m^3) and fissured, are highly heterogeneous and so no truly representative intact samples from which were obtained to determine the actual shear strength values by laboratory tests. Back calculation was performed, from one of the profiles of slide (2), where almost the whole surface of failure was located in the weathered marls, to determine shearing parameters of this soil.

18. Figure 6 is a cross-section of the cutting showing the level of the upper surface of the water table (about 2 m below the level of the natural ground), the three main soil layers involved in the study of this slide, and the position of the failure surface.

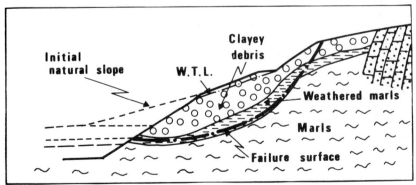

Figure 6. Cross section of slide (3) used for back calculation

Following a number of calculations, we obtained a safety factor F = 1.05 for shear strangth values c' = 0 and φ' = 29°, subsequently used for the design of the remedial measures.

19. The intact marls are dense (γ = 22 kN/m³), fissured, and interspersed with decomposed layers. Intact samples were taken rather easily and used for triaxial CD or CU tests and pore pressure measurements. The test results pose problems of interpretation. Firstly, the values of the shear strength parameters are quite dispersed : c' ranges from 50 to 250 kPa and φ' from 18° to 40° (extreme values in both cases). Secondly, they do not reflect the evolution of the mechanical properties at the base of the cutting, which results both from the release of stresses ans continued decomposition. In this case too, back calculation was accordingly performed on a profile taken in a part of the slope that had suffered disorders. This profile involves a failure surface a part of which passed through the intact marls.

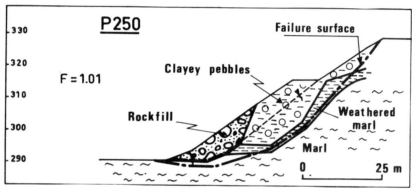

Figure 7. Reference profile (P250) for back analysis of shear strength parameters in marl

20. Figure 7 shows this reference profile. Following various calculations made for several pairs of values of c' and φ', we obtained F = 1.01 for c' = 20 kPa and φ' = 30°. These parameters were used in further stability analysis.

STUDY OF REMEDIAL WORKS

21. Three types of solution were examined with a view to the permanent stabilization of the slopes :
- a solution based on earthworks and drainage, consisting both of unloading the top of the slope and of draining the "ditches" of debris, bounded by the sill fo the "gutter" ;
- a solution based on the construction of a closed-frame engineering structure, a sort of tunnel at the base of the cutting, rigid enough and strong enough to withstant the active

earth pressure ;
- a solution based on the construction of a cast wall fixed in the intact marls and anchored at the top by tie-rods.

22. This last approach was quickly discarded because of the high cost of the work, with no guarantee of safety based on adequate experience of such works at unstable sites.

23. The second solution offered a guarantee of safety : it was checked that the stability of the reworked cutting slope would in fact be substantially improved by the installation of the closed frame (F approx. 2). On the other hand, the works would have been very expensive.

24. The study therefore focused on a solution of earth works and drainage. Figure 8a shows the profile P250, typically involved by a new design, on which are drawn the planned earthworks, compatible with the rebuilding of the access road to the "Tredez" property. These earthworks would increase the safety factor from 1.01 to 1.15. Since this value of the safety factor, $F = 1.15$, was insufficient, the earthworks had to be completed by drainage. The most effective drainage possible would lower the water table to the level of the decomposed marl-debris contact. Figure 8b shows that this increases the safety factor to 1.30.

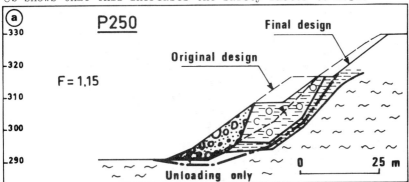

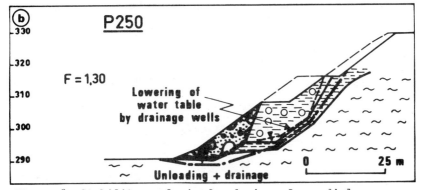

Figure 8. Stability analysis for design of remedial measures

25. Since the values of the shear strength parameters were determined by reference to observed slide surfaces, this value of the safety factor was regarded as adequate. Also, the design and the repair of the slides were based on these two measures, removal of soil and deep drainage.

EXECUTION OF THE REMEDIAL WORKS

26. The consolidation works were executed in accordance with the findings of the studies described above. The earth-moving work consisted of two measures. The first was the building of a platform 7 m wide half-way up the slope, at "bank 310", for the first unloading of the soil mass. The second was the building of a new slope on the upper part of the site, set back from the lower part. The slopes chosen were generally 1/1. Figure 9 is a drawing of the site after the earthworks. It should be compared to figure 1, which shows the situation before the remedial works. It can be clearly seen that the crest of the slope has been set back a bit and that the access road to the Tredez property was shifted upslope of "bank 310".

27. These earthworks were attended by two minor slides at the crest of the new slope and at each end of the cutting. These disorders were repaired by replacing the slipped materials with rockfill.

28. Drainage was provided by the following works :
. The digging of eight pumping wells 125 mm in diameter, each with a pump in the bottom. The operation of the pumps is discontinuous and is controlled by the water levels in the wells. These wells were drilled down to the intact marls, below the level to which the water table was to be lowered (debris-decomposed marls contact). The wells, the positions of which are shown on figure 9, are in two groups of four, one on the platform at "bank 310", the other on the Tredez road.
. The drilling of subhorizontal drains. A first set of three drains (80 m long, slope 20 %) was installed in the western part of the cutting to drain the end of the debris. A second set of short bored drains (10 to 20 m long) was installed starting from the motorway to complete the drainage of the water from the debris-decomposed marls contact.
. Runoff ditches. These ditches were placed on the Tredez road and on the platform at "bank 310", where they receive the seepage from the slopes and the localized and temporary seepage that appears on the slopes.

29. All of the arrangements are working well, and the piezometers indicate that the ground water level remains practically at the bottom of the layer of debris.

CONCLUSIONS

30. The slides of the large slopes of the "chez Tredez" cutting resulted from the combination of a number of

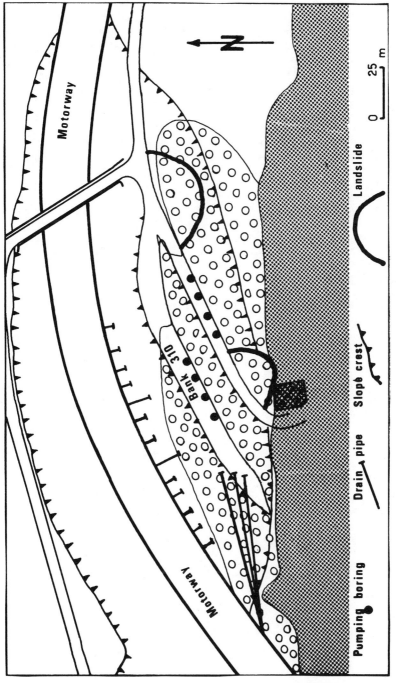

Figure 9. Map of the site after execution of remedial works

unfavourable factors.: the shape of the debris-marl contact, which promoted slides after earthworks, the weathering of the marls, and the gathering of water in the "gutter". The remedial solutions were studied using conventional stability calculations along non-circular surfaces placed along the failure surfaces corresponding to geological discontinuity. The values of the shear strangth parameters, which could not be measured in the laboratory, were determined by back analysis along the observed slide surfaces.

31. The remedial works used, with a safety factor of 1.30, include earthworks at the top of the slope and a number of drainage measures (pumping wells, vertical drains, ditches for surface water) that are working well. These slopes have withstood the unfavourable ground water conditions of the winters of 1983, 1984, and 1985 with no further slides.

ACKNOWLEDGEMENTS

The authors wish to thank ESCOTA, firm in charge of the motorway, for permission to publish this paper.

10. Water and frost—stability risks for embankments of fine-grained soils

A. EKSTRÖM, Chief Engineer, Swedish Geotechnical Institute, and T. OLOFSSON, Head of the Geotechnical Office, Swedish National Road Administration

SYNOPSIS.   In Sweden road construction is primarily performed during the winter period to maintain even employment
over the year. This circumstance must be considered in the
evaluation of the stability of embankments of fine grained
soil. When the embankment is more or less frozen during the
filling out improper compaction will be the result. Such an
embankment may be stable due to capillary forces as long as
it is not oversaturated. However, if it is oversaturated,
failure may occur by the gravity force or by dynamic influence. It is also important to provide for proper drainage
so that water will not be trapped in the embankment. The importance of these factors is described in some cases where
failures have occurred due to liquefaction, oversaturation
at thawing or trapped water.

ROAD No. 351. THE ROAD FAILURE AT ÅSELE IN OCTOBER 1983
   1.   On the 4th of October 1983, a road embankment carrying
Road No. 351 near Åsele in northern Sweden failed (Fig. 1).
The failure, a slide, occurred when the embankment was partially submerged during the course of impounding a reservoir
for a new hydro-electric power station on a nearby river.

Fig. 1 View of the failure, October 4th 1983.

2. The slide was triggered by a 3.3 tonne tractor-drawn vibratory roller carrying out repairs to the road surfacing that had occurred as a result of damage caused by the damming of the river. The tractor and roller slid into the reservoir with the embankment, as a result of which the driver lost his life. The tractor and roller were later found at the bottom of the reservoir, about 60 metres from the road.

Geological conditions and the design of the embankment

3. The embankment was constructed on firm till along the edges of two small lakes. The fill for the embankment was taken from adjacent cuts containing fine-grained till (Figs. 2 and 3).

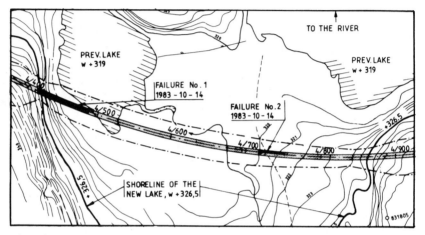

Fig. 2 Plan.

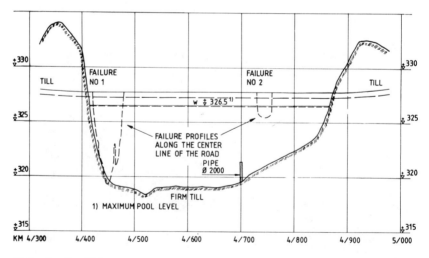

Fig. 3 Profile.

4. Because the embankment was to be constructed with relatively steep slopes (1:1.5) and to be saturated in the future, the method of construction was specified especially to ensure its stability (Fig. 4). The fill was to be compacted using the wet fill method[1], which implied that the work could not be performed when there was a risk of the material freezing.

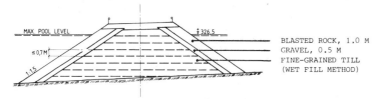

Fig. 4 Principal section for the construction of the embankment.

Impounding and the course of the slide (see Fig. 5)

5. The road was completed and opened to traffic in August 1978. Filling of the reservoir was started on the 15 August 1983.

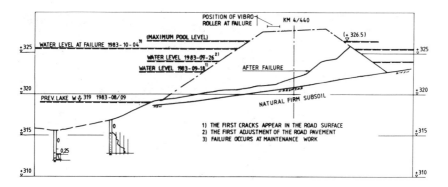

Fig. 5 Cross-section of the embankment after failure. The figure illustrates the schedule of raising the water level and the influence on the road embankment.

6. By the beginning of September the water had reached the toe of the embankment. The first signs of damage to the road, in the form of longitudinal cracks in the road, were noticed on the 18 September, when the water had reached a level of +322.1 (about 3 m above the toe). By the 26-27 September, the water level had been raised an additional 2 m and the first repairs of unevenness in the road were carried out. By the

[1] Nilsson, T. & Löfquist, B.: An earth and rockfill dam on stratified soil. The wet fill method. 5. Congr. Large Dams, Paris 1955. Trans. Vol. 1, Quest. Nr 16, pp. 403-413.

4 October, the water level had been raised to +325.5, corresponding to a level about 2 m below the level of the road. Damage to the road was then so extensive that the surfacing on the section km 4/400-4/500 had to be stripped, and the pavement levelled and compacted. The work was carried out with a grader and a 3.3 t vibratory roller drawn by a wheeled tractor.

7. The slide occurred suddenly, in the course of about 10 seconds, when the roller was being drawn alongside the railing on the river side of the road for the first time. No warning, in the form of increased cracking or settlement in the shoulder, was observed. The slide comprised the whole width of the road over a stretch of 60 m.

8. After the slide, filling of the reservoir was stopped until the machines had been recovered. Cracking and deformation of the road increased greatly and, on the 14 October, a second slide occurred on the section km 4/700-4/800 (Fig. 6). This slide, which occurred spontaneously, without the effects of traffic or other live loads, was shorter and narrower than the first one, removing only half the road over a stretch of about 40 m.

Fig. 6 View of the failure, October 14th 1983.

Investigation of the cause of the failures

9. After the slides, a comprehensive investigation was made of how the embankment had been built and the factors that could have contributed to its instability. The nature and rapid course of the first slide indicated spontaneous liquefaction of the fill.

10. To provide employment the earth works were started and mainly performed during the winter contrary to the specifications. Due to snow and frozen soil the compaction was poor especially in the slopes.

11. The investigations of the remaining embankment revealed in places very loose layers of the finer fractions of the till. The thickness of these layers amounted to 1 to 2 metres.

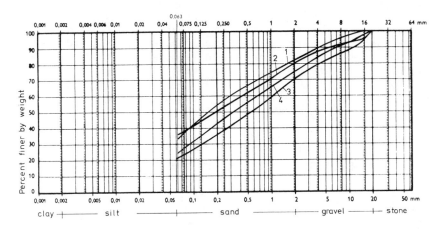

Fig. 7   Grain size distribution curves of till from the
embankment fill. (Material < 20 mm.)

12.   The grain size distribution (Fig. 7), water content and
strength properties of the till were determined in the laboratory . The shear strength was determined by direct shear tests
on samples of dry and saturated material, in a loose and dense
state:

| | Angle of internal friction, ϕ | |
	Loose	Dense
Dry	$\sim 27^0$	$\sim 35,3^0$
Saturated	$\sim 24^0$	$\sim 34^0$

13.   Stability calculations (c-ϕ analyses) based on the
laboratory results and the probings into the embankment indicated that the factor of safety of the embankment in the
saturated state was about 1.0, in other words, it was on the
verge of collapsing, without any external load such as that
from traffic. This conclusion was confirmed by the second
slide and the extensive cracking of the road surface. The stability of the embankment during the period 1978-83 was probably
due to the stability improvement obtained by "false cohesion"
of the fine-grained material of this type at natural moisture
content.

14.   The assumption of occurrence of "liquefaction" caused
by pore pressure elevation in the saturated fill during compaction is supported by a special study made by Prof. Rainer
Massarsch. The study shows that the penetration resistance
measured in the rest of the embankment was considerably below the values stated in literature as the upper limit at
which soil flow occurs under dynamic loading.

Restoration of the road

15. Reconstruction of the embankment by blasting combined
with partial excavation of the unsatisfactory fill and replacement with blasted rock would have been technically possible. For different reasons, however, it was decided to rebuild
along the shore of the lake formed by the reservoir. In conjunction with this work, the damaged part of the embankment
was demolished by blasting with surface charges placed on the
underwater part of the slopes. The bank "flowed away" when
relatively small charges were fired.

ROAD No. 341 AT NÄSÅKER IN ÅNGERMANLAND

16. The Åsele case shows that an embankment of moist, fine
cohesionless soil has sufficient stability even under heavy
traffic although its slope is steeper than the angle of re-
pose of the soil. But if the water content increases the embankment will gradually become unstable. The effect of an increased moisture content may be creep or flattening of the
slope due to earth flow or a slip. The river beds in North
Sweden are often eroded deep into alluvial sediments of sand
and silt and have steep river banks. When the water level is
dammed up upstream of hydro-electric power stations the banks
will be flattened out.

17. At Näsåker Road No. 341 is excavated in the north bank
of the Ångerman River for a distance of 700 m. The road descends
from the sediment plane at level +135 m to a bridge across the
river at level +90 m.

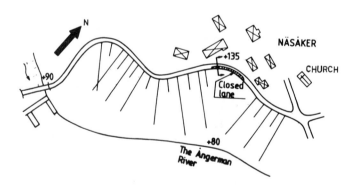

Fig. 8 Plan.

18. The valley consists of medium to coarse silt. The slope
is 155 m high with an inclination of 1:1.4 to 1:1.5. Considering the properties of the silt the slope is surprisingly steep.
The road has carried heavy traffic for more than 30 years,
during which time minor deformations have occurred, particularly in spring when the snow melts and the ground thaws.

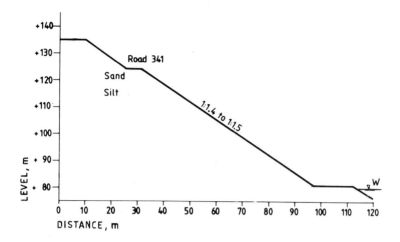

Fig. 9   Section.

19.   In May 1983, movements took place behind the crest
which were so large that one lane of the 6 m wide road had
to be closed for a distance of 100 m. The wearing course on
the closed lane had cracked severely due to settlements of up
to 0.2 m and the railing was displaced up to 0.3 m from the
road.

20.   Investigations into the cause of the deformations revealed that a water pipe had been broken sometime in early
spring. The leakage was discovered when the cellar in the
church was flooded. Soundings showed that the soil was loose
6-13 m below the road. There was only a very shallow ditch
along the cut of the road. The angle of internal friction of
the silt was 27 to 28 degrees. Stability calculations based
on friction only yielded a factor of safety of 0.9. Taking
into account the capillary force, T, the effective pressure,
p, in the soil will amount to

$$p = \frac{2T}{R}$$

where R is the radius of the meniscus. According to Taylor
the radius of the meniscus can be assumed to be a quarter of
the grain diameter. As the upper limit of coarse silt is
0.06 mm and T is equal to 0.075 N/m the fully developed capillary force gives a maximum incremental contribution to the
effective grain pressure of 10 kPa. At Näsåker this contribution increases the computed factor of safety from 0.9 to
1.45. By including the "false cohesion", p tan ϕ, in the stability calculation the deformation pattern fits the actual one
quite well. It is apparent that part of the false cohesion
will be lost at an increased moisture content so that the
slope stability will decrease and the slope will become more
susceptible to creep due to traffic vibration. To reduce the

risk of saturation of the soil by meltwater at Näsåker a minimum 1.5 m deep drainage ditch was excavated during the 1983/84 winter (Fig. 10). In the excavated ditch plastic drains were laid on a non-woven geotextile and filled with coarse gravel. This drain functioned satisfactorily during the 1984 spring when the quantity of meltwater was considerably above average.

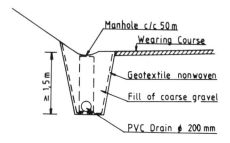

Fig. 10   Drainage ditch.

ROAD No. 95 IN CENTRAL LAPLAND
The slide of the embankment, km 19/280-19/510

21.   On August 13th 1974 at the village of Aborrträsk, 20 km south of Arvidsjaur, 60 m of a 9 m high by 230 m long embankment suddenly slid after several days of heavy rain (Fig. 11). No one witnessed the slide. Fortunately the slide was discovered soon afterwards and the road was closed before any accidents occurred. The slide comprised 7500 m³ embankment fill of glacial till.

Fig. 11   View of slide.

22.   Geological conditions. This region is a hilly dead-ice moraine area with plenty of tarns and bogs in its lowlying parts. The embankment which gave way crosses the outer edge of a bog between two hills of till (Fig. 12). The ground surface is sloping from the west towards the bog. From one tarn about 100 m to the west of the road a brook flows to the bog.

162

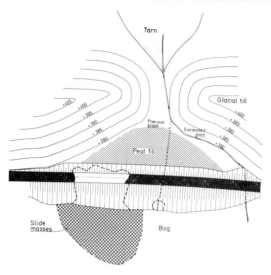

Fig. 12 Plan.

23. The embankment. The peat under the embankment had a
maximum thickness of 4 m. The peat was excavated and the embankment filled out with till from the adjacent hills. The
excavated peat was dumped on the west side of the embankment.
The brook was led via a drainage ditch at the edge of the
peat fill to a culvert through the embankment (Figs. 13 and
14). The embankment was completed early in 1972 and the road
was opened for traffic the same summer.

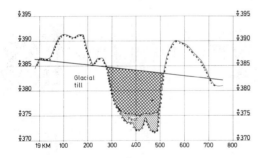

Fig. 13 Profile.

24. The cause of the slide. The displaced earth mass was so
saturated that it could only be entered on wooden planks. Test
pits in the till of the hillside to the west of the slide scar
revealed water bearing layers of stony coarse gravel.

25. The gravel layers were apparently underground outlets
into the bog from the lakes and tarns. When the peat was replaced with till of very low permeability (Fig. 15) the
gravel layers were closed. Thus high pore water pressures
were built up in the embankment fill. The snow blanket was

163

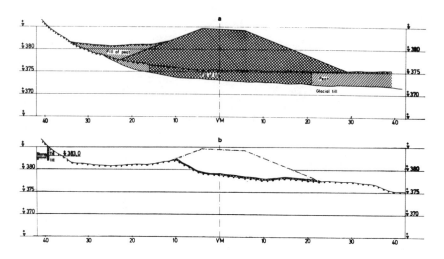

Fig. 14   Section 19/360 a) Cross-section of embankment
          b) Cross-section after the slide.

extremely thick in the previous winter, so that a major infiltration took place during the spring melt. In addition
heavy rain fell in the weeks before the slide.

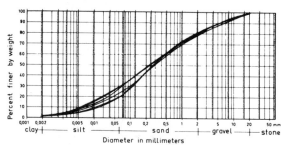

Fig. 15   Grain size distribution curves of glacial till.

26. **The restoration of the road.** A new ditch for the brook
was excavated down to the waterbearing gravel layers across
the embankment at the south edge of the slide scar.

27. A culvert was built on a gravel fill. To further relieve
the pore pressures under the embankment, pits were dug through
the till fill and refilled with stony gravel along the east
embankment toe. These pits were spaced about 10 m apart. The
slide scar was covered with an 0.5 m thick layer of gravel.
The embankment was then completed with dry glacial till. Pore
pressure measurements proved that the drainage functioned

satisfactorily as the pore pressures were kept within tolerable
values during the following spring and summer.

Slide in the slope of the embankment, km 11/380-11/420
28. This 10 m high embankment of till crosses a depression

in the glacial till. The toe follows the shore of a small lake (Fig. 16).

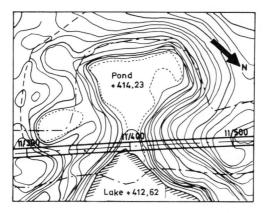

Fig. 16   Plan.

29.   Due to the lake the embankment was made with slopes as steep as 1:1.5. In June 1978 shallow slides occurred on the slope towards the lake. The slides reached the top of the embankment without affecting the road (Fig. 17).

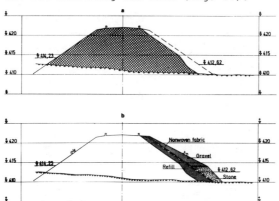

Fig. 17   a) Cross-section km 11/400 after the slide,
          b) Restoration of the embankment.

30.   The embankment had dammed the depression so that a pond had formed. The water level in the pond was 1.6 m above the level of the lake on the lower side of the road.

31.   The cause of the slides was seepage through the embankment from the pond to the lake and oversaturation of the frost active till at the embankment toe. During the thawing of the frozen till fill, earth flow occurred so that the embankment slope became unstable. In the scar some metres above the embankment toe the soil was moist, indicating the height of the capillary saturation zone.

32.   The slope was stabilized by a pervious fill of coarse gravel as support of the frost active till. The scars of the

165

slides were first made level with the slope. A working traverse of stone was filled along the embankment toe on nonwoven fabric and then coarse gravel was filled on the embankment slope.

CONCLUSIONS

33. The case histories on road embankments of fine-grained soil described above illustrate the necessity of considering the following circumstances for design and construction:

- properties of the fill, particularly angle of internal friction as a function of relative density, capillarity, frost activity and influence of the water content on the bearing capacity

- climatic conditions during construction

- influence on stability and settlements of submergence or seepage.

34. Construction during winter periods with more or less frozen soil will result in a loose fill with risk of settlements or earth flow at thawing. Precaution methods should be flat slopes or stabilizing slope fills of coarse pervious material.

35. Embankments that may be subjected to submergence or seepage should be designed according to the principles of earth dams so that sufficient drainage paths of the fill is provided for. Improper compaction can - as in the Åsele case - be disastrous if the embankment is submerged.

36. It is imperative that drawings and instructions for the construction are presented so that the motives of the design will be understood by the contractor and supervising staff. The soil expert's terms must be expressed so that the meaning is explicitly clear for the non-expert.

37. Finally it must be remembered that even the best designed and well prepared earth structure may run the risk of ending in a failure if the communication between the designer and the site people is lacking.

11. The use of stress measurements as an early warning method of failures in compacted embankments

H. ØSTLID, Construction & Maintenance Division, Norwegian Public Roads Administration

INTRODUCTION

In spite of the steadily increasing flow of papers, publications and various forms of scientific work in the field of
soil mechanics, the author strongly believes that **installation**
of measuring instruments and recording of data will have to
increase. This is necessary in order to check the validity
of our theories and also to furnish the necessary data to confirm the theory used on a particular project.
There is (in the author's opinion) a pronounced tendency in
our field of work to solve the problems theoretically and
than leave it at that.
This is very unsatisfactory, especially for the engineers in
charge on site, and we all should make an **effort** to see our
various tasks as a part of a finished project and not as a
verbal job in itself. When driving on top of a large embankment nobody is particulary interested in the complex soil
mechanics theories underneath, they want a cheap embankment
with an even and **comfortable** road surface. By the way, this
is in fact the object **of** the verbal exercise, to give our
clients safe and economic embankment constructions.
This paper intends to demostrate that comparatively simple
measuring equipment may give very clear warnings early enough
to prevent actual failures **from occurring.**

1.   SITE DESCRIPTION

The site is located at Hønefoss not far from Oslo and the
deposits in this area are all below the marine limit. The
materials consist of sand and silty clays in some places the
clay may be quick. The shear strength of these clays varies
from very low, 5-10 kN/m², to very high in the upper dry crust.
Erosion has created small valleys in this clay material;
valleys with depths of about 20 meters are quite commom. The
road embankment was constructed across such a valley giving an
embankment construction depth of about 22 meters.

2.   INSTRUMENTATION

The cross section of the embankment can be seen in fig. 1, and
the embankment is of the sandwich construction type.

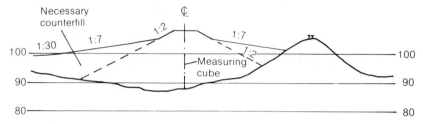

Fig. 1. Cross section of embankment with position of measuring
cube recording stresses acting on all 6 faces.

The instrumnet used to measure the triaxial total stresses is
described elsewhere, but very briefly it is a cube measuring
25 x 25 x 25 (cms) with earthpressure cells on each face.
This arrangement allows a check on the measurements, as there
are two measurements in each direction.
The cube was placed in the centerline of the embankment and
about mid-depth. Point of installation can be seen in fig.1.
In addition to the cube a number of electric piezometers were
installed in order to check the development of pore pressures
in the construction and also in the existing ground.
The cube itself had piezometers on three faces and by using
these data the effective stress could be estimated.

3. RECORDED DATA
The embankment was constructed over a period of about 12 months
and the cube was installed as the embankment reached half the
total length.
The recording started and has been going continously up to
the present date. These recordings together with the pore
pressure measurements and general settlement control will go
on for at least 10 more years.
After installation of the cube the construction at the embankment went on, but due to a misunderstanding on site the prescribed counterfill was not placed according to the plans and
consequently **an imbalance** existed in the cross section of the
embankment.
A very interesting development in the total stresses could in
fact be seen before it was clear that something was wrong on
site and this stress plot can be seen in fig. 2.
The total vertical stress was very similar to the stress along
the embankment and showed a **close** relationship with the actual
weight of the overburden.
The recording of porepressure also gave very reasonable values
and it is therefore some justification for accepting these
measurements as being of a reasonable quality.
As can be seen on fig. 2, the two curves depart very sharply
at about 180 days, and in fact this was when the transverse
movement started in the embankment.
Later measurements and observations have confirmed the movements to be about 60 cms on top of the embankment along the
centre line.

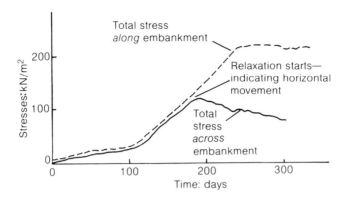

Fig. 2. The development of total horizontal stresses in the
embankment in two directions.

A number of other recordings were also made such as piezometre
values and settlement measurements.
These data will be published at a later stage.
The necessary counterfill was placed as fast as practically
possible, and the transverse movements stopped.
The stress   levelled off at about the values showed in fig. 2
and has stayed there up to this date approximately 4 years
later.

4.   CONCLUSIONS
1.   The continuous recording of horizontal and vertical total
stresses can be done for several years using this measuring
cube.   One cube has been working for 7 years.
2.   The development of horizontal stresses may be an indication
of both direction and magnitude of movements in an embankment
(or dam).
3.   Dramatic changes as shown in this case should not be expected normally, but the position of installation in the embankment is very important.
4.   A future installation using 5 cubes to give axis symmetry,
will be used.
This should further give inforamtion about the usefulness of
this type of measuring cube.

Discussion on Papers 8–11

MR F. HUGHES, Cementation, Piling and Foundations,
Rickmansworth

My firm put in the band drains referred to by Dr Arber in
paragraph 2 of his paper. Apart from some problems of entry
into the blanket due to its heavy compaction by traffic in
certain places the placement of the drains went well. An
Armco culvert provides a drainage route across the road in the
treatment area; it was necessary to place this in position
before the vertical drains were installed. Following a
successful inclined drain installation at the Selby railway
diversion we were able to place raked drains from either side
of this culvert.
Dr Arber, was the drain installation successful? I
understand that the stage II fill was placed early and it
would be interesting to learn whether the vertical drains
played a part in this?
The paper is a valuable contribution because the use of
pulverized fuel ash for embankments is now widespread. You
did not say whether the large Armco culvert caused any
cracking in the fill. This information could be of use to
others who may be thinking of using such a combination.

DR N. R. ARBER

You are referring to the Armco culvert which runs beneath the
embankment at ch 10,690. The crown of this culvert was above
original ground level and there was therefore a change in the
height of pulverized fuel ash (PFA) fill placed over the
culvert. Its settlement was monitored during placement of the
first-lift PFA and the subsequent consolidation period. The
rate and magnitude of the settlement and the shape of the
settlement profile were very similar to those recorded in HPG
35 at ch 10,710. No cracking of the PFA was observed in this
area of the embankment and this could be attributed, in part,
to the general uniformity of settlement.
The design requirement for the band drain system was to
enable 90% primary consolidation within 12 months. The back-

calculated apparent field Cvh was higher than the design value
and this combined with the band drains enabled this design
requirement to be met. In this respect the band drains were
considered to be a success. In the event, the second lift of
the embankment was not placed early. Had the first-lift PFA
not cracked it would have been possible that the second lift
could have been placed early.

DR A. B. HAWKINS, University of Bristol

Dr Arber, you explained the cracks in the PFA as resulting
from differential consolidation due to the changing thickness
of the alluvium. In Figs 2(b) and 6 you gave the rockhead
contours and isopachytes for the alluvium. Was the rockhead
determined on borehole, probe hole or geophysics, and could
the steep area indicated with slopes up to 1:2.5 actually be
steeper, i.e. be an old semivertical cliff line?
  Were such features as buried 'cliffs or islands' envisaged
at the precontract stage? What methods were used to locate
them and to determine their slopes?
  Little is given on the nature of the alluvium. If the
alluvium was laminated, then with the band drain spacing at
this site primary consolidation would be quick. If, however,
it was not laminated, then the rate of porewater dissipation
is reduced and hence consolidation is slower. For this reason
it is essential to obtain the correct geological
characteristics of the soil and its lateral consistency to
establish not only the consolidation rate but also the
significance of this when part of the embankment will be on
only a thin alluvial cover. What extra maintenance costs are
anticipated in this area, and were they foreseen at tender
stage?

DR N. R. ARBER

The design of the embankment was governed by the alluvium. As
a result, ground investigation boreholes did not generally
penetrate to bedrock, as mentioned in paragraph 6 of the
paper. Geophysics were not used to determine the rockhead as
it was not considered to be relevant to the design. The
alluvium isopachytes were orginally determined from borehole
information. These were updated during band drain
installation as the band drains completely penetrated the
alluvium and were pushed 0.5 m into the underlying sand and
gravel.
  Long-term settlements of the embankment due to secondary
compression of the alluvium were envisaged at the design
stage. It is not envisaged that there will be any additional
maintenance costs in this respect arising from the cracking of
the PFA.

172

MR I. L. WHYTE, University of Manchester Institute of Science and Technology

The development of smooth slickensided slip planes due to the interaction of earthmoving plant and soil was first reported in detail by Pavlakis (ref. 1) on observations made during the construction of the Pournari Dam. Such surfaces can occur at the junction interface between each compacted layer of clay fill under particular conditions of compactive effort (from any type of plant) and soil moisture content. The laboratory study suggested a possible mechanism for the formation of these smooth surfaces and indicated a limiting range of moisture contents, related to wetness index, under which they can develop. The main problematical features of smooth slickensided surfaces at the junction of compacted layers are

(i)   they can be of large area and extend over considerable cross-sections of clay fills
(ii)   the development of such planes is not detectable at the fill surface but only by careful pit excavation in the fill
(iii) they induce incipient slip planes into fills such that short-term strengths are highly anisotropic (with strength reduction along the surface possibly being reduced by about 50% or more compared with strengths from homogeneous samples) and long-term shear strength parameters may be reduced to residual values (particularly under higher effective stresses).

Although the development of such highly problematical shear surfaces in fills may only be within rather limited moisture content ranges (generally approaching upper limits of acceptability for compaction), should they occur and failure result, the back analysis of the slip cannot be explained using conventional parameters and can only be understood by strength anisotropy (short-term or total stress analysis) or partial development of residual strengths along critical slip path sections (long-term or effective stress analysis).

Reference
1.   PAVLAKIS, G.   In Clay fills, pp. 266–269.   Telford, London, 1979.

MR F. HUGHES, Cementation, Piling and Foundations, Rickmansworth

Mr Pilot, I was interested to see in your paper that pumps were installed in the vertical drainage holes. Pumps need maintenance. Was it not possible to use a combination of vertical and horizontal drains or galleries to give long-term drainage without pump maintenance and problems of power supply?

I have a comment arising from Mr Whyte's contribution on the problems of possible slide surfaces caused by rollers compacting layers of fill. Problems have arisen in the past with embankment slopes where access ramps have been travelled by works' traffic, much of which is rubber tyred. Such ramps must be well removed and scarified to prevent later slip movements on the old ramp surface. The same problem can arise with ramps out of cuttings.

MR G. PILOT

Horizontal drains were drilled in the slope but cannot keep water at a convenient level. Galleries were too expensive to build in unstable masses of soil, and that is why pumps were installed. It was not a problem to have a power supply on the site and, on another site of the motorway, the cost of maintenance of the pumps was found to be low.

MR I. L. WHYTE, University of Manchester Institute of Science and Technology

The problem of fills heavily trafficked by earthmoving plant and the possible development of incipient slip planes was not investigated during the laboratory model studies. I believe, however, that smooth slickensided incipient slip surfaces were encountered along haul roads and traffic routes across the clay core fill at Pournari Dam and that these were attributed to the action of earthmoving plant such as dump trucks. These plant-induced slip planes, I believe, were different from those induced by compaction plant in that they were of more limited extent across the dam (being limited to trafficked areas) but could occur not only at the interface within compacted layers but also within the compacted layer thickness. It does not surprise me therefore that a failure can develop along haul roads to embankments. The mechanism of formation of slip planes under rubber-tyred plant, however, could differ from that suggested for compaction equipment such as smooth or tamping foot rollers. To my knowledge, the action of tyred plant on clay fills and possible slip plane development has not been researched, and the work undertaken at the University of Manchester Institute of Science and Technology and reported here is the only detailed study made of the problem in relation to compaction plant. By necessity, the study was limited but I consider that the implications, particularly if greater use is to be made of the wetter and weaker clay fills, are such that more research is essential if 'unexpected' failures are to be avoided in otherwise well-engineered embankments. Such a study could include the influence of traffic routes and haul roads since a major factor in the development of incipient slip planes is 'over-compaction', a condition likely to be achieved along such areas.

174

MR T. OLOFSSON, Swedish National Road Administration, Borlange

The first case in our paper describes two slides in a five-year-old road embankment. The fill consisted of fine-grained glacial till, the most common soil in Sweden. The slides were caused by liquefaction when the fill was submerged during impounding a reservoir for a new hydroelectric power plant. The necessity of good compaction of the embankment fill was considered in the specifications for the embankment in the following way:

(i)   fill to be spread and compacted in layers of maximum depth 0.7 m
(ii) compaction by the wet-fill method developed by the Swedish State Power Board
(iii) fill and ground had to be unfrozen.

The road construction was carried out by the Labour Market Board who granted funds for the construction to relieve unemployment during the winter of 1976-77. The work began in January 1977. The work staff wanted a transportation road across the depression in which the road was to be built. As it was winter with temperatures between -20 and -30 ^{o}C it was concluded that the wet-fill method could not be used. Instead the fill was spread in layers 2-3 m thick and compacted only by the transportation vehicles, trucks and tractors. The embankment functioned without any problems before the submergence. The first slide was triggered by a vibratory roller. The energy released in the slide moved the 3.3 t roller and the tractor 60 m out into the 10 m deep reservoir. The second slide was due only to gravity.

It is apparent that the people in charge of the construction did not understand the motive and importance of the specified compaction method. This method is uncommon in road building. When special methods are prescribed it is important that the instructions and drawings for the job are presented in easily understandable terms.

The unstable embankment was removed by blasting in the water. The blasting was performed in sections about 50 m long along the road. The charges of 27 kg of explosives each were spaced at 5 m intervals on both sides of the embankment about 3 m below the water-table. At the first section 65 m of the embankment were blasted out so that the top of the fill was 3 m below the water-table.

The other cases in our paper deal with failures caused by trapped water and the effects of frost.

The second case describes creep in steep slopes of silt and fine sand, when the water content increases. The slopes are stable owing to capillary forces. At periods of increased infiltration the capillary forces decrease and an unstable state is obtained. By collecting the water from rain and melting snow in a drainage ditch along the road the damage to the road caused by creep can usually be controlled.

The third case describes the slip of a road embankment caused by high porewater pressure. The embankment passes a peat bog. The peat under the road was excavated and the embankment was filled out with silty glacial till. After the slip had occurred water-bearing layers of stony coarse gravel were discovered in the till. These layers had their outlets in the bog. Filling the excavation with glacial till of very low permeability was like closing a water tap. The embankment was repaired by providing drainage of the water-bearing layers.

The fourth case is a road embankment which cuts off the inner part of a depression in glacial till. In the confined depression water from rain and melting snow formed a pond, because the bottom was so impervious and there was no culvert through the embankment. After several years during a thaw shallow slips occurred in the embankment slope caused by earth flow in the embankment toe. Seepage from the pond had oversaturated the frost-active till. To stabilize the slope a supporting fill of coarse gravel was filled out.

DR A. D. M. PENMAN, Sladeleye, Chamberlaines, Harpenden

I should like to point out, in relation to Dr Østlid's cube, that the embankment dam engineers have been measuring total pressures for many years and prefer thin earth pressure cells to avoid the difficulties caused by compressibility of the fill material. Usually five or six cells are used, to enable the magnitude and direction of principal stresses to be measured.

Often the horizontal stress in a direction across the embankment is lower than that in the axial direction.

DR H. ØSTLID

I agree with your remarks about the measurements of total stresses in earth dams not being anything new. I am not even sure whether combining earth pressure cells, pore pressure transducers, inclinometers and temperature sensors in a single cube is new, although I have not seen it before.

The use of thin earth pressure cells is standard practice, but not without problems. As always, the installation procedure and technique is an important factor in the quality of later measurements.

These objections apply equally to the measurements with the cube; however, after using this cube for about 8 years and studying the results our confidence in the results has increased.

This is particularly true when the cube is installed in comparatively plastic clays. In clays with high strengths the measurements of stresses are very difficult both with thin earth pressure cells and with the cube.

MR D. J. KNIGHT, Sir Alexander Gibb and Partners, Reading

Dr Østlid's description of the use of total stress measurements within the body of a wide road embankment 30 m high constructed of soft clay is interesting, and it prompts comparison with similar measurements within soft clay cores of embankment dams constructed of material at a high natural moisture content well above optimum. Dr Penman has also referred to total stress measurements in embankment dams.
Three questions or comments arise from Paper 11.

(i) Would Dr Østlid say where the description of the triaxial total stress cell may be found?

(ii) What were the ratios of the total horizontal stresses, measured both along and across the embankment, to the total vertical stress?

At Monasavu Dam in Fiji, a rockfill dam 85 m high with a very wet, soft clay core, the ratio of total horizontal to total vertical stresses was generally between about 0.85 and 0.90. A set of three total stress cells was provided at each of several elevations up the dam height, to measure vertical, horizontal upstream and downstream and horizontal cross-valley stresses and gave consistent results, published in paper Q56/R68 of the Proceedings of the 15th ICOLD Congress in Lausanne, June 1985. Because of the softness of the clay, in which it was permissible to allow rock plums up to 400 mm size to be left within the fill (they were completely squeezed into the plastic surrounding clay), I do not think that Dr Penman's concern about the usual influence of a large instrument influencing readings is so significant. The high ratio measured at Monasavu demonstrated the attainment of very pleasing, similar level, all-round stress conditions. Another major soft, wet core dam is now under construction in south east Asia, to a height of 120 m.

(iii) It is noted that the counterweight fills were not built simultaneously with the soft clay, which would have prevented the concerns experienced.

At Monasavu Dam strict rules were laid down for core advancement above the general level of the shells, to ensure a good minimum level of support of the weak clay during construction.

DR H. ØSTLID

Results of stress measurements using this cube are reported in the Proceedings of the International Conference on Soil Mechanics and Foundation Engineering, Stockholm, 1981, and further information may be obtained by contacting the Norwegian Road Research Laboratory (NRRL) in Oslo.

The cube was constructed by the NRRL and Geonor, and the cube is designed specially for each case. Consequently, the cube is designed to fit the requirements as the case may be and again detailed information may be obtained from the NRRL.

The question about the variation in stresses in the three directions is the most important feature of the cube measurements, and very briefly our experience is the following.

The ratio of horizontal to vertical total stresses will vary between wide limits, say from 1.5 to 0.4. After the cube has been placed and compaction proceeds, the vertical stress increases regularly with the overburden height. However, the horizontal stresses may be very much higher than the vertical stresses owing to compaction confinement effects. As the height of the fill increases the horizontal stresses do not increase at the same rate and somewhere during the filling operation the ratio passes through a value of unity, i.e. the stresses may be roughly equal in all directions. After the fill has been completed the vertical stress changes very little, but the horizontal stresses continue to decrease. This is reasonable in a high fill of plastic clay: small horizontal deformations should be expected to take place. In the oldest installation that we have done the horizontal stresses stabilized after about two and a half years and the final stress ratio became 0.45. This seems to be very low, but it may be a result of a very long-term continuous horizontal spread of the whole fill which is not unreasonable in these very plastic clays. On future major road fills more cubes will be installed and the measurements will be recorded over a number of years.

MR T. OLOFSSON, Swedish National Road Administration, Borlange.

I agree with Dr Østlid that taking measurements of earth structures is useful for gaining experience in advancing our knowledge, but it is important that we consider the accuracy of the recorded data. An uncritical application of the data may lead to wrong conclusions.

During the reconstruction of a bridge approach the abutment moved laterally. Continuous measurements of the abutment were taken over a long period. The measurements began with a Geodimeter AGA 12 that has an accuracy of ±5 mm (Fig. 1). The results indicated that the abutment did not move. However, more accurate measurements were needed so a Mekometer ME 3000 with an accuracy to ±0.1 mm was used. The results from it indicated a large lateral movement which was not confirmed by the position of the bearings. The explanation of the incorrect result was that the Mekometer was too sensitive to temperature changes during the measurement. The heat of exhaust gas from vehicles was enough to disturb the accuracy severely.

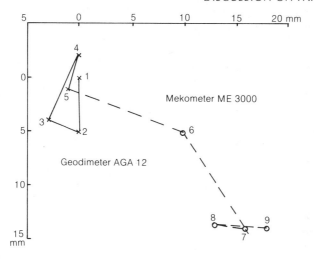

Fig. 1

PROFESSOR A. F. VAN WEELE, Foundacon, The Netherlands

TEMPORARY BUT IMMEDIATE STABILIZATION OF CREEPING SLOPES
Vacuum has been used in The Netherlands to stabilize deforming
slopes temporarily. The system only functions in soils with a
low permeability, as follows.

A single row of vertical sand drains (of the non-
displacement type) were installed at intervals of
approximately 30 m. Each drain was equipped with a small
plastic filter at 5-10 m depth and closed at the top with a
cement-bentonite plug of 2-3 m thickness.

All the filters were connected by means of a small plastic
hose to a vacuum system, consisting of two separate vacuum
tanks and one vacuum pump.

An underpressure was applied to one of the tanks, which was
connected to the filters in the vertical sand drains. These
drains transferred the underpressure to the surrounding soil,
in which the shear deformation took place.

This underpressure resulted in an immediate increase in
effective stress in all directions and thus to an increase in
shear resistance. This stopped further deformation
instantaneously.

The vacuum brings porewater into the tank and when it is
nearly full the vacuum is applied to the second tank, so that
the first tank can be emptied before the vacuum is re-applied.

The method enables adequate permanent measures to be taken
to stabilize the slope.

The system requires only a limited number of drains and, if
the necessary equipment is kept available on site for
emergencies, the vacuum can be applied within 12 hours of the
decision to install it.

It is only effective if the soil has a low permeability so
that the underpressure can develop before large quantities of
groundwater have to be evacuated.

12. The failure of embankment dams in the United Kingdom

J. A. CHARLES, MSc(Eng), PhD, ACGI, MICE, Head of Dams and Earthworks Section, Geotechnics Division, and J. B. BODEN, MSc, FICE, Head of Geotechnics Division, Building Research Establishment

SYNOPSIS. The history of embankment dam failures in the United Kingdom is reviewed, and an attempt is made to classify the types of failures and serious incidents that have occurred in the UK during the last 200 years. Some of the more important of the various categories of failure are illustrated by short case histories. Nearly 100 dam incidents drawn from published information are assessed, and an indication is given of the probable hazards that UK dams face as they continue to age in service.

INTRODUCTION

1. Whereas with most earthworks failures damage will be limited to an area in the immediate vicinity of the earth structure (except for the case of flow slides) the breaching of an embankment dam and the consequent release of impounded reservoir water can cause immense destruction over a wide area downstream of the dam. Many British dams are located in river valleys upstream of densely populated industrial regions and the structural stability and safety of these dams is of major importance. The majority of the dams in the United Kingdom (UK) are earth embankments. About half the total population of dams were built prior to the beginning of the present century and almost all of these dams were earth structures. Reservoir safety in the UK is thus intimately concerned with the behaviour and long term performance of old embankment dams.

2. Until 30 years ago most embankment dams in the UK were built with central puddle clay cores having a typical cross-section as shown in Figure 1. The clay was puddled by destroying the natural structure of the clay, remoulding and working in extra water to form a very wet clay with an undrained shear strength as low as 10 kN/m^2. Frequently a cut-off was formed by excavating a trench along the centre-line of the embankment and filling it with puddle clay. It was common practice with early puddle core dams to take outlet pipes or culverts through the earth embankment. Serious accidents and major disasters led to some changes

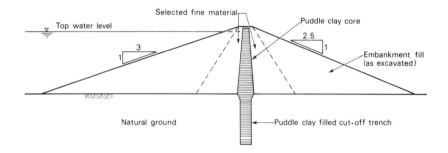

Fig. 1. Typical cross-section of an old embankment dam with
         a puddle clay core.

and developments in the design and construction of puddle
core dams during the nineteenth century. It became
recommended practice to place a selected fill, which was more
cohesive than the general embankment fill, on either side of
the puddle core. The practice of placing outlet pipes
through the embankment was largely superseded by the more
costly but much safer expedient of driving a tunnel through
the natural ground. The superiority of backfilling the
cut-off trench with concrete rather than puddle clay was also
recognised.

3. In this paper an attempt has been made to classify the
types of failures and serious incidents which have occurred
in the UK during the last 200 years. Some of the more
important of the various categories of failure have been
illustrated by short case histories. A total of nearly 100
dam incidents drawn from published information has been
assessed, and an indication is given of the probable hazards
that UK dams face as they continue to age in service.

CLASSIFICATION OF EMBANKMENT DAM FAILURES

4. Failure can be broadly defined as some incident,
occurrence or process whereby an embankment dam does not
perform that function for which it has been constructed,
namely to impound safely a reservoir of water. At the outset
two radically different situations should be distinguished.

(a) An embankment may fail during construction. Shear
    failure may cause instability in one or both of the
    embankment slopes and may also involve the foundations of
    the embankment. This failure situation is similar to the
    failure of any other type of earth embankment. As it
    occurs before the embankment has begun to act as a water
    retaining structure, reservoir safety is not involved at
    that stage.

(b) An embankment may fail when in service impounding a reservoir. The mode of failure is usually closely associated with the retained body of water and therefore differs significantly from the failure of other types of earth embankment. The consequences of the failure may be very severe if the embankment is breached and the impounded water is released. It is useful therefore to have two subdivisions in this failure situation:

(i) Ultimate limit state - catastrophic failure in which the embankment is breached and the reservoir water escapes.

(ii) Serviceability limit state - a serious incident or situation in which it is considered necessary to draw the reservoir level down, carry out other emergency measures and/or carry out major remedial works.

5. During construction, the embankment failure mechanism most likely to occur is that due to shearing along a slip surface. However, an embankment dam impounding a reservoir may fail due to a number of quite different causes. Shear failure could, of course, occur during impounding as well as during construction. In addition, the dam could be overtopped during a flood and fail by surface erosion caused by the reservoir water flowing over the crest and the downstream slope. The dam could also fail by internal erosion caused by excessive seepage and leakage through the embankment or its foundations. (Leakage could be so excessive that it prevents a reservoir filling, but generally it is a problem because it leads to internal erosion). Failure can occur due to a combination of these causes. Other less frequent causes also lead to degradation in dams; see for example Penman, 1980 (ref 1).

6. This classification of embankment dam failures as discussed above is summarised in the following table.

TABLE 1: CATEGORISATION OF EMBANKMENT DAM FAILURES

1.	Construction failures	
2.	In-service failures	
	Ultimate Limit State	Serviceability Limit State
	(a) External erosion (overtopping) (b) Internal erosion (seepage, conduit leakage) (c) Shear (slips, slides) (d) Other	

7.   In reviewing earth dam practice in the USA Middlebrooks (ref 2) listed the unsatisfactory performance of about 200 earth dams. His analysis of this data is summarised in the following table:

TABLE 2: EARTH DAM FAILURES (after Middlebrooks)

Cause	%
External erosion (overtopping)	30
Internal erosion (seepage, conduit leakage)	38
Shear (slips, slides)	15
Other	17

Failure mechanisms, falling into the various categories discussed above are illustrated by the following case histories drawn from some of the incidents which have occurred at embankment dam sites in the UK.

CONSTRUCTION FAILURES AND INCIDENTS

8.   Failures during construction are generally associated with shear instability and are similar therefore to the failure of any other type of earth embankment. The presence and shape of a core, especially a very wet puddle clay core has a considerable influence on the stability of the embankment. The core should consolidate with time and generally stability should improve. The end of construction or the rapid first filling of a reservoir are therefore likely to be critical periods for this form of construction. Serious incidents occurring during the construction of earth dams have had a major influence on the development of soil mechanics in the United Kingdom.

9.   Chingford. In 1937 a shear failure occurred during the construction of a reservoir at Chingford in Essex and was reported by Cooling & Golder (ref 3). A continuous earth embankment some 3.5 miles in length was being built to form a storage reservoir. The embankment had a central puddle clay core and a puddle clay cut-off trench as shown in Figure 2(a). The bank was built directly upon a layer of soft yellow clay. At the end of July 1937, with the embankment 8 m high at a section where the completed height would be 10 m, a 90 m long section of the downstream (outer) slope moved 4 m as shown in Figure 2(b). An investigation was carried out by the Soil Mechanics Section of the Building Research Station. Excavation was carried out to determine the location of the failure surface. It appeared to pass through the puddle core and then to follow a path contained within the layer of soft yellow clay. The undrained shear strengths of the yellow and puddle clays were measured by laboratory direct shear tests yielding values of 14 and 10 kN/m² respectively. A stability analysis was carried out

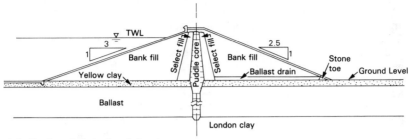

a) Section through the bank as designed

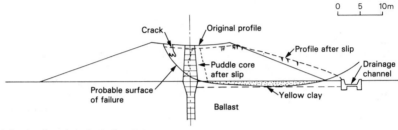

b) Section through the bank after slip

Fig. 2.   Shear failure during construction at Chingford
          (after Cooling & Golder, ref. 3)

in terms of total stresses and a factor of safety close to
unity was obtained. Chingford was one of the first dams
built in the UK using what in those days would have been
described as 'modern earth moving equipment'. Thus the
construction rate would have been faster than had previously
been common practice. It seems likely that the development
of high pore water pressures in the yellow clay due to rapid
loading by the embankment was a major factor contributing to
the failure.

   10. <u>Muirhead and Knockendon</u>. In September 1941 movements
were observed in the pitching on the upstream slope over the
central section of Muirhead dam (refs. 4 & 5). The
embankment was 21 m high at this stage and a further 5 m of
fill had still to be placed. The embankment had slopes of
1 in 3, a central puddle clay core and shoulders of boulder
clay. The Building Research Station carried out an extensive
investigation of the failure. An initial survey established
that the upstream slope had moved outwards up to 1.2 m and
that a berm on the downstream slope had moved 0.6 m. Movements were horizontal and the toe walls had not moved. When
0.5 m of fill was added further horizontal movements of about
0.3 m were monitored. It was believed that the embankment
had failed through the lower part of the shoulder fill. The
strength of this material was found to be very variable but

the average measured value of undrained shear strength was close to the value of 40 kN/m² which corresponded to limiting equilibrium. The final height of the dam was limited to 21 m and the upstream slope was stabilised by a substantial berm. At the time of the Muirhead failure a similar embankment was under construction nearby at Knockendon and the fill had reached about one fifth of the full height (ref 6). As a result of the events at Muirhead the section was modified by adding toe weighting to the upstream shoulder and by including a zone of stronger granular fill in the downstream shoulder. Standpipe piezometers were installed to monitor construction pore pressures in the fill and so check the rate of consolidation. The measured pore pressures were used together with the results of drained shear box tests to calculate the stability of the embankment.

11. <u>Usk</u>. The measurement of pore water pressures by the Building Research Station at the site of the Usk dam during the early 1950s had an important influence on the development of embankment dam design and construction techniques in the UK (ref 7). The Usk embankment dam was 33 m high with a central core of puddle clay. The cut-off trench was filled with concrete. The shoulders of the embankment were made of a boulder clay fill. A layer of silt was found to be present under the downstream shoulder and as a consequence a sand drain system was installed. Twin tube hydraulic piezometers were installed in the silt layer to check on the performance of the drains. Also three piezometer tips were installed at mid-thickness of the first season's fill in the downstream shoulder in July 1952. The tips in the silt layer in the foundation measured no significant pore pressures, thus indicating that the drainage system was effective. The pore pressures in the fill were however very large. Effective stress stability analyses indicated that the factor of safety would be unacceptably small if the dam was brought to full height with the average pore pressure ratio (r_u) greater than 0.5. The dissipation that occurred during the winter shutdown period appeared insufficient to ensure stability. Advice was sought from Professor Skempton of Imperial College. Fifteen steel standpipes driven into the fill confirmed the BRS pore pressure measurements. It was decided to place horizontal drainage layers within the embankment shoulders and it is believed that this was the first use of horizontal drainage layers of this type in an earth dam to control construction pore pressure.

12. The use of instrumentation and associated construction techniques such as those described above have considerably reduced the risk of embankment failure during construction. The current improvements in understanding of the interaction between construction rates and pore pressure development (and dissipation) in the embankment and its foundation should continue to reduce the probability of this type of failure in the future. Nevertheless the major upstream slip at

Carsington in 1984 has demonstrated that the risk of
construction failure has by no means been eliminated.

13. <u>Carsington</u>. At the beginning of June 1984 a 400 m
length of the upstream shoulder of Carsington dam slipped
some 15 m. Early details of the failure were reported by
Byrd & Middelboe in mid June (ref 8). At the time of the
failure, embankment construction was virtually complete with
the dam approaching its maximum height of 35 m. The design
of the Carsington scheme has been described by Davey & Eccles
(ref 9). A cross section of the embankment at its maximum
height is shown in Figure 3. Horizontal drainage blankets

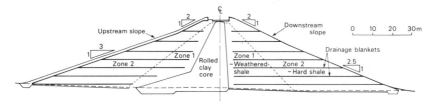

Fig. 3. Cross-section of Carsington dam (after Davey and
          Eccles, ref. 9).

were incorporated in both the upstream and the downstream
shale fill shoulders. Piezometers had been installed and
pore pressures were being monitored in the foundation, in the
clay core, and in the shoulder fill. Effective stress
stability analyses had been carried out. Byrd & Middelboe
(ref 8) have suggested that the failure surface probably
passed through the boot shaped clay core and a relatively
thin layer of surface clay in the foundation of the dam. At
first sight there appears to be some similarity between the
recent events at Carsington and the failure of Chingford 47
years previously. For example, at the time of the failure,
Chingford was about 8 m high and slipped about 4 m (ref 3);
whereas, Carsington was approaching its full height of 35 m
and slipped about 15 m (ref 8). As is often the case with
failures, it is possible that the recent events at Carsington
will make further important contributions to our fundamental
understanding of the behaviour of large earthworks of this
type. At the time of writing this paper (July 1984) the
investigations into the failure at Carsington (involving BRE
as well as many other geotechnical specialists) have just
commenced and therefore it is not appropriate to review
circumstances which are unique to the Carsington site.

14. <u>Recent developments</u>. Two recent developments in our
understanding of the stability of embankment dams during
construction may become part of design practice in the
future.
(i) It is now becoming recognised that in reality the shear
strength envelope for a soil may be significantly curved,

see for example Charles & Soares (refs. 10,11). However it is still current design practice to construct the failure envelope (determined by laboratory triaxial compression tests) as the best straight line approximation to the curve characterised by a cohesion intercept (c') and an angle of shearing resistance (ϕ'). The resulting values of c' and ϕ' therefore represent a geometric convenience rather than the true properties of the soil. Present evidence suggests that the real value of c' (due to electro-chemical bonding) is relatively small compared with that obtained from the linearisation of the curved failure envelope. Consequently this approach will over-estimate the strength of the soil at low normal effective stresses and this may be significant in stability analyses.

(ii) In wet clay cores the pore pressure ratio (r_u) may be close to unity ($r_u = u/\gamma h$ where u is the pore pressure at depth h and γ is the bulk density of the soil). It might be considered that the amount by which r_u is less than unity is an indication of the vertical effective stress in the core. However, a wet clay core is likely to arch between the stiffer embankment shoulders and the total vertical stress may be significantly smaller than γh. This means that the vertical effective stress may be even smaller than the r_u value appears to suggest. Limit equilibrium stability analyses do not model the arching of the core and little work has been done to examine their validity with high r_u values in a central clay core.

IN SERVICE FAILURES AND SERIOUS INCIDENTS

15. Unlike failures during construction, failures of earth dams when in service are usually intimately connected with the special function of the dam to impound a reservoir of water. The catastrophic failure of two dams in the nineteenth century led to early pressure for reservoir safety legislation. Bilberry dam failed in 1852 and resulted in the loss of 81 lives and Dale Dyke in 1864 led to 244 deaths. These are the two worst dam disasters that have occurred in Britain and brief accounts of these failures are given.

16. Bilberry. Bilberry dam was 20 m high and had a puddle clay core 2.4 m wide at the top and 4.9 m at ground level (ref 12). Excavation for the puddle clay filled cut-off trench began in 1839 and a spring was encountered in the bottom of the trench. The outlet works comprised a masonry culvert which had to cross the puddle trench. Serious problems soon became apparent. Muddy water came through the culvert in 1841 and in 1843 the leakage became worse and water burst through the culvert. Remedial works were unsuccessful and large settlements occurred. It was claimed that between 1846 and 1851 the bank settled 3 m. This settlement eliminated the freeboard and soon after midnight on 5 February 1852 the embankment was over-topped and breached. The resulting flood claimed 81 lives in the Holme

valley below the dam. It would appear that erosion of and
through the puddle clay was the cause of the settlement that
led to the catastrophe. It is noteworthy that the Bilberry
dam had: (a) a puddle clay filled cut-off trench, (b) a
highly permeable fill on either side of the puddle clay core,
and (c) a culvert through the embankment.

17. Dale Dyke. The original Dale Dyke dam was constructed
on the River Loxley, 9 miles west of Sheffield. It has been
described by Binnie (refs. 12,13). Construction work started
in January 1859. The cut-off was completed in 1861 and the
embankment was finished by April 1863. Impounding commenced
in June 1863 and by 10 March 1864 the water level was 0.7 m
below the crest of the weir. In the late afternoon of the
following day a crack was observed along the downstream slope
near the crest of the dam. At 23.30 a collapse occurred and
the dam was breached. In the resulting flood 244 lives were
lost. Extensive property damage was caused, including some
in the City of Sheffield. Figure 4 shows a cross-section of

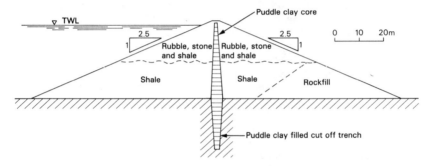

Fig. 4. Cross-section of old Dale Dyke embankment dam (after
Binnie, refs 12, 13).

the embankment as deduced from evidence given at the inquest.
The total height of the embankment was 29 m, and the maximum
depth to the base of the puddle clay filled trench was 47 m
below the crest. The top width of the core was only 1.2 m
and with batters on both faces of 1:16 produced a maximum
width at ground level of 4.9 m. The core within the
embankment was of course located directly above the
compressible puddle clay in the cut-off trench, whereas the
shoulders were founded upon the relatively incompressible
rock of the Millstone Grit Series. Differential settlement
would therefore be expected to take place between the puddle
clay core and the shale shoulders leading to arching across
the core. Under these circumstances hydraulic fracture may
have occurred and been the root cause of the failure. The
shoulders being relatively permeable (and not including
anything approaching the properties of a graded filter) would
be unable to contain the products of erosion of the core and
internal erosion would therefore be able to progress rapidly

through the body of the embankment. It is again noteworthy (as was the case with Bilberry dam) that the original Dale Dyke dam had: (a) a puddle clay filled cut-off trench, (b) permeable fill immediately on either side of the puddle clay core, and (c) outlet pipes laid in a trench beneath the embankment (although the available evidence suggests that in this case, the pipes did not play any part in causing failure).

18. At the inquest following the Dale Dyke disaster the jury stated that the Legislature ought to take such action as would result in government inspection of all works of this character and that such inspections should be frequent and sufficient and regular. However, no incidents involving loss of life occurred for another 60 years and no legislation was introduced. In 1925 two failures occurred which did cause loss of life. A small dam at Skelmorlie in Scotland failed and there was a disaster at Dolgarrog (Coedty dam) in North Wales. The latter failure was the more serious and is better documented; a brief description is given here.

19. Coedty. The Coedty dam was constructed in 1924 (ref 14). It had a maximum height of 11 m. Earth shoulders supported a central concrete core wall. On Monday 2 November 1925 the small concrete Eigiau dam failed. It is believed that this was due to a blow-out of the lower part of the dam wall at a point where there had been a seepage path for several years. The water thereby released surged into the nearly full Coedty reservoir below. The dam was overtopped, the material supporting the core wall on the downstream side was washed away and the wall collapsed. Sixteen people were killed in Dolgarrog by the resulting flood.

20. The reservoir failures of 1925 led to the Reservoirs (Safety Provisions) Act of 1930. Since this Act was brought into force, and periodic inspection by a qualified engineer became mandatory, no dam failures have occurred in Britain which have caused loss of lives. Dams have however been breached.

21. Warmwithens. Warmwithens dam was a clay fill embankment built more than 100 years ago near Oswaldtwistle, Lancashire. It was some 10 to 12 m in height and the reservoir it impounded lay in series above two other small reservoirs Cocker Cobbs and Jackhouse. During the period 1965 to 1966, the dam was raised to provide adequate freeboard and the old cast iron draw-off pipe was replaced by a reinforced concrete segmental tunnel driven through the dam itself. The tunnel contained a steel pipe for the water outlet. Moffat (ref. 15) has given an account of the failure. At 07.30 on 24 November 1970 an escape of water was detected and by 13.30 the dam was completely breached to foundation level. The water impounded by the dam was discharged into the two lower reservoirs. The embankment dam of Cocker Cobbs was overtopped, but it did not fail, and the water passed the spillways of the lowest reservoir, Jackhouse, without causing serious damage. Had a cascade

failure of the two lower dams taken place, the resulting flood could have caused serious damage in Oswaldtwistle. The breach occurred along the line of the outlet tunnel (Fig 5).

Fig. 5.   Failure of Warmwithens embankment dam (from Engineering Now, No. 4, 4 December 1970).

It seems possible, therefore, that seepage through, or along the perimeter of the abandoned cast draw-off pipe, or along the perimeter of the new tunnel could have played a part in causing the failure.

22.   Although complete failure and breaching of embankment dams fortunately have been relatively rare occurrences in Britain, there have been many serious incidents affecting dams in service. These incidents in some instances have warranted emergency drawdown of the reservoir and costly remedial works. Four incidents are now presented. They have been selected to give an indication of the wide range of problems that have developed with earth dams and the different ways those problems have manifested themselves.

23.   <u>Lower Lliw</u>. Construction of the Lower Lliw dam north of Swansea commenced in 1862 and was completed in 1867. The dam was 27 m high and had a puddle clay core. In 1873 water started to flow from downstream drains at a much increased rate and the water was turbid. A spring had burst through

the fissured rock below the puddle clay core. Erosion of the puddle clay led to settlement of the embankment. Remedial work involved an open cutting 50 m wide at the top and 15 m wide at the bottom to a depth of 11 m below the top of the embankment and a trench 9 m long and 6 m wide sunk from the bottom of the cutting to the rock, a total depth of 32 m below the top of the embankment. At a depth of 7 m in the trench a fissure 0.6 m wide was found in the puddle clay filled with the coarse material of the selected fill. The fissure extended down to the face of the rock. A drain was installed to take away the water from the spring which acted on the clay at the bottom of the trench. The trench was backfilled with puddle clay. In 1883 after two years of service, leakage again increased. Turbid water came from the drains and settlement occurred at the location of the remedial works. The Lower Lliw reservoir was then used from 1883 to 1975 with a top water level reduced by 5 m. The dam was eventually demolished and a new embankment completed in 1978 (ref 16).

24. <u>King George V</u>. King George V dam is a continuous embankment of some 4.5 miles in length. It has a height of 7.6 m and consists of clay fill with a puddle clay core. The embankment was built on unstripped grass and topsoil without any special provision for under-drainage of the downstream (outer) slope. The absence of drainage in this region could be expected to make the downstream shoulder more vulnerable to instability in the event of leakage through the clay core. In September 1939 is was decided to reduce the top water level by 1.5 m. This restriction was maintained until February 1945 when raising the water to its previous top level began. Water leakage appeared at the toe of the embankment as the original top level was approached. The possibility of the leakage indicating an incipient major failure was recognised and the water level was lowered and an investigation was initiated in association with the Building Research Station (ref. 17). The investigation revealed the presence of roots in the puddle clay down to the previous temporary top water level 3.1 m below the crest of the bank. This evidence together with extensive field observations and soil testing indicated that the passage of water was through the upper part of the core wall which had been subject to drying, shrinkage and cracking during the years 1939-45.

25. Refilling of a long empty reservoir impounded by an embankment dam having a puddle clay core wall should always be undertaken with the utmost caution. Recent research by the Building Research Establishment (ref. 18) has indicated the possibility of hydraulic fracture occurring during this operation. Especial caution is necessary in the case of a dam having any or all of the following features: (a) a puddle filled cut-off trench, (b) permeable fill on either side of the puddle clay core, and (c) outlet works passing through the body of the embankment.

26. <u>Lluest Wen</u>. On 23 December 1969 a horse fell into a
2 m deep hole in the crest of the Lluest Wen dam in South
Wales. The 20 m high dam had been built in 1892. Subsidence
had occurred previously in 1912 and 50 tonnes of cement grout
had been injected in the area of the valve shaft in 1915-16.
New investigations indicated that puddle clay was emerging
from an 0.15 m diameter drainage pipe where it terminated

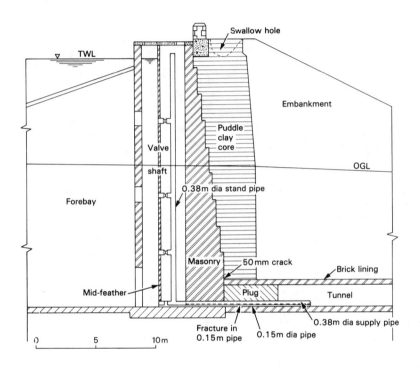

Fig. 6. Section at valve shaft at Lluest Wen embankment dam
(after Little, ref. 19).

near the plug in the draw-off tunnel (Fig 6). It was feared
that the dam would collapse and emergency measures were put
in hand. People living downstream were evacuated. The 0.38
m diameter draw-off pipe was inadequate for rapidly lowering
the reservoir water level and a large number of pumps, some
positioned by helicopter, were brought in to lower the water
level (Fig 7). Over 18 tonnes of clay/cement grout were
injected into a single hole in the neighbourhood of the valve
shaft where subsidence had occurred. An emergency cut was
made through the spillway lowering the overflow level of the
reservoir by 9 m. With the emergency over full grouting of
the puddle clay core was undertaken. Fifty tonnes of

Fig. 7.   Emergency measures at Lluest Wen embankment dam
          (from an original photograph by Mr Frank Herrmann,
          reproduced here by kind permission of The Times).

clay/cement grout were injected into the puddle clay core.
After grouting further investigation was carried out. The
core was found to consist of sandy silty clay with pockets of
silt and sand. The core had a series of cracks, many of them
open, iron stained by seeping water. About 75% of the cracks
were within 5° of the horizontal. Open water-worn cavities
were found in two drill-holes. The water content and
undrained shear strength fluctuated widely and erratically.
The core was very soft in the vicinity of the valve shaft.
In view of these findings it was decided that grouting alone
could not provide a satisfactory solution. A new plastic
concrete core was installed using the slurry trench method.
During the excavation of a 6 m diameter shaft at the valve
tower it was discovered that the brickwork of the draw-off
tunnel had not been bonded into the masonry at the back of
the valve shaft. Puddle clay had eroded through a 50 mm gap
and then through a crack in the 0.15 m diameter pipe. At the
time of the emergency there was a 0.06 cu m pile of puddle
clay at the downstream end of the 0.15 m diameter pipe.
Little (1977) (ref 19) noted how much hinged on a tiny detail
which might never have been detected until perhaps too late
but for the requirement for major remedial works.
     27. Withens Clough. Withens Clough dam consists of a
boulder clay embankment some 22 m maximum height built in
West Yorkshire between 1890 and 1894. It has a 2 m wide
vertical puddle clay core and cut off, the crest is 5 m wide

x 250 m long, and both shoulders have 1 in 3 slopes. The
foundation comprises boulder clay overlying grits and
mudstones. The investigation and remedial works carried out
at the dam have been described by Arah (ref 15). In the
original draw-off works the main low-level system consisted
of a 1.5 m diameter masonry culvert, closed by an iron
bulkhead in the centre of the dam from which ran a 460 mm
diameter cast iron supply pipe. The first control valve was
located at the downstream toe. This formed an intrinsically
unsafe system. With no upstream control valve there was a
serious risk of damage by erosion of the dam if the pipe or
the bulkhead failed. Deterioration of the culvert, where it
passed through the puddle clay core could create an
especially hazardous situation. Because of the superficial
evidence of deterioration in the condition of the dam,
detailed investigations were started in 1957. This revealed,
inter-alia, evidence of seepage through the core into the
downstream shoulder where very high pore water pressures were
recorded. In addition peaty silt tipped on the downstream
slope was found to conceal a previous slip. Remedial works
were carried out between 1971 and 1972, and included: (a) new
draw-off works constructed in tunnel around one of the
abutments, (b) old draw-off works backfilled, sealed and part
removed, (c) core and abutments grouted, (d) plastic concrete
diaphragm constructed along centre line of core wall, and (e)
peaty fill removed and old slip in downstream slope rebuilt
with free draining granular material.

ANALYSIS OF UK EMBANKMENT DAM STATISTICS
    28. Appendix 1 lists examples of unsatisfactory
performance of earth dams in the United Kingdom. It is
recognised that such a list will be incomplete and it is
difficult to define how serious an incident has to be to
merit inclusion. The examples are all ones about which there
is some published information and references are given.
Published information about early incidents is sometimes very
limited. For instance it appears that two dams near
Huddersfield, Diggle and Swellands, failed in 1810. The dams
are close to each other and none of the references mentions
two failures. The suspicion must be that the different
accounts refer to one failure. The danger of merely listing
failures found in previous lists without any other published
information can be illustrated by Middlebrooks data (ref 2).
He includes the failure of Dale Dyke in 1864 and also a
failure at Bradford in 1896. The reference for the latter is
an earlier list produced by Justin (ref 20) in which a dam
called Bradford, located in Sheffield, is reported as
failing in 1869. This however is clearly a reference to the
Dale Dyke (or Bradfield) failure in 1864. One failure has
thus been listed twice by Middlebrooks. For each dam listed
in Appendix 1 a brief description is given of the nature of
the unsatisfactory performance. Where the same problem has

recurred at a particular dam only one entry is made in the appendix. Where two different problems have arisen they are listed separately.

29. The data presented in Appendix 1 has been analysed. Firstly the proportion of the total number of failures and serious incidents that occurred during construction was examined. Where overtopping or seepage and erosion problems occurred due to impounding before embankment construction had been completed, these have been categorised as in-service failures. The construction category is thus restricted to shear failures and movements in embankments occurring before impounding commenced and consequently not involving reservoir safety.

TABLE 3: SITUATION IN WHICH FAILURE OCCURRED

Situation	No	%
Construction	17	19
In service	71	81

30. The in-service failures can be sub-divided into two categories; catastophic breaching of the embankment (ultimate limit state) and serious incidents (serviceability limit state).

TABLE 4: SERIOUSNESS OF IN-SERVICE FAILURES AND
         INCIDENTS LISTED IN APPENDIX 1

	No	%
Ultimate limit state	18	25
Serviceability limit state	53	75

31. The in-service failures and incidents affecting reservoir safety can also be analysed on the basis of the primary cause of the failure. Clearly in many situations there may be more than one cause, but the aim has been to identify the primary cause, eg if internal erosion over a long period reduces the freeboard and the dam finally fails by overtopping, the failure has been classified as due to internal erosion, if an inadequate spillway leads to overtopping during a flood and this causes a downstream slip, the failure has been classified as due to overtopping. The distribution of causes of failures derived by Middlebrooks from USA data is shown in brackets in Table 5. From this evidence it would appear that internal erosion is probably the greatest hazard which embankment dams face both in the UK and the USA. More detailed examinations of past failures and serious incidents together with appropriate investigations

and instrumentation of dams currently affected by internal erosion may yield a better understanding of the mechanisms involved and provide guidance for the design of more effective remedial works.

TABLE 5: CAUSES OF IN-SERVICE FAILURES AND SERIOUS INCIDENTS IN THE UK

Cause	No	%
External erosion (overtopping)	17	24 (30)
Internal erosion (seepage, leakage)	39	55 (38)
Shear (slips, slides)	10	14 (15)
Other	5	7 (17)

CONCLUSIONS

32. Although many old British embankment dams suffered shear failures during construction, it would appear that this hazard is of relatively minor importance to these dams as they age in service. This is probably because the wet central puddle cores in these dams have consolidated with time, so that the end of construction was usually the most critical period for shear stability failure.

33. There are three major hazards to UK embankment dams in service impounding reservoirs:

(a) Surface erosion when the embankment is overtopped.
(b) Internal erosion of the embankment or its foundations.
(c) Shear failure in the embankment slopes.

Internal erosion is probably the greatest hazard and also the least well understood.

ACKNOWLEDGMENTS

The work described in this paper forms part of the Research Programme of the Building Research Establishment and is published by permission of the Director. The main customer for the work is the Water Directorate of the Department of the Environment. The authors are grateful for the support, encouragement and assistance of colleagues both in the BRE and the Water Directorate, and to the dam owning authorities in the UK who have allowed BRE research to be carried out on their dam sites. The authors are especially grateful to Messrs M G Healey and J W Phillips for their guidance, and valuable criticisms and suggestions.

REFERENCES

1. PENMAN, A D M. Deterioration and failure of dams. BNCOLD News and Views 1980, pp 13-15.

2. MIDDLEBROOKS, T A. Earth dam practice in the United States. Transactions American Society of Civil Engineers, Centennial volume, 1953, pp 697-722.

3. COOLING L F and GOLDER H Q. The analysis of the failure of an earth dam during construction. Journal Institution of Civil Engineers, Vol 19(1), 1942, pp 38-55.

4. Building Research Station. Report of investigation on the stability of the earth dam at Muirhead reservoir, Dalry, Ayrshire, July 1942, 10 pp.

5. BANKS, J A. Construction of Muirhead Reservoir, Scotland. Proceedings 2nd International Conference on Soil Mechanics and Foundation engineering, Rotterdam, Vol 2, 1948, pp 24-31.

6. BANKS, J A. Problems in the design and construction of Knockendon dam. Proceedings Institution of Civil Engineers, Part 1, Vol 1, No. 4, pp 423-443.

7. PENMAN, A D M. Construction pore pressures in two earth dams. Proceedings of Conference on Clay Fills, Institution of Civil Engineers, London, November 1978, pp 177-187.

8. BYRD T and MIDDLEBOE S. Weak ground cited as Carsington fails. New Civil Engineer, 14 June 1984, pp 4-6.

9. DAVEY P G and ECCLES P G. The Carsington scheme - reservoir and aqueduct. Journal Institution of Water Engineers and Scientists, Vol 37, No. 3, June 1983, pp 215-239.

10. CHARLES J A and SOARES M M. Stability of compacted rockfill slopes. Geotechnique, Vol 34, No. 1, March 1984, pp 61-70.

11. CHARLES J A and SOARES M M. The stability of slopes in soils with non-linear failure envelopes. Canadian Geotechnical Journal, 1984.

12. BINNIE, G M. Early Victorian Water Engineers. Thomas Telford, London, 1981, 310 pp.

13. BINNIE, G M. The collapse of the Dale Dyke dam in retrospect. Quarterly Journal of Engineering Geology, Vol 11, No. 4, 1978, pp 305-324.

14. WALSH P and EVANS J. The Dolgarrog dam disaster of 1925 in retrospect. Quest, Journal of City University, Issue 25, 1973, pp 14-19.

15. British National Committee on Large Dams and University of Newcastle-upon-Tyne. Proceedings of Symposium on Inspection, Operation and Improvement of Existing Dams, Newcastle-upon-Tyne, September 1975.

16. HOWE, G R. Past and present of the Lower Lliw dam. Concrete, May 1977, pp 14-17.

17. BISHOP, A W. The leakage of a clay core-wall. Transactions Institution of Water Engineers, 1946, Vol 51, pp 97-131.

18. BODEN J B and CHARLES J A. The safety of old embankment dams in the United Kingdom - some geotechnical aspects. Municipal Engineer, Vol 111, No. 2, February 1984, pp 46-60.

19. LITTLE, A L. Investigating old dams. Quarterly Journal of Engineering Geology, 1977, Vol 10, No. 3, pp 271-280.
20. JUSTIN, J D. Discussion on The design of earth dams. Proceedings American Society of Civil Engineers, Vol 49, 1923, pp 1883-1900.
21. SCHOFIELD, R B. Benjamin OUTRAM (1764-1805), canal engineer extraordinary. Proceedings Institution of Civil Engineers, Part 1, Vol 66, 1979, pp 539-555.
22. HODSON G and F W. Loughborough Corporation Waterworks - the Blackbrook dam. Transactions British Association of Waterworks Engineers, Vol 7, 1902, pp 203-209.
23. HADFIELD C and BIDDLE G. The canals of North West England, Vol 2. David & Charles, Newton Abbot, 1970, 496 pp.
24. LINDSAY, J. The canals of Scotland. David & Charles, Newton Abbot, 1968, 238 pp.
25. International Commission on Large Dams. World Register of Dams (2nd update 31/12/1977), Paris 1979, 322 pp.
26. SWALES J K. Repair works in connection with the Belmont reservoir of the Bolton Corporation. Transactions Institution of Water Engineers, Vol 31, 1926, pp 203-240.
27. BATEMAN J F T. History and description of the Manchester Waterworks, Spon, London, 1884, 291 pp.
28. ARTHUR, C. History of the Fylde Waterworks, 1861-1911. Fylde Water Board, Blackpool, 1911, 383 pp.
29. FOX, W. Reservoir embankments, with suggestions for avoiding and remedying failures. Transactions Society of Engineers, 7 March 1898, pp 23-46.
30. LAPWORTH, H. The geology of dam trenches. Transactions Institution of Water Engineers, Vol 16, 1911, pp 25-66.
31. BOWTELL, H D. Reservoir railways of the Yorkshire Pennines. Oakwood Press, Blandford, 1979, 128 pp.
32. HENZELL, C G. Discussion on The use of grout in cut off trenches, and concrete core walls for earthen embankments. Transactions Institution of Water Engineers, Vol 27, 1922, pp 178-179.
33. International Commission on Large Dams. Lessons from dam incidents, Paris, 1974, 1069 pp.
34. WALTERS, R C S. Dam Geology, Butterworths, London, 1962, 335 pp.
35. McGAREY, D G. Emergency repairs to Weedon embankment on the Grand Union Canal, 1939. Journal Institution of Civil Engineers, December 1942, pp 128-132.
36. DAVIES, D G. The Harrogate dam failure. Journal Institution of Water Engineers, Vol 7, Part 1, 1953,   pp 57-79.
37. VAUGHAN P R, KLUTH D J, LEONARD M W and PRADOURA H H M. Cracking and erosion of the rolled clay core at Balderhead dam and the remedial works adopted for its repair. 10th Congress on Large Dams, Vol 1, Montreal 1970, pp 73-93.
38. British National Committee on Large Dams. Dams in the UK, 1963-1983. Institution of Civil Engineers, London, 1983, 215 pp.

39. KENNARD M F. Examples of the internal conditions of some old earth dams. Journal Institution of Water Engineers, Vol 26, part 3, 1972, pp 135-154.
40. British National Committee on Large Dams, 1982 Conference, University of Keele. Institution of Civil Engineers, London, 115 pp.
41. FERGUSON P A S, Le MASURIER M and STEAD A. Combs reservoir: measures taken followng an embankment slip. 13th Congress on Large Dams, Vol 2, New Delhi 1979, pp 233-245.
42. New Civil Engineer, 21 August 1980. Magazine of the Institution of Civil Engineers, London, p6.
43. New Civil Engineer, 9 June 1983. Magazine of the Institution of Civil Engineers, London, pp 16-17.

APPENDIX 1: UNSATISFACTORY PERFORMANCE OF EMBANKMENT DAMS IN UK

Dam	Location	Date built	Height m	Date of incident	Nature of incident	Refs.
Marsden	Huddersfield		18	1799	Partial collapse probably due to overtopping during floods.	21
Blackbrook	Loughborough	1797	11	1799	Large crest settlement associated with major leakage; embankment overtopped and breached.	22
Aldenham	Watford	1795	8	c1802	Slips in upstream slope following reservoir drawdown.	15
Diggle	Huddersfield			1810	Embankment burst during floods; 5 lives lost.	23
Swellands	Huddersfield	1810		1810	Failure probably due to under seepage; Colne Valley inundated and 5 lives lost.	21
Glen Clachaig	Crinan canal			1811	Following violent gale, embankment collapsed; canal closed for 1 year.	24
Whinhill	Greenock	1796	12	1815	Rebuilt after failure.	25
Whinhill	Greenock	1796	12	1835	Rebuilt after failure.	25
Belmont	Bolton	1826	25	1844	Slip in upstream slope after raising of embankment.	26
Darwen	Blackburn			1848	Failure during heavy flood, 12 or 13 lives lost.	27
Woodhead	Glossop			1849	Overtopped and breached by flood during construction.	12,27
Woodhead	Glossop			1851 1859	First embankment abandoned due to excessive leakage and indications of erosion.	12,27
Bilberry	Holmfirth	1845	20	1852	Settlement caused by internal erosion led to overtopping and collapse; 81 lives lost.	12
Torside	Glossop	1857	31	1854	Fractures in discharge pipes during first filling due to spread of base of embankment.	12
Grimwith	Pateley Bridge	1864	25	1861	Leakage caused by unequal settlement of culvert and valve pit.	12
Piethorn	Rochdale	1862	26	1862	Slip during construction.	12
Dale Dyke	Sheffield	1863	29	1864	Catastrophic failure of embankment during first filling of reservoir; 244 lives lost.	12,13
Doe Park	Bingley	1862	18	1864	Serious leakage through puddle core.	12
Stubden	Bingley	1862	20	1867	Serious leakage associated with culvert through embankment.	12
Grizedale	Preston	1866		1867	Serious leakage and erosion of puddle clay in trench.	28

Dam	Location	Date built	Height m	Date of incident	Nature of incident	Refs.
Damflask	Sheffield	1896	28	1869	Excessive leakage through jointed rock at one end of puddle trench during first filling.	12
Blenheim	Woodstock	c1760	9	c1873 & subs.	Settlement caused by internal erosion.	15
Lower Lliw	Swansea	1867	27	1873 1883	Serious leakage associated with erosion of puddle clay and settlement; recurred after remedial works.	12,16
Leeming	Keighley	1877	23	1875	Embankment settlement damaged culvert prior to reservoir filling.	12
Cowm	Rochdale	1877	16	1877	Serious leakage on first filling; grouted.	12
Den of Ogil	Forfar	1887	18	1881	Serious leakage on first filling through poor concrete in trench; some erosion of puddle clay.	29
West Hallington	Newcastle-upon-Tyne			1887	Slips during construction.	30
Monkswood	Bath	1896	15		Slips during construction.	29
Clubbiedean	Edinburgh	1850	17	1897 & subs	Limestone foundation dissolved causing leakage.	12
Butterley	Huddersfield	1901	31	1901	Leakage on first filling probably through foundation rock.	31
Upper Roddlesworth	Blackburn	1865	21	1904 1905 1908	Swallow holes in upstream side of clay core.	12
Eccup	Leeds	1897	25	1905	Excessive leakage greatly reduced by gravity cement grouting.	32
Thorpe Malsor	Kettering	1905			Partial failure during construction.	15
Upper Creggan		1851	10	1908	Failure due to overtopping.	15
Walshaw Dean Lower	Burnley	1907	24	1908	Serious leakage on first. filling through foundation rock	31
Walshaw Dean Middle	Burnley	1907	24	1908	Serious leakage on first. filling through foundation rock	31
Blaenant Ddu	Swansea	1878	24	1919	Taken out of service due to unacceptable seepage into mine workings, demolished 1978.	12
Cowlyd	Dolgarrog	1921	14	1924	Overtopping caused entire downstream shell to be scoured out exposing concrete core wall.	15
Castle Howard	York	1798	6	1925 & subs	Leakage and settlement over long period.	15
Coedty	Dolgarrog	1924	11	1925	Overtopped and breached during flood caused by collapse of concrete Eigiau dam, 16 lives lost.	14
Skelmorlie	Largs			1925	Overtopped and breached during flood caused by release of water from flooded quarry, several lives lost.	15
Alston No 1		1932		1927	Slips during construction.	33
Bartley	Birmingham	1931	20	1927	Slips during construction.	33
Bottoms	Macclesfield	1850	12	1929	Excavation of slip revealed fractured outlet pipe.	34
Broomhead	Sheffield	1934	31	1929	Leakage, movement of hillside.	33
Monkswood	Bath	1896	15	1931	Subsidence in downstream slope, cement grout injected into puddle core.	
Lower Rivington	Horwich	1857	18	1932	Slight depression near compensation shaft, erosion of puddle clay.	12
Blaen-y-Cwm	Ebbw Vale	1937	18	1936	Settlement of mine waste beneath embankment during construction.	33
Abberton	Colchester	1939	17	1937	Slip in upstream slope during construction.	33
Chingford	Lea Valley	1937	13	1937	Slip during construction, slip surface in downstream slope passed through puddle core and foundation soft clay layer.	3

Dam	Location	Date built	Height m	Date of incident	Nature of incident	Refs.
Weedon (canal embankment)	Grand Union Canal Northamptonshire		11	1939	Overtopped and breached.	35
Muirhead	Largs	1943	23	1941	Slip during construction when embankment 21 m high.	5
Bilberry	Holmfirth	1853	16	1944	Downstream slope eroded by valley run off and waves during storm	12
King George V	Chingford	1912	8	1945	Serious leakage on refilling after 2nd World War.	17
Holden Wood	Irwell Valley	1841		c1946	Swallow hole vertically above outlet tunnel, probably due to erosion of clay core.	15
Thorters	SE Scotland	1900	15	1948	Overtopped during flood.	15
Spott Lake	SE Scotland	1900	9	1948	Failed due to overtopping and abandoned.	15
Island Barn	Molesey	1911		1950	Settlement, tree roots had penetrated puddle core.	34
Harlow Hill	Harrogate	1870	9	1951	Major downstream slip.	36
Cwmtillery	Ebbw Vale			1954	Mining subsidence damaged culvert and puddle core.	15
Upper Roddlesworth	Blackburn	1865	21	1954	Sand toe partially washed away during heavy rain and slip occurred.	12
Blatherwycke	Northamptonshire			1959	Wetness where supply pipe laid in embankment.	34
Cowlyd	Dolgarrog	1921	14	1960	Excessive distortion of moraine fill shoulders.	15
Greenbooth	Rochdale	1962	34	1962	Deep seated slip movements commenced during construction.	33
Tittesworth	Stoke-on-Trent	1859	16	1960	Downstream slip during excavation for new dam.	19
Blithfield	Rugeley	1953	16	1962	Overtopping due to wind and waves caused downstream slip.	15
Balderhead	Barnard Castle	1964	48	1967	Seepage increased and swallow holes appeared during first filling indicating internal erosion of clay core.	37
Draycote	Rugby	1968	15		Minor slip during construction.	38
Covenham	Louth	1969	18		30 m length slipped during construction.	38
Auchendores		1880	10	1968	Damage caused by heavy spray from waves.	15
Lluest Wen	Maerdy	1892	20	1969	Swallow hole caused by erosion of puddle clay core.	19
Tamar Lake	Bude	1805	10	1970	Drainage layer and berm added to stabilise downstream slope.	39
Warmwithens	Oswaldtwistle	c1860	10	1970	Breached on line of new outlet tunnel.	15
Buckieburn	Denny	1902	26	1970	Slide on downstream slope during heavy rain and high winds.	15
Upper Creggan		1851	10	1971	Failed.	15
Banbury	Lee Valley	1903	c10	1971	Leakage through puddle core necessitated remedial work.	40
West Corrie	Kirkintilloch	1895	18	1971	Movement in downstream slope.	15
Withens Clough	Halifax	1894	22	1971	Internal erosion of core, major remedial work including new plastic concrete core.	15
Aldenham	Watford	1795	8	1975	Slips in upstream and downstream slopes.	15
Combs	Chapel-en-le Frith	1805	16	1976	Shallow slip in upper part of downstream slope.	41
New Pool	Church Stretton	1900	15	1977	Sudden increase in leakage; depression on crest.	40
Deep Hayes	Leek	1848	18	1977	Leakage led to permanent lowering.	40
Lockwood	Lee Valley		c10	1979	Leakage through clay core.	40
Winscar	Barnsley	1974	53	1980	Minor crack in asphaltic membrane repaired.	38
Snitterfield	Stratford on Avon	c1884	9	1980	Leakage through clay core.	40
Lower Glendevon	Perth	1924	35	1980	Earth tremors could have caused minor damage, reservoir emptied.	42
Greenbooth	Rochdale	1962	34	1983	Depression in crest, emergency drawdown followed by grouting.	43
Carsington	Derby		35	1984	Major upstream slip during construction.	8,9

13. Carsington Dam failure

PROFESSOR A. W. SKEMPTON, DSc,FICE, FEng, FRS, Imperial College of Science and Technology, and D. J. COATS, CBE, DSc, FICE, FEng, Senior Partner, Babtie Shaw & Morton

DESCRIPTION OF FAILURE

1.   Carsington Dam (Fig.1) is situated near the village of Hognaston in Derbyshire and has a maximum design height of 35 m. By the end of May, 1984 the placing of fill material was nearly complete. Piezometer levels at chainages 705 and 850 were as shown in Fig. 2 and the relationship between pore pressure and fill height is given in Fig. 3.

2.   Heavy rainfall in the period 1 to 3 June 1984 amounting to approximately 40 mm, brought earthmoving to a standstill. Due to the wet conditions earthfilling had not commenced on Monday 4 June 1984 when a crack was reported on the dam crest at 07.30 hours. Measurements made 40 mins. later recorded a 50 mm crack parallel to the dam centreline and 2 m into the downstream shoulder. The crack extended from chainage 675 m to 740 m with lesser widths beyond. At this time no noticeable differential vertical movement was apparent. Enlargement of the existing berm between chainages 650 m and 950 m commenced at about 15.40 hours on 4th June but by evening (21.00 hours) the crack width was 130 mm maximum, still with little vertical differential movement.

3.   By late afternoon on Tuesday 5 June the main crack was widening at 150 mm per hour. The original crack not only widened to 3 m maximum but on the upstream side between chainages 650 m and 800 m had dropped 3 m by the close of the day's operations. A second crack 2 m downstream of the original opened rapidly and the wedge of material between these cracks collapsed. Fissures also opened on the upstream slope and crumpling was seen at the toe. Cracks at the dam crest now extended south to chainage 1050 m, adjacent to the draw-off tower. Work on the berm enlargement continued throughout Tuesday.

4.   During the night of Tuesday/Wednesday the crest dropped 10 m exposing a striated scarp face. Parts of the

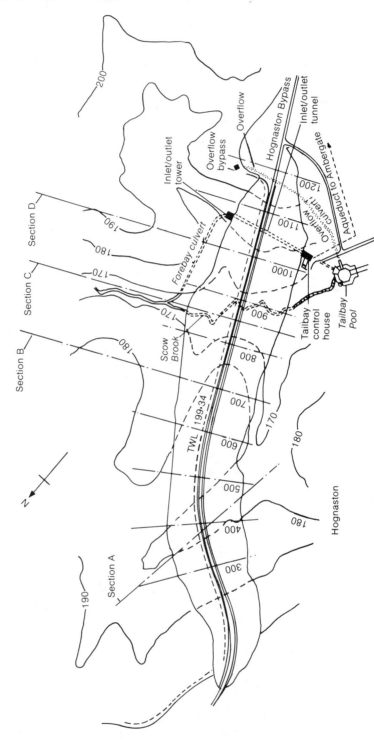

Fig. 1. Plan of the works (chainage and elevation in metres)

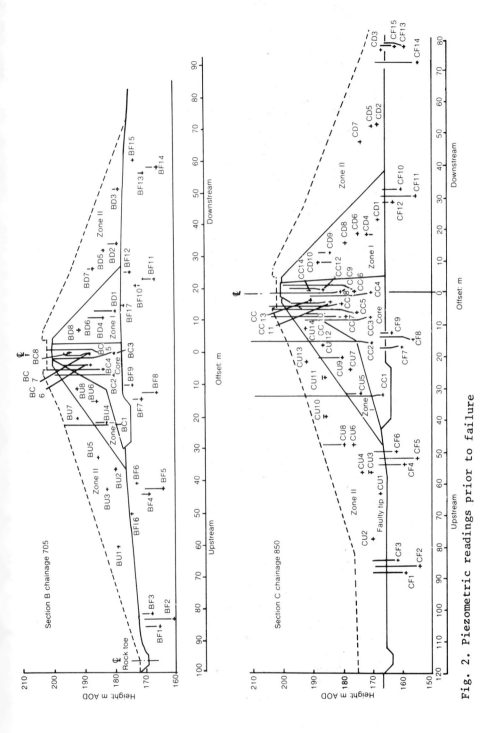

Fig. 2. Piezometric readings prior to failure

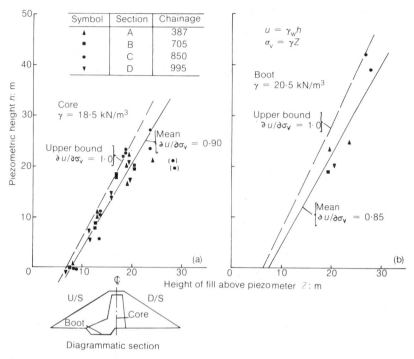

Fig. 3. Piezometer observations on 21.5.84

upstream toe had moved 13 m laterally and had lifted by
about 2.5 m at chainage 675 m and disruption of the
upstream slope was severe. Although movement of the dam
virtually stopped on 6 June, berm enlargement continued for
a few days (Fig. 4). Horizontal movements (Fig. 5)
illustrate the progression from about chainage 750.

INVESTIGATION AND ACTION FOLLOWING FAILURE

5. To relieve possible resultant loading on the draw-off
tower, 3 metres of fill was removed from the crest at the
south end.The vertical face of the failed section was also
removed in the interests of safety and stability. Much of
this material was placed as an upstream toe berm on the
unfailed northern section of the dam as a precaution.

6. An early exploratory trench at chainage 620 revealed
the slip plane in the foundation "Yellow Clay" and gave
some understanding of the distortion at the toe. Further
extensive exploratory works - notably trenches at chainage
720 and 825 - provided more information and allowed
sampling of embankment and foundation materials. From
survey and interpretation the geometry of the slip was
derived (See Figs 6 and 7).

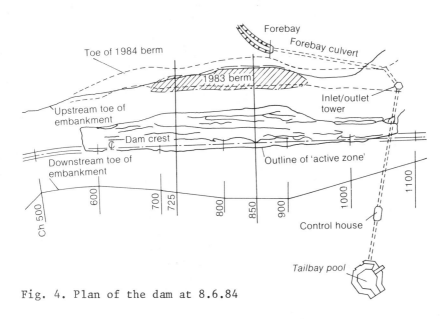

Fig. 4. Plan of the dam at 8.6.84

7.   From trial pits and trenches both under the dam and in ground beyond the upstream toe the succession of foundation strata was established as follows :-

SITE IDENTIFICATION		SYMBOL IN TRIAL PIT LOGS	CLASSIFICATION AND DESCRIPTION
Topsoil		TS	Topsoil
Subsoil		a_1	**Head deposits** Stiff brown and grey friable clay
Yellow Clay	Yellow Clay (a)	a_2	Firm and stiff light grey and yellow/ orange mottled clay. Both a_1 and a_2 contain some angular/subangular sandstone and limestone fragments, and rare rounded quartz pebbles
	Yellow Clay (b)	b_1 & b_2	**Weathered Bedrock** Soft to stiff grey, brown & yellow mottled clay with rare sandstone and coal fragments [Residual Soil]
Dark Clay		b_3	Soft dark grey and black clay with some very weak mudstone peds. [Highly brecciated and completely weathered mudstone]
Brecciated Mudstone		b_4	Dark grey laminated, highly weathered mudstone; very weak
Blocky Mudstone		b_5	Dark grey laminated, moderately weathered mudstone

8.   A Technical Steering Committee of representatives and advisers from the principal interested parties was convened at an early stage of the investigations and their observations and suggestions influenced the extent and content of the site investigation. A total of 1260 soil samples were taken and a comprehensive series of tests initiated.

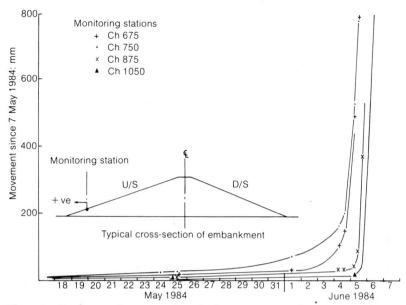

Fig. 5. Horizontal movement of the upstream face vs. time,
May–June 1984

GEOTECHNICAL PROPERTIES

Core Material

9. The core is made from the upper clay strata : chiefly
the "Yellow Clay" a_2 and b_1 with some of the "Dark Grey
Clay" b_3. Typical index properties are :

Liquid limit	74%
Plastic limit	32%
Plasticity index	42
Clay fraction	62%
Standard ⎰water content	28%
compaction ⎱unit weight	18.3 kN/m^3

10. On average the material has a water content of 34%
with a unit weight of 18.5 kN/m^3 and an undrained shear
strength of 65 kPa. The peak strength (effective stress)
parameters are (Fig. 8)

$$c' = 15 \text{ kPa} , \quad \emptyset' = 21°$$

11. During construction there has been only a small degree
of consolidation in the core. In undrained triaxial tests
the peak strength is reached at an axial strain of 3 to 4
per cent and the post-peak strength falls by about 10 per
cent at axial strains greater than 5 or 6 per cent. The
corresponding parameters are about $c' = 5$ kPa , and $\emptyset' =$
21

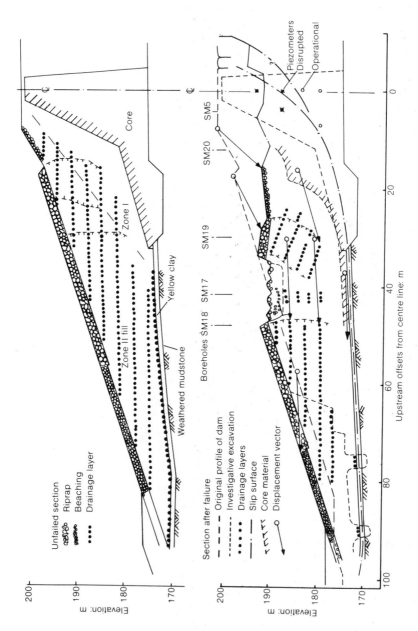

Fig. 6. Investigative section at chainage 725

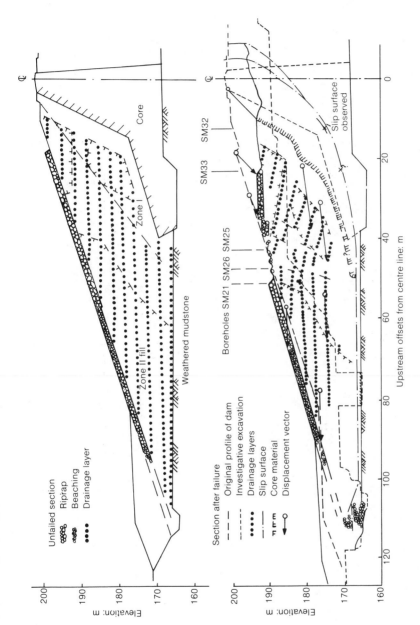

Fig. 7. Investigative section at chainage 825

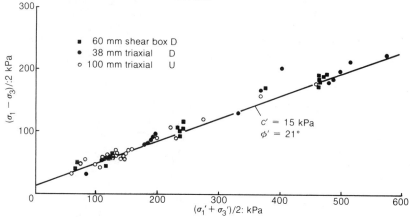

Fig. 8. Core material: shear strength (peak)

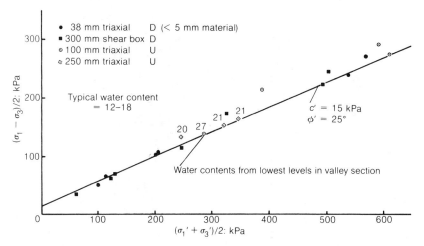

Fig. 9. Zone II fill: shear strength (peak)

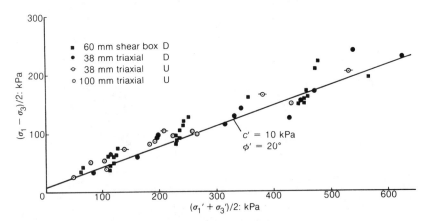

Fig. 10. Yellow Clay (a$_2$ and b$_1$): shear strength (peak)

12. The compaction equipment tended to cause rutting or small scale bearing capacity failures with consequent local shears on which the strength is at or close to the residual value ($c' = 0$, $\phi' = 13°$).

Zone II Fill

13. This material, forming the outer shoulders of the dam, is composed of the dark grey weathered and brecciated mudstone. Typically the water content is between 12% and 18%, the unit weight being about 21 kN/m^3 and the peak strength parameters (Fig. 9) are

$$c' = 15 \text{ kPa} \quad , \quad \phi' = 25°$$

A post-peak displacement of 30 mm reduces the parameters to

$$c' = 0 \quad , \quad \phi' = 22°$$

and the residual strength, as measured on slip surfaces, is given by

$$c' = 0 \quad , \quad \phi' = 15°$$

Yellow Clay

14. This is the name given to the head deposit a_2 and the uppermost layer of completely weathered mudstone b_1. Index properties of both strata are similar, with the following typical values

water content	43%
liquid limit	79%
plastic limit	34%
plasticity index	45
clay fraction	64%

15. There are no indications from the piezometers that any significant excess pore pressures existed in the foundation clays, so drained tests are relevant. The peak parameters of the Yellow Clay (Fig. 10) are

$$c' = 10 \text{ kPa, } \phi' = 20°$$

In shear tests the peak is reached at displacements equal to about 10 per cent of the sample thickness (Fig. 11), and the post-peak strength falls sharply.

16. The Head (a_2) was formed by periglacial solifluction during the last (Devensian) glaciation. This downslope

movement was accompanied by shearing, and in pits
excavated in the natural ground upstream of the dam,
above the valley bottom, the shears are seen near the
junction of beds a_2 and b_1 typically at a depth of about
1.3 m. They are gently undulating, with smooth surfaces.
In a total length of 62 m examined these pre-existing shears
were found to occupy 36 m (not including overlap), in ground
sloping at 4° to 8°.

17.   Tests on these shear surfaces show a peak strength
corresponding to $c' = 0$ and $\emptyset' = 16°$ reached at a small
strain, and an almost immediate drop (after not more than
5 mm displacement) to the residual strength $c' = 0$, $\emptyset' = 12°$.

18.   From trial pits examined through the upstream shoulder
of the dam at Ch.620 and Ch.720 it is clear that the main
slip passes through the Yellow Clay which , at these
sections, is between 0.5 m and 1 m in thickness under the
basal drainage layer.

<u>Dark Grey Clay and Brecciated Mudstone</u>

19.   In the trial pits at Ch. 620 and Ch.720, and also in
the pits in the natural ground, the Yellow Clay is
underlain by Dark Grey Clay b_3 and/or Brecciated Mudstone
b_4. Neither of these strata are sheared in nature or under
the dam. Tests on the clay b_3 show peak parameters of

$$c' = 20 \text{ kPa}, \quad \emptyset' = 20°$$

and for the b_4 material

$$c' = 10 \text{ kPa}, \quad \emptyset' = 23°$$

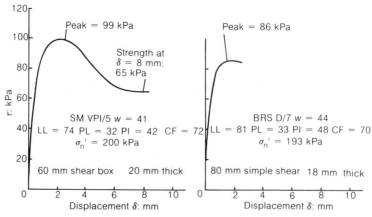

Fig. 11. Yellow Clay: drained shear tests

STABILITY ANALYSIS CH. 725

Method

20. The section at Ch. 725 (Fig. 12) is chosen as representing the initial failure which developed between Ch. 670 and Ch. 740. Pore pressures on the slip surface are taken from the (average) piezometer observations made a few days before failure took place. Expressed as a pore pressure ratio, $r_u = \gamma_w h / \gamma_z$ the values used in analysis are

Core	$r_u =$	0.42
Boot	$=$	0.53
Yellow Clay	$=$	0

21. Factor of safety is defined in the usual manner as

$$ F = \frac{\text{available shear strength}}{\text{shear stress}} $$

the correct value at failure being F = 1.0. To calculate the strength and stress the sliding mass above the slip surface is divided into a number of vertical-sided blocks or 'slices' and it is assumed that the forces between the slices are inclined at an angle $\delta = 10°$ to the horizontal; this angle having been found to be a reasonable average from a preliminary finite-element solution of the stress distribution within the dam. Given a value of δ the calculations become statically determinate.

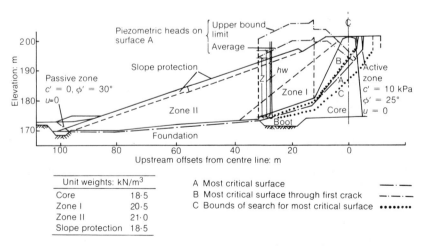

Fig. 12. Multiple wedge stability analysis: chainage 725

22.   In the following Table figures for F are given to 2
decimal places only for the sake of comparison.   In
practice a probable variation of ± 10% should be allowed.

Factors of safety   Ch. 725

Core	Foundation	Core c' ϕ'	Foundation c' ϕ'	F $(\delta = 10°)$
peak		15 21°	10 20°	1.41
'critical state'		0 21	0 20	1.21
'critical state'	peak, with 50% shears	0 21	5 16	1.08
	50% shears 20% at resid.	0 21	4 15	1.02
residual		0 12	0 12	≪1.0

Results

23.   The principal results of stability analyses at Ch. 725
are given in the above Table.

(i)   With peak strengths in the core and Yellow Clay, F =
1.4.

(ii)   A more conservative approach, sometimes adopted where
progressive failure is a possibility, is to use an
approximation to 'critical state' strength obtained by
putting $c' = 0$ and retaining the peak ϕ' values.   This
procedure gives F = 1.2.

(iii) Settlement gauge observations in the core show,
shortly before the slip movements began, a vertical
compression up to about 5 per cent (Fig. 13).
This is sufficient to take the strength past the peak, i.e.
there must have been some progressive failure in the core
preceding the slip.   Taking into account the presence of
occasional local shears, the parameters $c' = 0$ and and
$\phi' = 21°$ are not unreasonable as representative of the core.

(iv)   For the Yellow Clay, however, $c' = 0$ and $\phi' = 20°$
certainly overestimate the strength, for the following
reasons :-

216

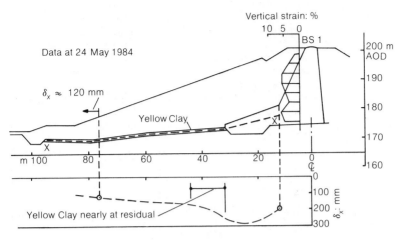

Fig. 13. Chainage 725: horizontal movements on line XX and vertical strain in the core

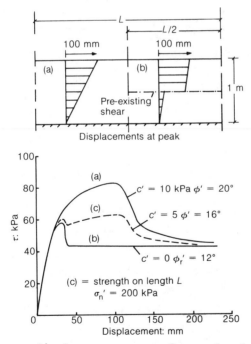

Fig. 14. Stress-displacement curves for a clay layer with a pre-existing shear occupying 50% of the total length

(a) Before failure, horizontal movements of at least 100 mm are known to have taken place in the upstream bank, and much of this would be accommodated by shearing displacement in the Yellow Clay. In a layer of this material 1 m thick such a displacement would just about mobilise the peak strength in previously unsheared clay. But where there are pre-existing shears the strength at this displacement would already be at the residual (see Fig. 14).

Thus if on average 50 per cent of the Yellow Clay contains pre-existing shears the field peak strength is

$$c' = 5 \text{ kPa}, \qquad \phi' = 16°$$

These parameters associated with $c' = 0$ and $\phi' = 21°$ in the core, give F = 1.1.

(b) In the foundation near the 'boot' and extending for a distance equal to roughly one-fifth of the length of the foundation clay between the boot and the upstream rock-fill toe, the displacements just prior to failure probably approached 200 mm (Fig.13). This would be sufficient to reduce the strength almost to the residual. Hence the average strength parameters in the clay are approximately

$$c' = 0 + \tfrac{4}{5} (5-0) = 4 \text{ kPa}$$

$$\phi' = 12 + \tfrac{4}{5} (16-12) \approx 15°$$

These values, taking into account progressive failure as well as the influence of pre-existing shears, give the correct value of F = 1.0.

24. Once failure had started and the displacements reached, say, 500 mm the strength throughout the length of the slip surface would be at residual. The factor of safety would then have been far less than 1.0 and very large movements were bound to take place.

25. A more sophisticated treatment, using finite-element analysis, is at present being undertaken by David Potts and Peter Vaughan.

STABILITY ANALYSIS CH. 850

26. This section represents the dam as built in the valley bottom, where the soft clays have been replaced by alluvium and this in turn was removed before construction. The dam here is therefore founded on brecciated and/or weathered

mudstone, b_4 and b_5. The slip surface passes through Zone
II fill at a level about 1 m above the basal drainage layer
(Fig. 15).

27. Pore pressures used in stability analysis correspond
to the following ratios :

Core	r_u = 0.52	
Boot	= 0.71	
Zone II fill	= 0.13	

The latter figure (for Zone II) represents a piezometric
height, as measured, equal to ground water level 4 m above
the slip surface.

28. With the unreduced peak parameters (c' = 15 kPa , ϕ' =
21° in the core and c' = 15 kPa , ϕ' = 25° in Zone II
fill) the factor of safety is 1.7. Reducing the core
parameters to c' = 0 and ϕ' = 21° (as before) and
taking c' = 0 and ϕ' = 25° in Zone II, F = 1.5.

29. However, it is known from field measurements that the
upstream shoulder of the dam at Ch. 875 had moved probably
as much as 180 to 200 mm prior to failure. A considerable
degree of progressive failure might, therefore, be expected
in the Zone II fill.

30. It is hoped to quantify the problem by means of a
finite-element analysis, but meanwhile the parameters c' =
0 and ϕ' = 22° would seem to give a realistic allowance for
this effect. With these, and c' = 0 and ϕ' = 21°
in the core, the factor of safety is 1.35.

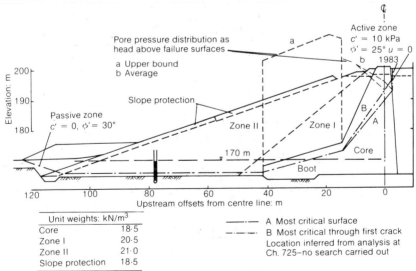

Fig. 15. Preliminary multiple wedge stability analysis:
chainage 850

31. Quite possibly the value of $\emptyset'$ in Zone II should be somewhat lower to fully allow for progressive failure ; but there is little doubt that to achieve a factor of safety of 1.0 an appreciable transfer of load is required from the initial slip further north.

32. Dr. Vaughan is investigating this problem ; he finds that load transfer can reduce the factor of safety at Ch. 850 by 30 per cent (based on reasonable, not in any way extreme, assumptions). Granted this result, the observed southerly spread of the failure to the valley section can be explained.

ACKNOWLEDGMENTS

The laboratory tests have been carried out by Soil Mechanics Ltd., the Building Research Station, Georesearch Ltd. and by Dr. D.J. Petley at Warwick University. Logging of all the trial pits was done by Soil Mechanics Ltd. and the geology of the area was reassessed by the British Geological Survey. Readings of pore pressures and settlements have been provided by G.H. Hill & Son while those of upstream movement have been made available by Shephard, Hill Limited.

We are grateful to the Severn Trent Water Authority for permission to present this paper.

Discussion on Papers 12 & 13

DR J. A. CHARLES

The government's decision to implement the 1975 Reservoirs Act
has focused attention on reservoir safety. The majority of
British dams are earth embankments and reservoir safety is
therefore primarily concerned with the long-term performance
of old embankment dams. The history of embankment dam
failures in the United Kingdom has been briefly reviewed in
Paper 12 and some case histories have been presented. These
have included both construction failures and failures in
service when impounding a reservoir. The latter category is
obviously the important one with regard to public safety.
However, although failures during construction may not involve
reservoir safety, they can be major events necessitating
expensive remedial works.

Two aspects of stability analysis, which may be particularly
relevant to the stability of some embankment dams during
construction, have been highlighted in Paper 12 (paragraph
14).

(i)   The effective stress Mohr failure envelope of some soils
        may exhibit significant curvature. In this situation
        the effective cohesion c' and the effective angle of
        shearing resistance ϕ' merely describe a straight line
        approximation to the actual curved envelope. This is
        illustrated in Fig. 1. The curved envelope is shown in
        Fig. 1(a). The parameters c'_L and ϕ'_L, derived from
        three triaxial compression tests carried out at low
        stresses, are shown in Fig. 1(b). The parameters c'_H
        and ϕ'_H, derived from tests at high stresses, are shown
        in Fig. 1(c). The two sets of parameters are
        significantly different. The use of the parameters c'_L
        and ϕ'_L would lead to an overestimate of shear strength
        at high stresses and the use of the parameters c'_H and
        ϕ'_H would lead to an overestimate of strength at low
        stresses. An approach that is sometimes advocated is to
        put $c' = 0$. The effect of this is illustrated in Fig.
        2. The method is conservative and in effect a factor of
        safety is applied to the shear strength of the soil as

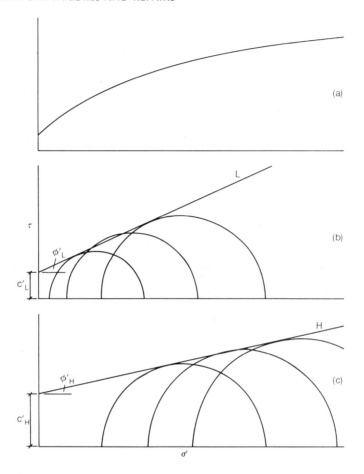

Fig. 1

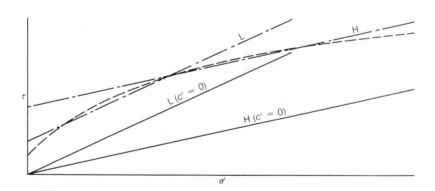

Fig. 2

determined from the laboratory triaxial tests. However, the magnitude of this factor is arbitrary and not clearly defined. It depends on the stress range of the laboratory tests in relation to the stress range applicable to the field situation which is under analysis. Where a failure envelope exhibits significant curvature it is sensible to carry out a stability analysis which takes this into account (references 10 and 11 of Paper 12). If this is not done, it becomes even more important to ensure that the stress levels used in the laboratory testing are relevant to the field situation.

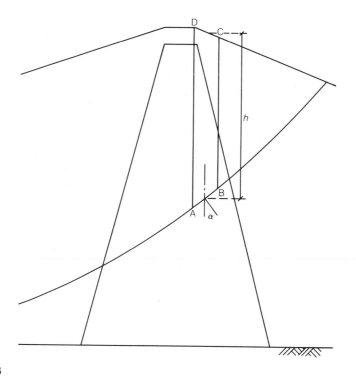

Fig. 3

(ii) In most slope stability problems, the pore pressure. ratio $r_u < 0.5$ ($r_u = u/\gamma h$ where u is the pore pressure at depth h in soil of bulk density γ). However, in the wet clay cores of embankment dams r_u may be as large as 0.7 or 0.8 during embankment construction. The amount by which the r_u value is smaller than unity may be an indication of the arching of the core between stiffer shoulders of the embankment and does not necessarily result from consolidation of the clay core. Fig. 3 illustrates a limit equilibrium stability analysis by the method of slices applied to an embankment with a central wet clay core. The analysis is in terms of the

223

effective stress and the shear strength that can be mobilized along the base AB of the slice ABCD is a function of the normal effective stress σ' on the slice base. The calculated magnitude of the normal effective stress depends on an assumption about interslice forces, e.g. in the Fellenius method where interslice forces are ignored $\sigma' = \gamma h(\cos^2\alpha - r_u)$; in the Bishop semirigorous method where the resultant of the interslice forces is assumed to act in the horizontal direction

$$\sigma' = \frac{\gamma h(1-r_u) - (c'/F)\tan\alpha}{(1 + \tan\alpha\tan\phi')/F}$$

With both expressions it can be seen that there are problems where there are high values of r_u and in such cases the calculated values of σ' may be very small or even negative. The magnitude of σ' depends on the assumptions made about interslice forces and these assumptions do not bear any resemblance to the actual situation in which the clay core may arch between stiffer embankment shoulders. Estimates of available shear strength based on these conventional calculations therefore may be unreliable. In cases where cores are placed wet and little consolidation is anticipated during the construction period, with consequent high r_u values at the end of construction, it might be preferable to represent the shear strength of the clay core by its undrained strength when analysing the stability of the embankment at the end of construction. There are serious problems with the determination of the appropriate undrained shear strength, but in this particular situation with very high r_u values it could be the most satisfactory approach despite the difficulties.

MR T. BULMER, L. G. Mouchel and Partners, Farnborough

The problems of puddle clay deterioration described on the King George V reservoir also occur on a number of other reservoirs in the Lea Valley area and are a characteristic of the native clay used in the puddle clay cores. Reservoirs of similar vintage in the Thames Valley are less prone to this type of deterioration.

A number of remedial measures have been carried out to reservoirs so affected over the years. Replacement of the upper part of the core has been tried but generally this deteriorated in a similar fashion after a few years.

In 1972 the Banbury reservoir puddle core was successfully repaired by forming a vertical grout screen (50 mm wide) in the core by repeatedly driving a single H-pile overlapping and with a tube attached through which a specially formulated grout was pumped as the pile was withdrawn. The grout was designed to simulate puddle clay when set.

The technique has also been successfully employed at the Lockwood reservoir and most recently at the King George V West Warwick reservoirs where proving tests are currently in hand.

The grout screen technique and grout formulation have been described by Ray and Bulmer at the BNCOLD conference at Keele University in 1982 (reference 40 in Paper 12). The technique can now be regarded as an established method of repairing leakages of the nature described in the paper.

DR J. A. CHARLES

You have referred to some interesting remedial measures for puddle clay cores of embankment dams which have suffered some form of deterioration. Remedial measures for old embankment dams is a most important subject and it is very helpful when case histories are published.

DR J. R. GREENWOOD, Department of Transport, Bedford

EFFECTIVE STRESS STABILITY ANALYSIS
In his presentation Dr Charles described problems with the commonly used Swedish and Bishop methods of effective stress stability analysis when applied to the clay core of a dam where 'slices' of the analysis have large base angles α.

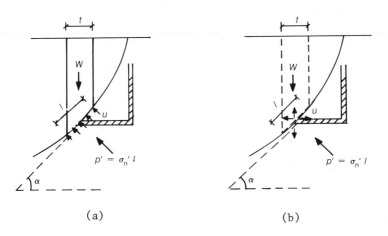

Fig. 1. Comparison of methods: (a) Swedish method; (b) simple method.

Referring to Fig. 1(a), using the Swedish method and the usual notation, the effective normal stress on the base of the slice is given by equation (1). I believe that the problems encountered by Dr Charles and others relate to the fundamental error in this equation (and the Bishop equation) of resolving total forces rather than effective forces to determine the effective normal stress on the slip surface. If effective forces are resolved, as in Fig. 1(b), equation (2) is obtained.

$$\sigma_n{}' = \frac{W \cos \alpha - ul}{1} \qquad (1)$$

$$\sigma_n{}' = \frac{(W - ut) \cos \alpha}{1} \qquad (2)$$

In equation (2) the porewater force is correctly deducted from the total weight of the slice to give the effective weight which is then resolved normally to the base of the slice. It is generally erroneous to resolve the total force as equation (1), as this contains the porewater pressure which acts 'all round' and therefore cannot be resolved. This point was picked up by Turnbull and Hvorslev (ref. 1) but has been ignored since then.

Equation (1) implies that the slice is surrounded by an impermeable membrane (Fig. 1(a)) with the water force acting outside that membrane along the base of the slice. This particular case would be relevant only where the water pressure is concentrated for a short time along an inclined fissure or drainage channel without penetrating the surrounding soil. In the majority of slope stability problems, the water pressure is present throughout the region of the slip surface acting as an all-round neutral pressure and therefore effective forces should be resolved. Case (b) is consistent with the principles of effective stress; case (a) is not.

The simple stability equation is derived from equation (2) (ref. 2)

$$F = \frac{1}{\sum W \sin \alpha} \sum [c't \sec\alpha + (W - ut)\cos \alpha \tan \phi'] \qquad (3)$$

or if the curved failure envelope is used as suggested by Dr Charles (ref. 3)

$$F = \frac{1}{\sum W \sin \alpha} \sum A[(W - ul \cos\alpha)\cos\alpha]^b l^{1-b} \qquad (4)$$

where $1 = t/\cos \alpha$.

With the factor of safety defined as the shear resistance along the slip surface divided by shear stress along the slip surface (ref. 4) equation (3) (and (4)) can be used for both circular and non-circular slip surfaces. It gives sensible results for all failure surfaces whether shallow or deep and its simplicity enables the engineer to develop an understanding of the analysis that he is carrying out.

Equation (3) may also be derived by consideration of the Mohr circle diagram for the soil elements at the base of the slice giving the full stability equation (ref. 2)

$$F = \frac{1}{\sum W \sin \alpha} \sum [c't \sec \alpha + (W-ut)(1+K \tan^2\alpha)\cos\alpha\tan\phi'] \qquad (5)$$

where K is the ratio of horizontal to vertical effective stresses at the failure surface.

It is seen that when horizontal stresses are ignored (i.e. K=0) equation (5) becomes identical with equation (3).

Equation (5) is a useful aid when studying the effects of horizontal stresses (see Technical Note No. 3) or soil reinforcement but for routine analysis K cannot readily be determined and as its effect on the factor of safety is small it is usual to make the slightly conservative assumption K=0.

References
1. TURNBULL, W. J. and HVORSLEV, M. J. Special problems in slope stability. Proc. Am. Soc. Civ. Engrs, 1967, vol. 93, SM4, 499-528.
2. GREENWOOD, J. R. A simple approach to slope stability, Ground Engng, 1983, vol. 16, no. 4, 45-48.
3. CHARLES, J. A. and SOARES, M. M. Stability of rockfill slopes. Géotechnique, 1984, vol. 34, no. 1, 61-70.
4. LAMBE, T. W. and WHITMAN, R. V. Soil mechanics, pp. 363-365. Wiley, New York, 1969.

DR J. A. CHARLES

I cannot agree with your assertion that there is a fundamental error in these methods due to resolving total forces rather than effective forces to determine the effective normal stress on the slip surface. In limit equilibrium stability analyses by the method of slices, each slice must be in equilibrium under the action of the total forces. For a dry slope or a submerged slope effective forces can be used in the equilibrium equations because in the former case total and effective forces are equal and in the latter case forces due to porewater pressure can be ignored since hydrostatic conditions exist within the slope. However, in the more general situation where porewater pressures are related to steady state seepage or to the dissipation of excess construction pore pressures, the forces due to pore pressures cannot be neglected and total forces should be used in the equilibrium equations. The differences between your approach and the Fellenius and Bishop methods may not be large in many cases and may be less important than the assumptions about interslice forces which also affect the calculated values of normal effective stress.

MR P. W. MOTTRAM, John Burrow and Partners, Lincoln

In 1978, on the instructions of R. H. Cuthbertson & Partners, Nuttall Geotechnical Services carried out an investigation into the internal conditions of Torduff Reservoir, Colington, an earth-filled dam 120 years old.

The purpose of the investigation was to determine the cause of leakage through the dam. The work was carried out in two phases. Phase 1 consisted of six boreholes sunk through the puddle clay core and then continued into the bedrock. Permeability tests were carried out in the boreholes and piezometers were installed in four of the holes. This phase of the investigation revealed no possible reasons for the leakage. Phase 2 consisted of trial pits. One of the trial pits, 7 m deep, was dug in the location of a depression that had formed on the downstream side of the dam. This revealed an outlet pipe tunnel excavated through the bedrock. The tunnel was backfilled with puddle clay and contained a cast iron pipe resting on wooden sleepers. A void approximately 900 mm wide and 300 mm high was noted at the roof of the tunnel. This was the cause of the leakage and probably resulted from consolidation of the puddle clay. The void was grouted up and no further problems have been noted at the dam.

DR J. A. CHARLES

One problem with the investigation of old embankment dams is that the problem may be related to quite a minor detail of design or construction or to some unrecorded event in the long history of the dam.

MR R. A. SMITH, Cambridge University Engineering Department

EMBANKMENT FAILURES OF SUPPLY RESERVOIRS FOR THE HUDDERSFIELD NARROW CANAL
The authors of Paper 12 correctly recognize that their list of unsatisfactory performances of embankment dams will be incomplete and that 'published information about early incidents is sometimes very limited'. They suspect that the references to failures of Diggle and Swellands in 1810 may refer to a single failure - examination of original sources suggests that this suspicion is correct, but the failure was to another reservoir!

Two accessible modern references are Hadfield and Biddle (ref. 1) and Schofield (ref. 2). Nearly identical contemporary accounts are given in The Times (ref. 3) and The Gentleman's Magazine (ref. 4), differing only in the location of the reservoir: 'Driggle on the top of Stanedge' (The Times) and 'the reservoir at the top of Standedge', an account duplicated in a local history (ref. 5), which ascribes the collapse to 'Swillers' reservoir. However, both accounts make

it clear that the flood water advanced into the Colne Valley
running eastwards from the Pennine watershed towards
Swaithwaite and Huddersfield (Fig. 1). This confirms my
feeling that the collapse could not be of the Diggle reservoir
which is well below the western watershed of the hills, in a
district where I have previously lived and have not been aware
of local folklore of such a flood calamity.

Further searches into the minute books of the Huddersfield
Narrow Canal Company, who owned the reservoir for supply
purposes, at first appear to complicate the issue: according
to the minutes of the annual meeting of the Huddersfield Canal
Company, 27 June 1811 (ref. 6)

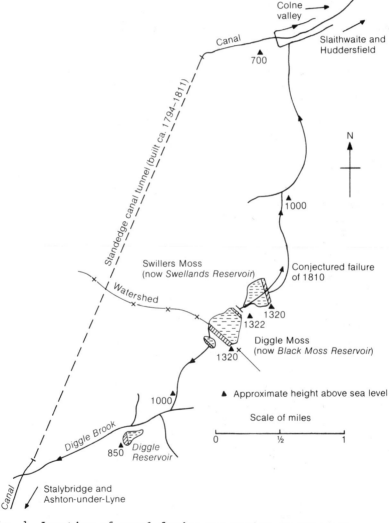

Fig. 1. Location of canal feeder reservoirs on Standedge
watershed

'Early in the morning of the 29th November last, owing to continual rains, the black earth at the tail of the Diggle Moss Reservoir gave way, and let off the water in such a quantity as to cause the loss of five peoples lives and do other damage to the amount of near four hundred and fifty pounds.'

and from the Canal Committee, 4 March 1814

'The water which anciently flowed into Diggles Brook and which by the making and breaking down of Diggles Moss Reservoir had been diverted into the River Colne to be restored to Diggle Brook.'

However, meetings some five years later clarify the situation: from the Canal Committee, 24 June 1819

'Engineer reported suitable site for additional reservoir on Diggle Moss to east of present reservoir.'

and on 1 August 1821

'£1000 instead of £500 to be spent on Swillers Moss reservoir.'

These reports confirm the date stone of 1821 still clearly visible on the Swellands masonry work and clearly indicate that the failure must have been on the top of the north-east bank of Diggle Moss reservoir, now called Black Moss reservoir, see Fig. 1. Both these reservoirs are situated in a slight depression on a peat plateau approximately 1320 ft above sea level, Moss being the local name for such high moorland areas and Black being a particularly apt description of the peat. Black Moss has both south-west facing (main) and north-east facing (subsidiary) embankments, and the geography is such that a failure of the 'black earth tail' would cause 'water which anciently flowed into Diggle Brook' to be diverted into the River Colne.

An excellent description of the cause of failure has been given by Sutcliffe (ref. 7)

'Such was the inattention of the engineer, that he laid the foundation of the bank upon nothing but the moss and ling that covered the soil. When this reservoir was filled to the depth of about seven feet, the water found its way under the foot of the bank, and floated it away, as if it had been only a deal plank!'

This particular account ends on a happy note, uncharacteristic of the times, concerning the treatment of the flood orphans by the Canal Committee:

'But to the credit of the committee, they supplied the

distressed family with every thing necessary, and became a father to the fatherless.'

This was not the only bad experience of the Canal Committee caused by their reservoirs; the incident at Marsden in 1799 is the first of Charles and Boden's list, cited from Schofield (ref. 2). Schofield also states that Swineshaw Common reservoir was so badly damaged in 1799 that it was abandoned for good, a statement which appears to be incorrect. This reservoir in the Swineshore Valley above Stalybridge, about six miles SSW of the Swellands Reservoir, was also built to supply the Huddersfield Narrow Canal. Bateman (ref. 8) first investigated this valley for the possibilities of supplying the Manchester and Salford Waterworks Company and stated that 'the Canal reservoir was destroyed by the failure of the embankment due to neglect to provide an overflow'. A modern account (ref. 9) dates this failure as 10 March 1803, which appears to be confirmed by the following Canal Company minute, at the meeting of 25 May 1803.

'Mr Rooth having agreed with Mr Stelfox the occupier of Staley Mill to pay him eighty four pounds for damages done to said Mill by breaking down of the Swineshaw Reservoir.'

The site of the canal reservoir was eventually sold for £500 by the then owners (London and North Western Railway) to the Ashton and Stalybridge Waterworks Committee in 1864 (ref. 10). A new embankment was built which still impounds the Higher Swineshaw Reservoir.

The details of accounts such as these are necessary, as is the use of primary reference material and, if possible, local knowledge, if accurate accounts of these early failures are to be unearthed. That such work is valuable in the light of concern about our ageing embankment dams should not be in doubt. It is clear that the list of Charles and Boden is only the beginnings of an accurate catalogue of unsatisfactory performance of UK embankment dams.

References
1. HADFIELD, C. and BIDDLE, G. Bursting of Diggle reservoir on 29 November (1810) with the loss of five lives. In The Canals of North West England, p. 328. David and Charles, 1970.
2. SCHOFIELD, R. B. Benjamin Outram (1764-1805), canal engineer extraordinary. Proc. Instn Civ. Engrs, Part 1, 1979, vol. 66, 539-555.
3. THE TIMES, 11 Dec. 1810, p. 3, col. 2.
4. THE GENTLEMAN'S MAGAZINE, 1810, quoted in Whitehead, 1942.
5. WHITEHEAD, L. B. Bygone Marsden, pp. 73-75. Hotspur, Manchester, 1942.
6. PRO RAIL 838 Minutes of Huddersfield Canal Company and Reports of Annual Meetings. Available from the Tameside

Local Studies Centre, Stalybridge.
7. SUTCLIFFE, J. A treatise on canals and reservoirs, pp. 119-120. Rochdale, 1816.
8. BATEMAN, J. F. T. History and description of the Manchester waterworks. Spon, London, 1884.
9. ANONYMOUS. Brushes Valley - a Stalybridge beauty spot. Staley Booklet No. 6, p. 23. Whittaker, Stalybrige, 1930. Available from the Tameside Local Studies Centre, Stalybridge.
10. KELSALL, J. Manuscript on the Ashton under Lyne and Stalybridge Corporation Water Works, Swineshaw. DDWW/86, p. 5, Tameside Local Studies Centre, Stalybridge.

DR J. A. CHARLES

Your conclusion that the failure did not concern either Diggle of Swellands but was to another resevrior is intriguing, and you correctly conclude that the list of incidents and failures in Paper 12 is only the beginning of an accurate catalogue of unsatisfactory performance of UK embankment dams.

DR A. D. M. PENMAN, Sladeleye, Chamberlaines, Harpenden

Several modern embankment dams have been designed to have clay layers under their upstream shoulders and while probably improving hydraulic conditions in the foundations this feature appears to increase the risk of slip movements.

Nature had provided a soft clay layer under Chingford Dam which led to the well-known failure in 1937 that was investigated by Professor Skempton. Charles and Boden (Paper 12) have shown a section through the dam (their Fig. 2): the significance of the clay layer in soil mechanics terms was not appreciated by the designers at that time.

A dam 40 m high built on 22 m of saturated sand has been discussed in Technical Note 6. A black clay was used for core and cut-off, as shown by the figures in the Note. A modification to the original design caused a blanket of the black clay 7 m thick to be placed at formation level to connect the two. This provided a path for a slip surface which caused failure of the upstream shoulder during construction when the dam had reached a height of 35 m.

A clay blanket was placed under the upstream shoulder of the 52 m high Kielder Dam: its behaviour has been discussed by Millmore and McNicol (ref. 1). The dam was well instrumented and the pore pressures that developed in the clay blanket during construction were somewhat higher than had been predicted. The calculated factor of safety against slip of the upstream shoulder had fallen to 1.28 when the dam height reached 45 m. A toe berm was therefore constructed to raise the factor of safety to 1.8 so that fill placement could continue. This ensured a factor of safety on completion of

1.5.

It is apparent that clay layers, clay blankets or clay aprons extending from a core under an upstream shoulder are likely to increase the risk of slip failure: is it a practice that should be encouraged? Would Dr Coats like to comment in relation to Kielder Dam?

Reference
1. MILLMORE, J. P. and McNICOL, R. Geotechnical aspects of the Kielder Dam. Proc. Instn Civ. Engrs, Part 1, 1983, vol. 74, 805-836.

PROFESSOR A. W. SKEMPTON and DR D. J. COATS

At Kielder Dam the foundation was generally Boulder Clay overlying rock of the Lower Carboniferous series whose mass permeability we took some care to assess. We considered that a positive deep cut-off such as may have been achieved by grouting was not necessary provided that the seepage path could be made extensive. For this a clay blanket between 5 m and 6 m thick under the upstream shoulder was placed and this clay had an undrained shear strength of the order of 100 kPa. Whether or not we would adopt such an arrangement again would depend on the foundation's permeability and the availability of suitable materials. You have pointed out that the pore pressures in the clay blankets under the upstream shoulder do not dissipate rapidly and great care must be taken to monitor them in relation to factors of safety. The shear strength of the clay in the blanket is also, of course, of significance.

MR M. F. KENNARD, Rofe Kennard and Lapworth, Sutton

In my view Paper 13 need not have been written as the failure need not have occurred because the design of Carsington Dam was inadequate, as had been stated before June 1984. The design parameters were too strong, the stability analyses were not critical, and early warning signs were not critically examined.

The embankment fill at Carsington was specified to be placed in three seasons (1982, 1983 and 1984). At the end of the first season, the engineer reduced the design parameter of ϕ' from 27° to 23° without a change in slope. At the end of the second season I, with Dr Bromhead who carried out a rigorous series of analyses, wrote an independent report to the Contractor to meet the Engineer's request for the Contractors' figures. Amongst other points these conclusions stated the following.

(i)   The quoted factor of safety for the upstream slope could not be achieved. Using the Engineer's original parameters and pore pressures, lower values were

obtained for a slip surface in the fill, and less than
unity, for certain pore pressure assumptions, for a non-
circular surface in the foundation.

(ii)   Computer analyses by the Engineer had assumed
unrealistic surfaces in the fill and had not included
the foundation.

(iii) The Engineer's reassessed parameters of $c' = 20$ kN/m^2
and $\phi' = 23°$ can be considered to be unrealistic and
values of $c' = 0$ and $\phi' = 21°$ may be appropriate, as
confirmed by Professor Skempton. My view based on work
done by Mr Hoskins in my office in August 1983 on the
test data then available was expressed before the
detailed and extensive investigation described in the
paper.

(iv)   'More information on foundation properties is required.'

(v)   'Further analyses are necessary and a revised design is
essential, so that the embankment can be completed in
1984 with safety and confidence.'

Two cases can be illustrated to show the effects of c' and
ϕ' on the factor of safety for the 1:3 upstream slope. Case A
is based on the Engineers' design pore pressure assumptions
and case B is for the actual lower pore pressures at failure.

It does not matter precisely on the case A assumptions:
different workers would make different assumptions at the
design stage. With any ϕ' value, a non-circular slip surface
in the foundation is on the wrong side of $F = 1$ with the
original pore pressures assumed for the fill.

With case B and $c' = 0$, as given in the paper for the core,
but not for the foundation, which surprises me, an average ϕ'
$= 20°$ gives a factor of safety of about unity. A more
rigorous design as described in the paper is a better
approach, but mine is sufficient for illustration.

The Engineers' quoted factors of safety for case A (or
similar) are far too much to the right.

The cross-section was not changed when ϕ' was changed from
$27°$ to $23°$ despite a drop in the factor of safety of 0.24.
The effect of c' makes a difference of about 0.2 in the factor
of safety, which is very significant, and approximately 0.06
for each degree change in ϕ'.

The paper by Skempton and Coats states that the details of
the slip are complex. A non-circular slip in shale fill, a
rolled clay core of asymmetrical shape and a thin clay
foundation with problems of residual strength and progressive
failure would of course be complex. However, adequate
stability analyses with reasonable design parameters and pore
pressures and a conservative factor of safety of 1.5 at the
end of construction need not be very complex and would have
produced a safe design, especially with more information on
the foundation as I had concluded.

The slip surface given in the paper is almost identical with
the critical surface analysed by Dr Bromhead in our report,
and with the sketch in New Civil Engineer within ten days of

the slip based on a rough sketch I made for the reporter.

If symposia like this one are to reduce the risk of failures in the future, as already said by Mr Garrett, then lessons must be learned from the papers and discussions and future designers (not only design engineers but also advisers and contractors and administrators) must not ignore the past.

The lessons from this classic case, which I restate should never have occurred, include the choice of value of ϕ' for design; the use of $c' = 0$ in analyses; the reasonableness of design pore pressures; the adequacy of site investigation, including the geological implication of the foundations and the use of the critical information; the need to review the design as work progresses; the means of reviewing the design of a major dam by a review board or panel of experts; the relationships and trust that is needed between engineers, contractors, advisers and employers.

PROFESSOR A. W. SKEMPTON and DR D. J. COATS

We must confine our discussion of the failure of Carsington Dam in June 1984 to the mechanism of failure determined following extensive site investigation. We greatly appreciate the Severn Trent Water Authority's permission to present the paper but we have no authority to extend our remarks beyond the mechanism of actual failure. We are, therefore, unable to comment on your contribution to the discussion.

DR D. W. COX, Polytechnic of Central London

I should like to ask Professor Skempton and Dr Coats whether they have considered the possibility of the collapse of the shale fill structure on wetting owing to rainfall etc. accumulating at the base of a particularly dry shale layer. The resulting collapse would set up a high pore pressure confined to that layer with little possibility of drainage. Constant total stress, rather than constant rate of strain, tests would be required to prove this either way.

PROFESSOR A. W. SKEMPTON and DR D. J. COATS

The zone II shale fill at Carsington Dam did have rather large compressions, for example about 3% during the 1982-83 winter shut-down period, but piezometers in the fill showed no significant excess pore pressures at any stage during construction. However, it is evident from inspection on site that some deterioration takes place after removing this material from the borrow pits, and we now have tests showing that ϕ' decreases from about 28° for compacted shale freshly dug to 25° for shale fill in the dam (Fig. 9 in the paper).

This 'weathering' effect is not directly relevant to an

analysis of the failure since, as recorded in the paper, at Ch. 850 where the slip surface passes through the base of zone II fill, Fig. 15, the factor of safety is still around 1.35 (neglecting load transfer) even if strength parameters of the fill are reduced to $c' = 0$ and $\phi' = 22^\circ$. Such a reduction is an attempt to allow for the possible effects of progressive failure; it amounts to an assumption that 30% of a potential slip surface was already at residual strength.

MR D. A. B. BAKER, Balfour Beatty Engineering Ltd, Sidcup

The geotechnical analyst's task in back analysis such as the Carsington failure is always enviable as he knows the failure mechanism and the factor of safety to be attained. Even so, as we have seen, much reasoned juggling of parameters is required to predict failure correctly. This emphasizes the need for the designer to be cautious in selecting the design parameters and factors of safety. At the same time, there is a danger of designers over-reacting and producing unnecessarily uneconomic designs. Perhaps the most important lessons here are the re-emphasized need for designers to pay more attention to the possible presence of what Terzaghi termed 'minor geological defects' and the need for further study of post-peak strength behaviour leading to formulation of rational design guide-lines for the selection of strain-related shear strength parameters for various engineering applications.

On the subject of 'reasoned juggling' of parameters, I note that despite a reasonably logical basis for evolution of the foundation shear strengths based on field evidence of 50% of pre-existing shears and strain-related arguments for 25% of the failure surface being at the residual strength, selection of the critical state parameters for the core is still rather arbitrary. Bearing in mind that Dr Vaughan has found that load transfer can reduce the factor of safety at Ch. 850 by 30%, does the converse not imply that there could have been a significant end restraint effect on the first section which failed and hence that strength parameters could be even lower than assumed in the back analysis?

PROFESSOR A. W. SKEMPTON and DR D. J. COATS

Your remarks about 'juggling' of parameters to explain the failure at Ch. 725 are rather disparaging. Precise values can be debated, but no juggling is required to show that the strength of the Yellow Clay with its pre-existing shears cannot be very different from $c' = 5$ kPa and $\phi' = 16^\circ$, results derived from careful field studies and laboratory tests. Nor can it be doubted that some reduction in strength, below the peak, must be allowed in the core in view of the presence of compaction shears and strains of the order of 5% before failure. Reducing c' to zero, without changing φ', would not seem to be extravagant, yet the combined result of these two effects is to bring the factor of safety to within 10% of the correct value, using observed pore pressures.

Perhaps you would like to see a smaller φ' value in the core, and the answer to whether there may not have been end restraint on the section at Ch. 725 which would lead to lower parameters than those assumed in the back analysis is that there could not have been such an end restraint. Differential movements between various parts of the dam immediately before the initial failure were relatively small and in fact the movements near Ch. 850 were slightly more (not less) than at Ch. 725. Once the failure had started at Ch. 725 large differential movements developed, leading to a transfer of load to the sections further south which then in turn started to fail. This sequence of events is clear from the excellent field observations made before and during failure.

MR D. J. KNIGHT, Sir Alexander Gibb and Partners, Reading

We have just had the description of the sliding failure of an embankment dam during construction. The following details are for a failure by piping erosion of a dispersive clay dam during fast impounding. Such clays are not experienced in the UK as they are in many areas overseas.

The photographs shown here and the details of description of the event have been given to me by an engineer involved with the project. He has kindly given me permission to make known this information at this symposium. Such graphic records, showing the progression of a failure as it occurred, are not only very uncommon but are very instructive to dam designers. In showing these details I wish to make it absolutely clear that neither were Sir Alexander Gibb and Partners connected in any way with the design and construction of the dam, nor did any of their staff witness the event.

The dam is situated in the east of Thailand near the border with Kampuchea for supplying water to a refugee-holding camp and for irrigation of farm land. It is a homogeneous earth fill with a maximum height above the river bed of about 10 m with a reservoir capacity of about 10 million m^3. Dam construction was finished in July 1981, after which the reservoir was unfilled for two years. Then in September 1983, heavy rain caused the reservoir to impound quickly. In the early morning of 7 October 1983, at 7.0 a.m., a leakage was found at the downstream toe of the dam, which progressed to show a pipe flow characteristic (Fig. 1.). At this time, on the upstream side opposite the leakage, a vortex was formed (Fig. 2) within the reservoir 1 m below retention level, representing 5 million m^3 of water or about 50% of the reservoir's capacity.

Matters then developed quickly (Figs 3 and 4). By 2.30 p.m. a clear bridge had appeared (Fig. 5) over the erosion tunnel,

Fig. 1. Pipe flow forming

Fig. 2. Vortex in reservoir opposite the leakage

Fig. 3. 11.15 a.m. 7 October 1983

Fig. 4.   11.30 a.m.

Fig. 5.   14.30 p.m.: formation of bridge

Fig. 6.   14.50 p.m.: close-up of bridge

Fig. 7.   14.52 p.m.

Fig. 8.   15.10 p.m.: start of collapse

Fig. 9.   15.25 p.m.: collapse

Fig. 10. 15.30 p.m.: complete breach

Fig. 11. 8 October 1983, 9.00 a.m.: reservoir completely drained

Fig. 12. One of many sinkholes found on the slope of the embankment

241

through which upstream daylight could be seen. The next sequence (Figs 6-8) shows the onset of final collapse at 3.10 p.m. The collapse occurred with a large explosive crash (Fig. 9) of the crest at 3.25 p.m., and at 3.30 p.m. the dam was completely breached, with a 'thin slice of cake' type of gap (Fig. 10). By 9.0 a.m. on 8 October 1983 the reservoir had completely drained (Fig. 11), and many sinkholes (Fig. 12), typical of tunnel erosion within a dispersive clay, were seen.

The dam was repaired to a flatter downstream slope, but I do not have more details. The design preventative is, of course, provision of adequate filters.

MR I. ELLIS, Fondedile Foundations Ltd, Yiewsley

I should like to describe briefly an example of the use of reticulated Pali Radice structures to solve slope stability problems.

Light but mobile drilling equipment only is needed to drill and install the Pali Radice piles and the scaffolding erected to support the rigs and for providing access for placing the shuttering and concrete of the capping beams.

Netlon mesh and topsoil is spread over the whole slope to hide the capping beams.

New trees and shrubs are then planted and the slope can be completely covered with vegetation within a period of 9 months (Fig. 1).

Fig. 1

14. Keuper Marl as a road foundation — predicting its strength

A. J. TURNER, BSc, BA(Law), MSc, MICE, Barrister, Senior Geotechnical Engineer, and K. L. SEAGO, BSc, MICE, MIHT, MIWPC, ALArb, Associate, Ove Arup & Partners

SYNOPSIS

A road constructed on a sub-grade of Keuper Marl experienced failure in areas of cut following the trafficking of the road base by construction traffic. Laboratory tests carried out on the sub-grade material subsequent to failure showed the material strength to be very sensitive to moisture content. This paper attempts to relate strength in terms of C.B.R. to a non-dimensional Condition Index, based upon moisture content and Plasticity Index, so that comparison with similar materials can be made.

A.   INTRODUCTION

1.   The strength and performance of a road pavement is dependent on the strength of the sub-grade foundation. Guidance on the strength of various sub-grade soils has been given in Road Note 29 published by Road Research Laboratory in 1970 (1). In general, for cohesive soils, the strength, measured in terms of the California Bearing Ratio (C.B.R.), is indicated to increase with decreasing Plasticity Index (PI) for cohesive soils down to a PI of 10%. Site experience shows, however, that soils of low plasticity are very susceptible to wetting and a sub-grade formation can quickly deteriorate to a point where it becomes unsuitable.

2.   Problems can occur where heavy inundation of the sub-base layers of a pavement causes unseen deterioration of the road sub-grade, with the result of premature failure of the pavement. This effect has been examined by Black and Lister (2) who show that, although Road Note 29 (1) predicts with reasonable accuracy the strength of intermediate and highly plastic soils, it can seriously under-estimate the strength of low plasticity soils where an increase in moisture content has occurred.

3.　　Soils of low plasticity are commonly found in the Midland regions of England, either as in-situ sedimentary formation, or as drift deposits. Such materials include Keuper Marl of the Triassic series and Old Hill or Etruria Marl of the Upper Coal Measures. Although called Marl, neither is a true marl.

4.　　Keuper Marl has been extensively studied by Chandler (3) who has suggested a classification of in-situ material into six grades, based upon the materials structure. In general, the upper layers of any exposed formation will be weathered and structureless and would fall within Class 5 or 6 of Chandler's (3) classification. Glacially transported material can resemble completely weathered Keuper Marl and can be difficult to distinguish from the in-situ material.

5.　　The following section describes the failure of a road sub-grade formed of Keuper Marl. The failure occurred after the pavement had been subject to loadings not envisaged in the road design. It is interesting to note, however, that the road sections which suffered failure were areas constructed in cut and that in general areas constructed at the original ground level or on fill remained undamaged.

B.　　CASE HISTORY

6.　　The road was constructed in a steeply sided valley with a small stream discharging to the southwest. The land was in agricultural use where gradients permitted, both for cropping and grazing. The estate road onto the site follows the eastern perimeter and from this road are spurs leading to the accommodation units. The road is in a cut along the majority of the eastern perimeter of the site and also, to some extent, along the western boundary.

7.　　A site investigation was carried out with trial pits to examine the nature of the material. A number of these were related to an assessment of the sub-grade to the road. The investigation showed that the sub-grade would consist of Keuper Marl with a sand content which varied across the site. The ground to the west of the stream had a high sand content in comparison with that to the east of the stream. Laboratory testing indicated that the Plasticity Index of the material at sub-grade level varied between 20% and 30%, and that C.B.R. values varied between 0.6% and 25%. The low values of C.B.R. occurred exclusively in the wetter materials which lay more than 1m below formation level.

8. The predicted traffic loading was low relating to use by private cars. No information was available in detail, and it was agreed that the design should be to Category 1 of Table 1 in Road Note 29, the 3rd Edition 1970. This led to a number of accumulated standard axles of 180,000. From Table 3 in Road Note 29, the C.B.R. for the sub-grade design was chosen at 4%, with the principle that sub-grade drainage would be provided to ensure the watertable remained more than 600mm below formation level.

9. Based upon the above parameters, and taking note of the need to keep costs to a minimum, a light form of construction was chosen as follows:-

    Sub base   - 230mm of type 1.
    Road base - 75mm of bitumen or tar macadam.
    Surfacing - 55mm single course dense bitumen macadam.

10. For reasons of programme and economy, it was decided to divide the project into two parts:-

    a) A civil engineering contract to create earthwork platforms, construct the road and provide surface and foul water drainage.
    b) Building works and associated landscaping.

11. It was decided that the road would be completed to top of roadbase only in the civil engineering contract and surfaced at the end of the subsequent building contract. The surface of the roadbase was to be sealed by the application of sealing grit.

12. The civil engineering contract started in the autumn of 1973, and was completed in July 1974. Work was hindered during the winter by wet ground conditions and the preparation of the sub base and formation to the roads took place in the drier weather of March, April and May, 1974. The resident engineer's records show that the road formation in the cut areas was in extremely good condition and very hard. The type 1 sub base was placed and compacted as the works proceeded and again no difficulty was experienced in compaction. The roads were used by the contractor to run on during this stage of the works and in this period there was an average amount of rainfall. As the works came to a completion, the contractor cleaned the roads, provided new sub-base as required and laid the tar macadam roadbase and the sealing grit.

13. The building works started in the autumn of 1974 and almost immediately the building contractor noticed damage to the roads under heavy construction vehicles making deliveries of materials to the site. The damage

245

Photograph 1 Trial hole through sub-base.

took the form of deflection under single wheel roads with "crazing" of the roadbase, see photograph 1. The area of crazing then began to spread. No problems occurred in areas of fill and, with few exceptions, all the damage was concentrated into the ares of cut, particularly along the eastern site boundary. Although the actual area of damage was small in percentage terms of the complete area of the road, it was considered necessary to re-seal the road in the areas of cut to prevent ingress of water and softening of the sub-grade.

14. At the end of the building contract, repairs to the road were carried out representing some 5% of the total road area. The surfacing was then applied and subsequently damage ceased, with the exception of occasional local patches requiring immediate repair to prevent the ingress of water and the spread of the damaged area.

C. ANALYSIS OF THE FAILED AREAS

15. The concentration of the sub-grade failure to areas of road constructed in cutting suggests that the failure resulted from an increase in moisture content of the sub-grade consequently leading to a reduction in strength.

16. In order to examine the behaviour of the material with respect to its moisture content, samples were taken of

the sub-grade in the areas of damage. The Plasticity Index of the soil in these areas varied between 13% and 24% and the results are shown in Fig.1. The samples of soil were compacted at varying moisture contents using the 2.5kg rammer into a C.B.R. mould. The C.B.R. of the sample was measured at each moisture content. As could be expected, the C.B.R. values measured decreased with increasing moisture content.

17. The natural moisture content of the sub-grade was measured at the time of taking the samples and also at two occasions subsequently. The results of these tests are summarised in Table 1.

Table 1. Variation of In-Situ Moisture Content

		Natural Moisture Content %		
Sample	Depth (m)	6.7.76	23.9.76	16.12.76
1	0.60	9.2	12.0	14.0
2	0.60	17.0	20.0	13.0
3	0.50	16.0	17.0	17.0
4	0.45	6.4	11.0	11.0
5	0.45	6.6	9.4	13.0
6	0.50	11.0	11.0	13.0

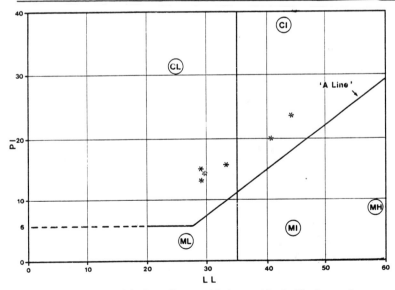

Fig.1 Atterberg limits for soil beneath failed road pavement.

18. Black and Lister (2) have shown that a relationship exists between the Consistency Index of the soil, defined as

$$\frac{LL - W}{LL - PL,}$$ (1)

(where: LL - Liquid Limit
W - Natural moisture content
PL - Plasticity Limit),

the Plasticity Index and the strength of the soil measured either as C.B.R. or shear strength. It can immediately be seen that the term LL - PL is, in fact, the definition of Plasticity Index and the expression can be re-written as:-

$$\frac{LL - W}{PI}$$ (2)

19. It can be seen from an examination of the test results that a more rapid decrease in strength was found to the wet side of the optimum compaction moisture than content occurred on the dry side. In order to examine this effect the Liquid Limit in the above expression of the Consistency Index was replaced by the optimum compaction moisture content. This dimensionless "Condition Index" can therefore be written as:-

$$\frac{O.M.C. - W}{PI}$$ (3)

20. The results of this is shown in Fig.2. It can be seen that although there is a degree of scatter in the results, there is a reasonable trend of decreasing C.B.R. with an increase in the dimensionless Condition Index. The values of C.B.R. recommended in Road Note 29 are plotted as are the in-situ moisture contents measured within the sub-grade.

21. It can immediately be seen that as the moisture content of the sub-grade increased the strength rapidly decreased.

22. The limits recommended by Road Note 29 are plotted on Fig.2. The change in moisture content required to cause a soil with a PI of 15% to fall below the lower limit is 1.5% from optimum moisture content.

23. Such a change in in-situ moisture content was shown in soil samples, taken at various dates from the site, to be exceeded. The limits of soil strength proposed by Black and Lister (2) are shown on Fig.2. The difference in moisture content from optimum to achieve the suggested lowest strength is 6% for a soil of PI of 15%. This difference in moisture content is compatible

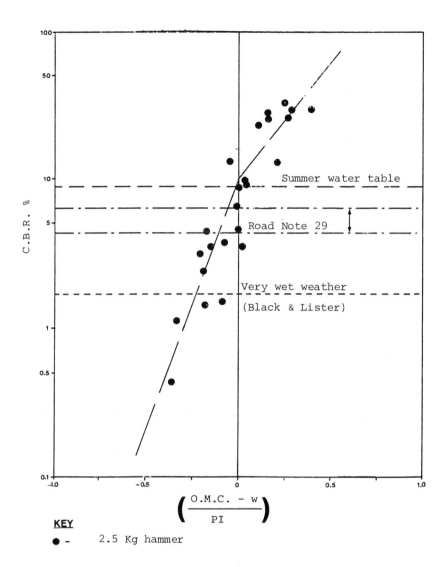

KEY

● -   2.5 Kg hammer

Fig.2 Strength of soil beneath failed road pavement.

with the in-situ moisture contents measured in soil samples taken from the area of the failed section of the road.

D.   EXAMINATION OF MATERIAL FROM OTHER SITES

24.   Following the damage experienced to the previously mentioned road it was decided, in similar developments in the Midlands area, to investigate further the

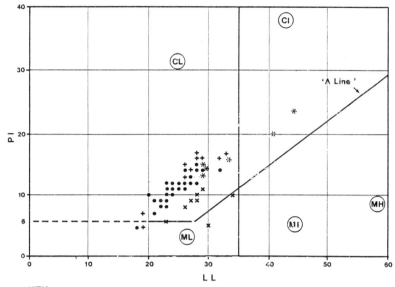

+ -

x -

• -   Soils from other locations.

* -   Soil beneath failed road pavements.

Fig.3   Atterberg limits of soils from comparative
        locations.

behaviour of similar materials as a road sub-grade.
Samples of the soils at the proposed sub-grade horizons
were taken to examine the soils strength relationship
to increasing moisture content. The Classification
Limits of the soil, a glacial drift deposit of Keuper
Marl origin, are shown in Fig.3, together with those
from the failed sub-grade, where it can be seen that
they are of similar classification.

25. Unlike the samples taken from the failed sub-grade, it
     was decided to examine the strength/moisture content
     relationship using a vibrating rammer to compact the
     samples to simulate the effect of a vibrating roller.

26. The results of the tests are shown on Fig.4, plotted
     with results for the soil from the failed sub-grade.

27. All results can be seen to show a similar rapid
     decrease in strength with an increase in the
     non-dimensional Condition Index. By taking a value of
     PI equal to 15%, a value of C.B.R. = 6%, can only be

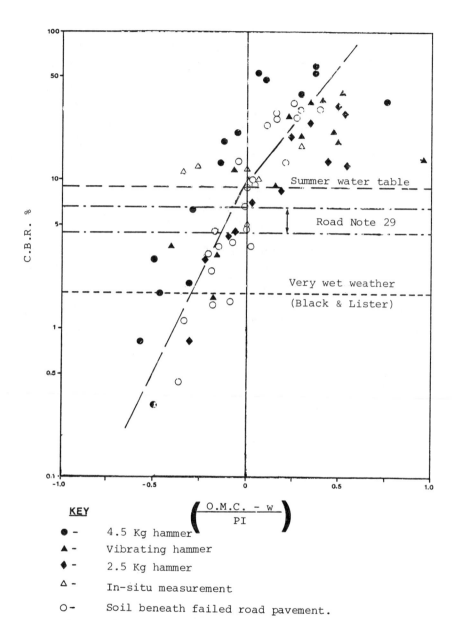

KEY

$\left(\dfrac{O.M.C. - w}{PI} \right)$

● -   4.5 Kg hammer

▲ -   Vibrating hammer

◆ -   2.5 Kg hammer

△ -   In-situ measurement

○ -   Soil beneath failed road pavement.

Fig.4   Strength of soils from other locations compared with that beneath failed road pavement.

expected with an in-situ moisture content less than 2% wet of optimum.

E.   CONCLUSIONS

28.   The foregoing results and case history show that the choice of sub-grade strength must be carefully considered and the strengths suggested by Road Note 29 be viewed with care. This is especially so if the road is constructed at a time when either bad weather may reasonably be expected, or in areas of cutting where a relief in vertical stress would be expected to result in an increase in the moisture content of the road sub-grade.

REFERENCES

1.   TRRL. Road Note 29 "A Guide to the structural design of pavements for New Roads". H.M.S.O. 1970.

2.   BLACK W. and LISTER N.W. "The strength of clay fill sub-grades: its prediction and relation to road performance". Clay fills. Institution of Civil Engineers, London 1978.

3.   CHANDLER R.J. and DAVIES A.G. "Further work on the Engineering Properties of Keuper Marl". C.I.R.I.A. Report 47.

15. Earthworks formation failures

N. McFARLAND, Senior Engineer (Material), Ove Arup Partnership

SYNOPSIS. The paper describes the types of formation failures encountered in the writer's involvement in the initial construction and more recently re-construction on major motorway contracts over a period of twenty years. Formation failures attributable to problems associated with design, control, materials and workmanship are reviewed with reference to specific cases. Possible remedies are suggested for failures on existing motorways and recommendations to prevent recurrence of these failures for future highway construction.

1. INTRODUCTION

1.1. The development of a major arterial motorway network system (Fig.1) in this country began in the late nineteen fifties and was substantially completed by the mid nineteen seventies.

1.2. Based on a design life of twenty years, calculated on a standard axle loading, (Fig.2) the early sections of the motorways constructed over the period 1957-1968 have been subjected to significant increases, both in the growth in numbers of commercial vehicles using the motorways and an axle loading substantially greater than that at design stage.

1.3. A damaging factor due to the increased growth and loading has considerably reduced the design life of the construction, necessitating extensive major restrengthening on motorways throughout the country over the past eight years.

1.4. I have been closely associated in the construction of numerous major contracts on the M1 London-Yorkshire, the M6 Midlands Links motorways and subsequently on the major restrengthening, often on the same stretches, over the past 20 years. This has been invaluable in assessing the possible reasons for various forms of pavement failures, the most practical and economical remedies and, hopefully, most important, prevention of recurrent failures in future.

1.5. The performance of a motorway is to a large extent determined by the strength of the underlying formation. In this country the standard of earthworks construction is dependent on many factors, not least of which is the re-use of infinitely variable, naturally occurring materials, in many cases excavated and placed under extremely difficult

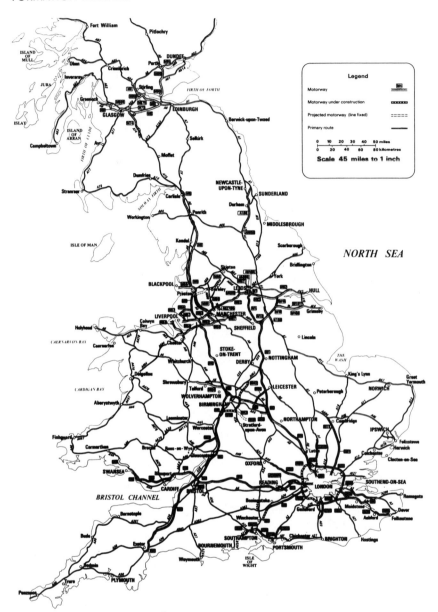

Fig. 1. Motorway network system

climatic conditions.   The fact that subsequent formation
failures occur is therefore not totally surprising.

1.6.   The origin of formation failures can be inbuilt as
early as the planning stage of a motorway contract when
political decisions and cost effectiveness can restrict the
final choice of route to one which presents minimal disruption to the public and is generally through areas of land

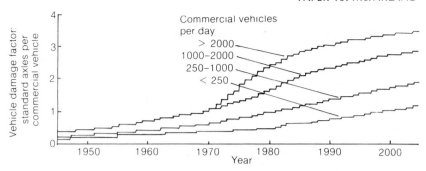

Fig. 2. Estimated changes in vehicle damage factors for four levels of traffic: 1945 to 2005

which are unsuitable for any other forms of development, such as housing or industrial building.

1.7.   In general at design stage the earthworks form some 20% or even more of the total contract value and unlike the structural or pavement element, which seldom vary to any great extent, the earthworks values have, in many cases, been known to have doubled or sometimes trebled the earthworks tender figure.   Over a number of contracts in the Midlands area of the United Kingdom the average increase in earthworks value was in the region of 40%.   This indicates the variability with which the designer has to contend.

1.8.   A study of earthworks design balances on current major motorway contracts indicates a definite shift in the assessment of suitable/unsuitable volumes, generally result- ing in a situation of a surplus of suitable material on completion of earthworks excavation.   With the availability of selectively better quality materials in this situation, the likelihood of a high standard of earthworks construction is greatly enhanced.

FORMATION FAILURES
2.   Failures Due to Classification

2.1.   The criteria for classifying suitability of materials for earthworks construction is normally specified in the contract in terms of moisture content.

2.2.   For cohesive soils the maximum permitted moisture content is expressed in terms of Plastic Limit of the material (e.g. 1.2 x P.L.) and for non-cohesive soils, the maximum permitted moisture content is expressed in terms of the Optimum Moisture Content and is usually within the range Optimum Moisture Content + 1.5 to 2.5%.

2.3.   Although pre-contract site investigation and advanced trial hole information are indicative of the suit- ability cf the earthworks materials, problems arise when dealing with materials which fall into the category of marginally suitable or marginally unsuitable.

2.4.   A daily rate of 10,000m³ is an easily achievable target for the large scraper fleet used on major earthworks

operations on motorway contracts and even the most efficient,
well equipped site laboratory encounters problems in achieving
sufficient representative test results to permit the Engineer
to control suitability of earthworks materials on an end
product test results basis.

2.5. In order to permit smooth progress of the works,
minimise delays and subsequent costs, the Engineer in many
cases has to make on site decisions based on experience and
rely on subsequent test results to substantiate his decisions.

3. Examples of Failures

3.1. Incorrect classification based on limited pre-site
investigation information or on site decisions made on classification involving the use of marginal materials have been
found to cause failures. This may have been the result of
attempting to achieve the most economical use of available
materials within the works, or in many cases, the result of
on site decisions taken to achieve the designer's assessment
of balance of materials.

3.2. During construction, movement, cracking and rutting
of the overlying granular or pavement construction is
indicative of formation failures, due to the use of incorrect
materials. Site trafficking usually highlights areas where
failure is likely to occur.

3.3. Formation failures after completion and during the
life of the road are again indicated by local depressions,
longitudinal cracking or crazing of bituminous construction.
In the case of concrete pavement, again local depressions,
cracking or crazing of the surface.

3.4. Remedial Treatment. During the various stages of
construction the remedial treatment generally presents minor
problems, mainly associated with responsibility and cost.
Availability of plant and materials are at hand and the treatment involves removal of the unsuitable material to a sound
base and reconstruction with suitable materials.

3.5. Remedial work to formation failures on trafficked
motorways is a much more complex problem, involving extensive
signing and coning and is generally carried out by motorway
maintenance organisations. The restrictive working conditions
on this type of work limits treatment to that of a temporary
nature involving corrective measures within the upper layers
of pavement construction. The in-depth corrective treatment
is undertaken in a programme of major restrengthening of the
motorways, in which case the treatment of defective sub-grade
is undertaken.

4. Prevention of Recurrent Failures

4.1. To achieve the most economical use of the available
existing earthworks materials on a major motorway contract,
through the existing methods of materials classification, it
is unlikely that formation failures will not occur.

4.2. The earthworks balance on completion of a number
of current Midlands contracts have resulted in quite large

volumes of suitable surplus material being removed to tip. Whilst not gaining the most economical use of existing materials, this form of design assessment allows for the selective use of better quality materials design and, hopefully, fewer formation failures.

5. Failures Due to Transition Areas Cut to Fill

5.1. The weather-affected materials which exist at cut/fill transition areas are generally a source of weakness in the earthwork construction.   Either due to inadequate treatment being specified at design stage or insufficient treatment, either in depth or extent, at construction stage, formation failures occur at this location.

5.2. Examples of Failures. During construction the limits of weak weather-affected material at the cut/fill transition areas is generally indicated by the extent of rutting by heavy earthworks plant such as loaded dump trucks, scrapers or lorries delivering imported materials to the site.   Evidence of failures of this type on live motorways is in the form of local depressions in the upper layers of the pavement construction.

5.3. Remedial Treatment. During construction the weak weather-affected material at cut/fill transitions should be removed to a firm base or to a specified depth above adjacent drainage invert.   Backfilling either with suitable excavated site material or imported granular material should be to a standard to ensure uniformity of formation strength spanning this area of weakness.

5.4. On trafficked motorways the remedial treatment is generally limited to superficial treatment within the upper layers of pavement construction.   The extent of remedial treatment is determined by the visual evidence of the surface failure.

5.5. Prevention of Recurrent Failures. With the knowledge that transition areas cut to fill are a source of weakness, the designer should take this into consideration and provide in the contract for special treatment at these locations and, in so doing, minimise the risk of formation failures.

5.6. The site supervision should ensure that the irregular limits of cut/fill transition weaknesses are defined and the full extent of the area is adequately covered by the specified treatment.

6. Formation Failures Due to Drainage

6.1. Possibly the main source of formation failures is attributable to the ingress of water which adversely affects the performance of the cohesive soils at and below formation level.

6.2. An essential stage of earthworks construction is an efficient functional drainage system as early as practicably possible and most certainly prior to formation preparation and acceptance.

6.3. Examples of Failure. A blockage in the central reserve drainage system of the M6 Midlands Links immediately north of Junction 1 Interchange resulted in considerable ingress of water through the Type 1 granular sub-base which softened up the cohesive boulder clay at formation level and resulted in the collapse of the composite pavement construction in the area of the slow and middle lane. A number of instances of drainage blockages, sometimes due to faulty workmanship and sometimes due to the weakness of porous concrete pipes collapsing, have been the cause of failures at formation level due to softening up of the materials to very low C.B.R.'s in the region of 1 - 3.

6.4. Remedial Treatment. Locate the faulty drainage, remove and replace where necessary and, again, rectify the defective construction within the depth of drainage invert level. In some cases on the M6, where the softening up of cohesive material extended below adjacent invert level, the remedial treatment consisted of a composite construction. The soft material was partially removed and a nominal layer lean mix concrete placed and compacted directly on the soft material within drainage invert level. Whilst the lean mix was still "green", the overlay of bituminous layers of construction were placed and compacted. This form of treatment was of a temporary nature, implemented in order to eliminate the long curing period required for lean mix concrete. Subsequent observation of the area showed no signs of failure due to this method of treatment, but as stated, this was a temporary method of remedial work, used to reduce the hazard of lane closures on a heavily trafficked motorway to the travelling public.

7. Prevention of Recurrent Failures.
7.1. A recent development in the routine maintenance of motorway drainage systems has been the use of the mobile "Jetter" system of cleansing and flushing drains. This consists of a large mobile water tanker equipped with a lengthy reel of rubber hose which is propelled through the drain by the force of water jets from a steel head on the end of the rubber hose. The efficiency and rate of progress of routine drainage checking by this system ensures the drainage is functional, locates any blockages which could result in formation problems due to water ingress and is especially effective when used in the difficult area of central reserve drainage checking.
7.2. The introduction of perforated plastic pipes in lieu of the brittle porous concrete pipes in motorway french drainage systems appears to have reduced the number of blockages due to pipe breakages.

8. Formation Failures Due to Granular Sub-Base.
8.1. One of the main functions of granular sub-base is to provide a uniformly interlocked stable platform immediately above the earthworks formation and, at the same time, allow

for rapid dispersal of water below the impervious pavement construction.

Although in many cases in compliance with specification requirements, granular sub-base manufactured towards the finer side of the grading envelope has a greatly reduced permeability and is, when in-situ in pavement construction, in a saturated condition. This layer of construction is continually stressed by traffic loading and unless consideration has been taken at design stage on the strength of the underlying formation, softening up of the formation can result in failures.

8.2. Examples of Failures. Many instances occur during construction when granular sub-base placed and compacted in dry conditions becomes unstable when exposed to inclement weather conditions. Phasing of the construction operations often results in large areas of granular sub-base being exposed to the weather elements over the winter period and material which has retained water generally has had to be removed.

8.3. On live motorways the evidence of formation failures due to wetting up from the granular sub-base is shown in the form of depressions, cracking or crazing at finished road level.

8.4. Remedial Treatment. Granular sub-base that has become defective due to excessive increase in moisture content should be treated either by allowing the material to dry out to an acceptable standard or by complete removal and replacement of the defective material.

8.5. On live motorways the treatment usually consists of removing the material to a sound base and reinstatement with new material, preferably on the coarser side of the specified grading limits.

8.6. Prevention of Recurrent Failures. Possibly a revised grading specification of Type 1 granular sub-base is required which would perform as a stable interlocked construction platform and in the long term provide a sound sufficiently permeable layer to permit the movement of water without detrimental effect to the underlying formation.

9. Formation Failures Due to Composite Construction

9.1. The use of lean mix concrete in composite pavement construction provides an excellent load-spreading platform during construction, minimises the damage from heavy site trafficking and, in the long term, acts as an excellent load distributor at a high level in the pavement construction. In order to maintain the minimum strength requirements in the manufacture of lean mix concrete, strengths well in excess of the 28 day specified cube strength are common. Contraction of the concrete results in transverse cracking of the lean mix concrete at regular intervals, in some instances as regular as 10 m intervals. In due course this cracking is reflected through the bituminous overlay, allowing ingress of surface water and inevitably a softening up of the formation.

9.2. Examples of Failures. The reflective cracking on
the existing surface of the M6 Motorway (Fig.3) in Warwickshire
was so extensive as to warrant consideration over and above
other methods used to determine the nature of the new
construction.

9.3. Remedial Treatment. Generally results in the
removal of the fractured overlying pavement construction and
if necessary the removal of defective lean mix concrete.
The reinstatement generally consists of bituminous pavement
construction to new levels.

9.4. Prevention of Recurrent Failures. A design requirement is needed which would increase the overlay of bituminous
construction and reduce the extent of reflective cracking, or,
alternatively, a change in the specification of the bituminous
overlay construction materials to provide composition of construction layers which will resist the tendency of reflective
cracking from the movement of the lean mix concrete.

10. Formation Failures Due to Settlement

10.1. A major source of formation failure occurs in the
area of backfill to structures where, during construction, the
compaction of earthworks materials in the restricted working
area is an extremely difficult operation.

10.2. Embankment construction on major motorway contracts
in the area of structures is generally undertaken in two
distinct stages. Firstly, the bulk embankment filling is
completed well in advance of the area of backfill to the
structure and has a time period in which natural settlement
and consolidation can take place prior to formation and pavement construction. Secondly, the infill area between the
structure and the main embankment is generally constructed
rapidly to formation level, then followed immediately by
pavement construction before the same degree of settlement
and consolidation has taken place as in the main embankment.
This results in differential settlement at formation level
and is a continual source of failure.

10.3. Examples of Failures. Throughout my period of
imvolvement on motorway construction, numerous variations in
the material specification and methods of placing and compacting backfill to structures have been exercised to minimise
the problem of settlement in the backfill area at underbridges and to date there has not really been a fully successful
ful solution to the problem. One of the current structural
designs which includes cantilevered wing walls has, in fact,
aggravated the problem in achieving an acceptable standard
of compaction in this area. Contractors have, in many
instances, elected at their own cost, to concrete the void
under cantilevered wing walls in order to achieve a reasonable
standard of compaction in the backfill area.

10.4. Although action is taken at design stage on
structural foundations to minimise the degree of settlement
where weak materials are known to exist, little or no consideration is given to minimise the degree of settlement under

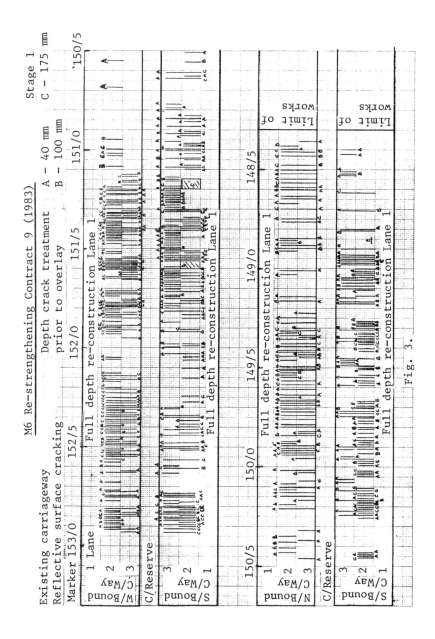

M6 Re-strengthening Contract 9 (1983)

Existing carriageway   Depth crack treatment   A – 40 mm   Stage 1
Reflective surface cracking   prior to overlay   B – 100 mm
                   C – 175 mm

Fig. 3.

261

embankment immediately behind the structure on the same weak materials.

10.5. A number of contracts on the M1 and M6 Motorways included in the design of backfill to motorway underbridges a 600 mm layer of rockfill overlayed by 125 mm blinding concrete and a 450 mm thick reinforced concrete floating slab as a means of bridging the area from rigid structure to adjacent embankment.

10.6. During restrengthening contracts, checks have been carried out at the location of floating slabs and in most instances settlement had occurred, in some cases voids as large as 150 to 200 mm were found immediately adjacent to the bridge abutment. On early contracts in Warwickshire a decision was taken to remove the floating slabs and reinstate the backfill area with rockfill. Where this was done subsequent settlement has taken place within a reasonably short period of trafficking. On later restrengthening contracts, the decision to remove the floating slab was reversed and instead a treatment involving grouting of the void beneath the floating slab was considered a better solution.

10.7 <u>Remedial Treatment</u>. Where depressions have occurred in the surfacing due to settlement, the remedial treatment has generally been to remove materials within the upper layers of pavement construction and reinstate to correct levels with bituminous pavement construction.

10.8 <u>Prevention of Recurrent Failures</u>. The problem of settlement in the backfill areas to structure has not been satisfactorily resolved. Consideration could be given at design stage for adequate treatment to the existing ground on site of embankments, especially at structures where the structural designer has had to specify piling or other forms of treatment to minimise structural settlement. The use of specialised compaction plant and quality inert granular materials placed and compacted to a high standard of workmanship would, to some degree, reduce the extent of settlement but not necessarily eliminate the problem.

11. CONCLUSIONS

11.1. The true stalwarts in the front line of any major motorway contract are the Resident Engineer and the Contracts Manager and not necessarily in that order.

11.2. On their shoulders lies a great responsibility which, if properly exercised, has a significant influence on the standard of the final completed product.

11.3. A sound working relationship between the Resident Engineer and the Contracts Manager is considered an essential element in maintaining smooth progress of the work, within an agreed programme, and inevitably leads to a high standard of achievement on completion.

11.4. Administration of a team of engineers, some of whom may be gaining site experience for the first time, providing technical advice and guidance on problematical situations are a few of the qualities required of the Resident Engineer.

11.5. The principal task of the Contracts Manager is to execute the works to the satisfaction of the Engineer and in so doing, gain the maximum financial benefit for his company. The profitable contracts generally reflect the standards of achievement.

11.6. The current established policy within the construction industry is to sub-contract practically every facet of the work and the nucleus of the main contractor's senior staff act as administrators. This form of contracting on major motorway contracts introduces problems, demanding a much greater degree of control on behalf of all supervisory staff.

11.7. Finally, it is unlikely that the problems resulting in earthworks formation failures will ever be fully resolved. Hopefully, with sufficient exchange of information on most likely causes of failures, preventative measures can be taken to limit the number and extent of formation failures on future motorway construction.

11.8. This, in turn, should minimise the necessity to undertake highly dangerous maintenance and restrengthening works which are extremely disruptive to the road user and at present succeed in reducing an efficient motorway network system to a series of minor roads interconnected by short stretches of high speed motorways.

Discussion on Papers 14 & 15

DR C. S. DUNN, Travers Morgan and Partners, East Grinstead

A considerable amount of information on Keuper Marl has been published by Chandler, Davis, Birch and others at Birmingham University but there are relatively few published case histories of failure. Mr Turner has therefore provided a useful study of a pavement failure on Keuper Marl. The problem with Keuper Marl is that like so many heavily overconsolidated clays of low plasticity it often gives the appearance of rock when first excavated. Everyone is surprised when it softens to such a weak state when it is unloaded and exposed by excavation and subjected to trafficking under poor drainage conditions. Keuper Marl generally has a high clay mineral content but this is present in aggregated beds and it actually behaves as a soil with a much lower clay content and the plasticity indices give a good indication of that behaviour.

It is recalled during construction of a section of the M5 motorway that the Keuper Marl embankment exposed to the weather was so softened by a period of continuous rain that a cow crossing the construction sank up to its knees in mud. This same material when dried during prolonged dry weather became so hard that it was not practical to pulverize the soil when attempting to stabilize it with cement.

The susceptibility of the Keuper Marl to softening is dependent on its weathering classification and most problems can be expected with the grade 5 or 6 materials referred to by Mr Turner. The less weathered and indurated undisturbed materials will not be affected by stress relief and poor drainage as easily, which probably explains some contradictory accounts of the behaviour of the marl.

On the more general subject of pavement design over clays, the important variable which must be determined to produce an appropriate pavement design is the equilibrium moisture content. Insufficient attention has been given to this. The Transport and Road Research Laboratory many years ago carried out research on soil suction and produced several excellent papers on the subject and this led to a clear understanding of the phenomenon of suction and the concept of equilibrium

moisture content. This work was not universally understood by
engineers and Road Note 29 did not clearly suggest how
equilibrium moisture content should be determined. If the
likely depth of winter equilibrium water-table below formation
can be determined then it should not be difficult to determine
the corresponding subgrade moisture content. Fig. 1 shows a
diagram of an apparatus which may be used to model or
reproduce the stress and drainage conditions of subgrades.

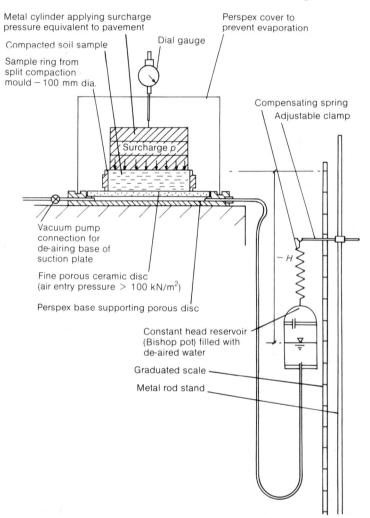

Metal cylinder applying surcharge
pressure equivalent to pavement

Perspex cover to
prevent evaporation

Compacted soil sample

Dial gauge

Sample ring from
split compaction
mould — 100 mm dia.

Surcharge p

Compensating spring
Adjustable clamp

Vacuum pump
connection for
de-airing base of
suction plate

— H

Fine porous ceramic disc
(air entry pressure > 100 kN/m²)

Perspex base supporting porous disc

Constant head reservoir
(Bishop pot) filled with
de-aired water

Graduated scale

Metal rod stand

Fig. 1. Equilibrium moisture content apparatus

A sample of subgrade soil is compacted in a special Proctor
mould which splits down after compaction to leave a central
ring of compacted soil which when carefully trimmed produces a
sample 25 mm thick and 100 mm in diameter. This ring with

sample included is placed on and in intimate contact with the surface of a suction plate about 150 mm in diameter (a ceramic disc having a high air entry value) sealed into a Perspex base. The sample is subjected to a surcharge pressure (by solid metal discs sitting on top of it) equal to the expected pressure of pavement on subgrade formation.

De-aired water which has previously saturated the suction plate is placed under a negative pressure by suspending a reservoir of water (on an extensible spring (Bishop pot) open to atmospheric pressure) at a level below the top of the sample corresponding to the expected winter water-table level below the formation level. In Mr Turner's case this would be 0.6 m. This representative sample is subjected to approximately the stress conditions and porewater pressure expected in the field. The sample must be permitted to reach equilibrium over a period of 2-4 days. The sample will either consolidate or swell and volume changes can be measured by a dial gauge. For low permeability samples the time to equilibrium may be too long for practical purposes and it may be necessary to use samples thinner than 25 mm. Having left the sample to equilibrate, its moisture content is then determined and this moisture content will be the most appropriate one at which to determine the California bearing ratio if the long-term conditions are the most appropriate to design.

MR W. H. LEWIS, Sir Owen Williams and Partners, Birmingham

My contribution is in relation to the section of Paper 15 that deals with formation failures due to settlement adjacent to structures.

As Mr McFarland says, the problem is the differential settlement which occurs between the structure and the embankment leading up to the structure.

The Department of Transport's decision in 1968 to prohibit the use of floating slabs, made known in Technical Memorandum T6/68, dated 14 October 1968, means that some alternative method of overcoming the problem has to be found. The solution adopted in the designs prepared by Sir Owen Williams and Partners and used in a number of schemes is to control very carefully the type of material used for filling behind structures and also to control the standard of compaction given to the material.

Figure 1 shows an underbridge abutment and the adjacent embankment. The normal embankment construction is terminated at least 30 m from the back of the abutment.

Zone 2 consists of selected materials from cuttings which are not susceptible to significant consolidation settlements after compaction. They are compacted to the standard specification requirements for earthworks.

The zone 1 materials are well-graded granular materials laid in layers 150 mm thick and compacted to the requirements for

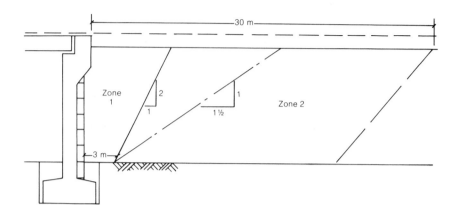

Fig. 1

granular sub-base (clause 803 of Specification for Road and Bridge Works).

The 2:1 slope between zone 1 and zone 2 is used when the two materials are laid concurrently. The flatter slope, of 1:1.5 is used when zone 2 materials are laid before zone 1 materials. The 3 m minimum dimension at the base of zone 1 is to ensure adequate room for compaction plant.

Any soil slope left for a period of time is trimmed back and benched before adjacent fill is placed. All topsoil beneath the zone filling is removed as is any suspect material below ground level.

The grading envelope for the zone 1 material is shown in Fig. 2. A number of specifications were examined before the envelope was chosen, including Missouri Standard Specification 1968, South Dakota Division of Highways 1968, the Staines Bypass 1963 (Road Research Laboratory No. 48) and the Department of Transport's Specification for Type 2 Granular Sub-base.

Although settlement of the backfill is minimized in this way, it is considered that some differential movement between the surfacing founded on the backfill and that on the structure is inevitable and must be accommodated.

The present requirements on Department of Transport schemes are that tenders are invited on the basis of alternative forms of pavement construction. If a rigid type of construction is adopted for the contract a fully flexible road construction is used on the approaches to the bridges for a distance of 30 m from the bridge abutments.

This form of construction is capable of accommodating considerable distortion without significant structural damage and facilitates the carrying out of remedial works to correct surface levels. Settlement due to consolidation of the ground beneath the embankment can be overcome by preloading the

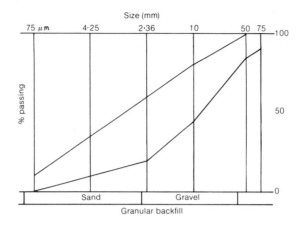

Fig. 2

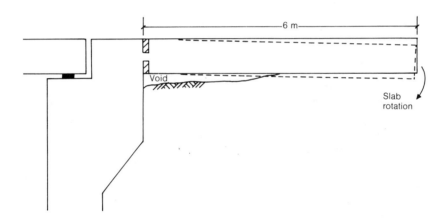

Fig. 3

ground, or by delaying the casting of the structure deck.

Mr McFarland mentions the floating slab treatment and mentions also the voids which form beneath the slabs. Voids which form beneath the slab (Fig. 3) are generally found close to the back of the abutment because this is where the greatest settlement occurs. The embankment further away from the abutment settles less, so the slab rotates about the hinge connection to the abutment and spans over the voids. The voids only become a real problem if they migrate towards the outer end of the slab, causing the slab to rotate an unacceptable amount and forming a step in the carriageway.

Remedial measures to deal with this condition, such as grouting, are comparatively simple and could be made less

269

disruptive to the traffic using the road if grouting tubes are cast into the slab at the construction stage. The operation can then be carried out from the outer edge of the carriageway (Fig. 4). Care would have to be taken to ensure that the grout did not find its way into any drainage measures behind the abutment.

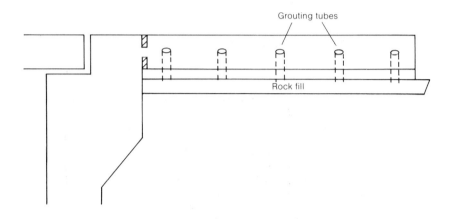

Fig. 4

The floating slab method therefore could be successful especially if combined with zone filling.

Mr McFarland, would you like to comment on this suggestion?

MR N. McFARLAND

Although I fully agree with Mr Lewis on the suggested design for earthworks construction in the difficult restricted area of backfill to structures, it is the implementation of a complex earthworks operation which contributes to the problem of settlement. The various structural designs and the variable earthworks outlines encountered in themselves complicate the situation. I consider that the main cause of settlement in this situation is the failure of the Contractor to produce the very high standard of workmanship essential in minimizing the extent of settlement in the critical area of earthworks construction and the failure of the supervising authority, in many instances, to ensure that the site control on materials quality and standards of earthworks construction are fully in accordance with specification requirements. I am an advocate of the floating slab construction which, I consider, did reduce the problem of settlement immediately behind the structure abutment.

Settlement has occurred at an underbridge on one of the current contracts in which I am involved and again it is due to

(i)   granular fill not to specification requirements
(ii) low standard of compaction
(iii) poor supervision.

Only when fully inert materials are specified, effectively placed and compacted under strict supervision will the problem of settlement in the area of structural backfill be reduced to an acceptable minimum.

MR A. R. DAWSON, University of Nottingham

In their Fig. 4, Turner and Seago draw a somewhat gratuitous line through a number of points for which the ordinate is the California bearing ratio (CBR). It is well known that the measurement of the CBR is highly test and test method dependent. Using their data, a typical road construction could have a design life between 6.8 and 16.5 million standard axles, the actual life depending on the value of the CBR. These values are obtained for the same soil at the same 'moisture condition index' (for a nominal life of 10 million standard axles on a 5% CBR subgrade).

The new Transport and Road Research Laboratory report LR 1132 tries to overcome this problem by qualitatively describing the soil in terms of 'good', 'average' and 'poor' construction conditions. This at least recognizes the difficulties of quantifying the soil's properties. However, this recognition should be a spur to research whose aim must be to determine simple and reliable means of assessing and predicting relevant properties (principally resilient stiffness) at working ('equilibrium') conditions. The penetrometer mentioned by Mr McFarland may be a first step in the direction.

None of the authors questioned the need for a granular sub-base. If it were possible to construct the bound layers without a granular working platform (and perhaps improved subgrades may enable this) then very little loss in road life would result. However, it is probable that the subgrades beneath many roads are wetted by water supplied under capillary action through the sub-base. Run-off from the carriageway can enter the sub-base at the road margins with the present system of connected sub-base and lateral drains. If this connection were severed, and this is most easily effected by removing the sub-base, dryer subgrades would probably result, which in turn would lead to roads with a longer life.

While engineers continue to believe that granular materials are necessarily drainage layers regardless of how low a hydraulic gradient acts, and while they believe that a lateral drain can draw down the water-table by 600 mm beneath the carriageway in a heavy clay over a distance of more than 10 m wet subgrades are likely to remain a problem.

MR A. J. TURNER and MR K. L. SEAGO

The problem with the qualitative approach of LR1132 to the
question of road construction conditions is that these will
not be known by the designer when designing the road or
preparing the bill of quantities or to the Contractor when
ordering the materials. Although on major road contracts,
such as motorways, where a certain degree of site 'redesign'
can be achieved in the vast majority of road construction, we
suggest that the only course of action is to adopt the
pessimistic approach of 'poor' construction conditions. In
practice, therefore, there is no real choice for the designer.

DR D. W. COX, Polytechnic of Central London

Miniature compaction trials in shallow trenches adjacent to
test pits are a useful method of simulating the effects of
compaction (to Department of Transport or other standards) and
subsequently of measuring the shear strength, California
bearing ratio, bearing capacity, rutting etc. of various soils
used, including the effect of wetting, and hence of
determining both suitability and trafficability.

MR C. WOOD, Department of Transport, London

In view of the large number of failures described could Mr
McFarland give his view of the failure rate for road
formations? The failures may be attributable to

(i)   failure by the Engineer to define limits for re-use of
       earth material adequately and to direct the best
       material into the formation properly
(ii)   failure of the specification to reduce the air voids in
       the top layer of an embankment
(iii) failure of the Contractor in not complying with clause
       20 (taking proper care of the works).

If these are in some way reasons for failures in road
formations then it is proposed in the next (6th) edition of
the specification to give designers a choice in suitability
parameters and to include both cohesive and granular suitable
material to aid direction of fill. Lime and cement
stabilization of formations and proper workmanship for
preparation to receive sub-base will be included.
  Designers will be encouraged to include construction
drainage and to produce a cross-section for earthworks to
improve drainage.
  Mr Turner has demonstrated the value of the prediction
method for California bearing ratio values to be used in
pavement design. This is good for materials which respond
rapidly to moisture changes but does not really meet the need

to predict correct equilibrium strengths for heavy clays.

MR N. Mc FARLAND

I have no numerical values of the precise failure rates of
road formations due to specific causes, but from my lengthy
practical experience of both new motorway construction and
motorway restrengthening I can express my opinion on the
points that you have raised.

Of the many contracts of new motorway construction in which I
was involved, those where the Specification permitted the
Engineer to direct the selectively better quality materials
from cuttings for the top 2 ft of construction of embankments
to a low air void content (5%) were effective in achieving a
high standard of uniform formation construction, capable of
withstanding the intensive stresses of traffic loading. The
introduction of the Method Specification for earthworks
compaction resulted in many cases of an acceptance of a
formation compacted to a high standard but, at the same time,
consisting of a high air void content. This form of
construction was vulnerable to failure due to reduction in
strength when an equilibrium moisture content is reached in
material placed at low moisture contents. I cannot recall
actual failures due to failure of the Contractor to take
proper care of the works and, in my experience, it has been
incumbent on the Engineer to ensure that where the Contractor
has failed adequate rectification measures have been taken at
the Contractor's expense to achieve a high standard of
acceptance at formation level.

I endorse the proposal of a choice in suitability parameters
for both cohesive and granular suitable material to aid
direction of fill, which provide both Contractor and Engineer
with the essential qualities to produce uniformly high
standards of formation.

Lime and cement stabilization have not been a commercially
viable method of earthworks construction throughout the
contracts in which I have been involved. Small experimental
trials have been undertaken and shelved, mainly because of the
cost effectiveness of the operation. I consider that the
proper workmanship for preparation has always been available
to the Engineer/Contractor but has not always been effectively
enforced.

I intended the theme of my paper to be that the main
contributory factor to the formation failures was the failure,
either at design or construction stage, to provide adequate
drainage to protect earthworks. Effective advanced cut-off
drainage and positive permanent drainage are essential to
long-term performance of earthworks construction.

In conclusion, the evidence of 7 years' experience in
restrengthening contracts on the Warwickshire section of the
M6 Midlands link motorway showed that the main attributal
factor to the extensive failures of the slow lane was the

damage factor of a heavy vehicle axle loading, about three times greater than that at design stage.

DR R. T. MURRAY, Transport and Road Research Laboratory, Crowthorne

Could either Mr McFarland or Mr Turner comment on the value of geotextiles in reducing formation problems?

MR N. McFARLAND

My experience on the use of geotextiles in the reduction of formation problems has been rather limited. Having been present before the introduction of geotextiles and having been accustomed to the use of locally available good quality rockfill, the treatment to the poorest of ground conditions consisted of punching in large-size rockfill to a maximum layer thickness of 1.0 m.

I have indirect experience on the use of Terram 140 placed on a wet subgrade consisting of a soil structure of uniform sand with little or no fines. This took place on the M69, Leicester section cut 32, southbound carriageway, in October 1975 (Fig. 1).

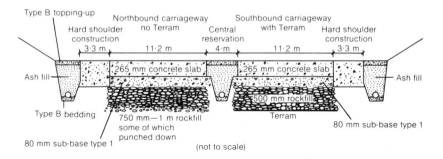

Fig. 1. Subgrade–sub-base separation in roads, M69, Leicester section: a section of dual three-lane motorway with 11 m carriageways, 3.3 m hard shoulders and 4 m central reservation, designed for 40×10^6 cumulative standard axles (40 years); soil structure, uniform sand with little or no fines

The water-table at the time of excavation was around formation level and it was not practicable to install carriageway drainage in advance of treatment. The Terram 140 was installed on the wet sand and overlaid with a 500 mm layer of rockfill (125–375 mm).

The operation was considered to be successful in reducing

the amount of rockfill required and the estimated saving on a 300 m length of one carriageway was in the region of £5000–10 000.

On the Solihull section of the M42 motorway, which was constructed 1974–1976, Terram 70 was used for capping French drains before overlaying with 100 mm of topsoil. In this instance the silt from the topsoil clogged the Terram, creating excessive ponding. The treatment in this case was considered to be a failure and the type B filter media were exposed at intervals of 5 lin/m to permit rapid dispersal of surface water.

Terram or similar geotextiles are in current use on sections of the M42 motorway. On the Lickey End section, Terram is being used in slope counterfort drainage to act as a filter and to prevent migration of particles of soil into the filter media.

On the Water Orton section of the M42 motorway, Terram has been effectively used to prevent the migration of very poor silty subgrade into the 500 mm layer of rockfill below formation level in cuttings.

This is the extent of my knowledge on the use of geotextiles and my main concern in their use as a separating membrane under rock at formation is what the life of the geotextile material is in this situation and whether there could be any significant detriment to the stability of the construction, should the geotextile rapidly deteriorate.

MR A. J. TURNER and MR K. L. SEAGO

The use of geotextiles to prevent the road base from being 'punched' into the road formation has proved to be successful. It prevents the road base from becoming contaminated with fines which would lead to failure of the road base.

However, it has come to our notice that designers are being encouraged to believe that the use of geotextiles can result in a saving of sub-base thickness. From the failures of pavements that we have observed where a lower thickness of sub-base has been used in combination with a geotextile, we consider that this approach should be viewed with caution.

It is difficult to make general statements concerning geotextiles as the various types available, such as woven and non-woven fabrics, each offer differing characteristics and performances. We would therefore welcome further research and guidance in this matter.

MR D. F. GAMLEN, Hereford and Worcester County Council, Worcester

Having been involved in the construction, in 1960, of the M5 motorway in Worcestershire, together with its subsequent maintenance and reconstruction, I was extremely interested in

Mr McFarland's paper and support many of the points he made.

About half the formation of the M5 motorway in Worcestershire is Keuper Marl, and as Dr Dunn has said the original intention was to stabilize the marl with cement. However, in situ stabilization of the imported gravel was used instead, and excavation for the reconstruction works has shown that in general this material is still in excellent condition (although the site problems of longitudinal joints between machine passes can be seen). There have therefore been few foundation failures - incidentally in the pavement design extra sub-base was allowed at cut/fill lines.

Little settlement occurred behind bridges; a wedge of graded backfill was used and particular care taken in the supervision of this work (an all-important aspect).

The Transport and Road Research Laboratory has reported on the differences between the slope stabilities of different ages of embankments, and I suggest that this reflects the early Department of Transport specifications which required 10% air voids in the main embankment but 5% in the top 2 ft. I hope that the new Department of Transport specification will resurrect this approach, and also require slope compaction.

Mention has been made of the use of video cameras in boreholes. This technique has been used very successfully to check the state of the drainage in motorways, preparatory to refurbishment.

MR N. McFARLAND

Great lengths of the earthworks construction on the M6 motorway section in Warwickshire consisted of Keuper Marl, both in cutting and embankment, as did much of the cutting and embankment earthworks construction of the M1 motorway on contracts between Crick and Doncaster. During both construction and subsequent reconstruction, my experience on the use of Keuper Marl has been such that, when properly excavated, placed and adequately compacted, this has produced an excellent impervious formation which has remained sound, on opening up after 12 years of intensive stressing from excessive traffic loading, well in excess of design criteria.

I agree that in situ stabilization of marl on major motorway contracts is not a realistic or practical proposition. I have been involved in small-scale lime stabilization trials on wet sand and gravel and although, again, not a realistically commercial nor practical function on a major motorway contract, the stabilized material, when inspected during restrengthening work, was in excellent condition.

I am pleased to hear that adequate treatment in the form of extra depth sub-base was successful in limiting problems at cut-fill transition areas on M5 construction works.

The bridge backfill settlement continues to be a problem and again on current work a failure has occurred after completion of pavement construction at an underbridge. Investigation shows that the failure is due to poor quality granular fill (outside specification requirements) that is poorly compacted and a collapse due to the ingress of water from open-jointed porous pipes.

This highlights the lack of adequate supervision, poor quality control on materials and faulty drainage design, which permits the ingress of surface water to susceptible earthworks construction.

As suggested in my paper, a wedge of inert granular fill placed and compacted to a high standard is an essential factor in limiting differential settlement in the restricted area of backfill to the various types of structural construction, especially when located in areas of existing weak earthworks materials.

I again concur with you on the use of an end product air void criterion for earthworks construction, e.g. 10% maximum in the main embankment and 5% in the top 2 ft of construction.

Even with the best will, the Method Specification on earthworks compaction is no guarantee of a high standard of compaction achievement in minimizing the extent of differential settlement in the restricted area of structural backfill, especially when this is on an area of existing weak ground conditions.

The use of an end product air void criterion and the choice of direction of selective materials would contribute greatly to a high standard of earthworks construction.

Great difficulty is encountered under the current economic pressures whereby contractors winning contracts at commercially impractical prices are financially restricted in the purchase of quality plant and materials that are essential for achieving high standards of construction, rather than marginally acceptable standards.

DR A. D. M. PENMAN, Sladeleye, Chamberlaines, Harpenden

Attention has been drawn during this session to the serious reduction in strength that can be caused by increasing the water content of a clay sub-base, leading to formation failure. Emphasis has been put on correct drainage methods and the desirability of ensuring that maximum groundwater levels are kept at a safe distance below formation.

The effect of cyclic loading does not appear to have been considered. Triaxial tests carried out some time ago at the Building Research Station on samples of overconsolidated clay showed that the application of shear stress reduced the pore pressure and when temporary drainage was allowed water moved into the sample. Cyclic repetition of this test increased water content and reduced strength until after sufficient reversals the sample failed at a stress much lower than its original strength. These tests were used to illustrate the mechanism of clay pumping observed on railways on heavy clay where cyclic loading reduced the clay to mud which worked up through the ballast and led to displacement of the track.

The size of modern continental trucks and trailers seen on our motorways gives the impression that cyclic effects must reach a clay formation, although the remarks of Black (ref. 1) are noted. He said then that investigations did not indicate a measurable increase in moisture content with time in clay soils that could be attributed reliably to this mechanism. Presumably the damage caused by contractors' plant running on lean mix reported to this symposium can be related to cyclic loading. Is the stiffness and weight of road construction over a clay formation always sufficient to reduce the cyclic loading effects from modern traffic to harmless levels?

Reference
1. BLACK, W. Discussion on Road subgrades. In Clay fills, pp. 245-246. Telford, London, 1979.

16. Major gabion wall failure

S. THORBURN, FEng, FICE, FIStructE, FASCE, Director, Thorburn Associates, and I. M. SMITH, DSc, MICE, Professor of Civil Engineering, University of Manchester

SYNOPSIS. The paper describes the failure of an eight metres high gabion wall constructed at the foot of a very steep rock slope. The historical background to the failure is discussed comprehensively and problems associated with the design and construction of gabion walls are highlighted. The results of a load test on a typical gabion basket is presented and recommendations are made for future designs.

INTRODUCTION

1. A landslip occurred in January 1978 adjacent to the Trunk Road A7 near Byreburnfoot, Dumfries, and removed the roadside wall and verge.At this location the trunk road had been constructed immediately adjacent to a very steep downslope comprising near vertical rock faces above the River Esk. The slope was 19 metres in height and had been formed by the erosive power of the river incising Coal Measures strata.The landslip exposed four metres of compact clayey silty sands beneath the road pavement and fifteen metres of a succession of sandstones with subordinate siltstones and claystones.

2.Post Carboniferous tectonic activity had formed a number of rock joint systems which together with the original bedding planes had provided ready access for weathering agencies. Weathering of the rocks together with growth of tree roots into open or clay-filled surface joints had resulted in the deterioration of the near vertical rock faces and many loose blocks of massive sandstone were in a precarious state of balance due to weathering of basal beds of weak and fragmented siltstones and claystones. Rock overhangs had developed and masonry retaining walls had been built on adjacent slopes as protective measures. The egress of groundwater on the rock faces had enhanced the continuous process of weathering and degradation.

3.Consideration was given to either (i)stabilising the near-vertical rock faces by rock anchors and surface shotcrete in combination with the construction of a gabion wall along the river bank or,(ii) forming a rockfill slope between the top of the same gabion wall and the road level to cover and protect the very steep natural slope. The decision was taken to adopt the latter solution and work commenced in July 1978.

4. The gabion wall was completed on 30th September 1978 and the rock fill forming the new downslope was placed by the 18th October, 1978. On the 26th October 1978, slip movement was observed in the rock fill and on the 27th October 1978 more extensive and serious slips occurred with substantial movements of the top of the gabion wall. The progressive outward wall movement continued until the gabion wall collapsed into the river on the 30th October.

5. The Scottish Development Department initiated a comprehensive investigation of events leading to failure of the wall involving theoretical analyses and a full-scale load test on a rock-filled gabion basket at Dundee University.

6. The paper describes the situation leading to failure; presents the results of the load test on a typical gabion basket; discusses the problems associated with the design and construction of gabion walls and makes recommendations for future designs. A description is given of the remedial works eventually constructed to support the trunk road.

DESCRIPTION OF SLOPE PRIOR TO CONSTRUCTION OF GABION WALL

7. An investigation of the landslip which occurred on 31st January 1978, was commissioned by Dumfries and Galloway Regional Council, and a report was received from a Site Investigation Contractor in June 1978.

8. The following descriptions of the condition of the rock strata in the vicinity of the landslip were provided by an engineering geologist employed by the Site Investigation Contractor. The landslip had revealed a relatively fresh rock surface beneath depths of about 4.0 metres below road level with a sandy soil overlying bedrock. The superficial condition of the rock exposed by the landslip was relatively good with very few discontinuities except locally where fissuring of the rock, and weathering agents had caused surface deterioration. The potential for failure of the rock strata on each side of the landslip was apparent from a visual survey of the site. Groundwater seepage from the soil or rock strata exposed by the landslip was not observed. The rock strata beneath the road consisted of weathered sandstone with subsidiary siltstone and mudstone layers to a depth of the order of 16.0 metres. Cores of the rock were generally badly broken with very low RQD values. The intact sandstone rock material was moderately strong with finer grained strata being weak to moderately weak. Weathering effects were pronounced throughout the profile apart from the finer grained strata in which these effects were not readily identified. No loss of air during drilling at any of the boreholes was reported by the geologist. Groundwater was encountered at all boreholes at varying depths and was present below a level of approximately 45 metres AOD. The regional dip of the bedding of the rocks varied but the geological sheet and site observations adjacent to the slip had indicated a dip of approximately 18° in an east 30° south direction. At the bore positions the true dip was approximately parallel to the road line at an

angle of 14° and thus the apparent dip perpendicular to the road at the position of the slip could be considered negligible.

9. Examination of the rock face had indicated that differential weathering would cause undermining of the upper sandstone and eventually lead to collapse in the form of dislodgement of surface rock blocks. The potential for failure was judged to be such that if the slope was not protected, a sudden landslip could occur which would endanger the entire trunk road.

10. It had been considered imperative by the Site Investigation Contractor that remedial measures should be executed immediately, otherwise the road had to be closed. Toe and slope protection were recommended with the former consisting of a gabion wall or similar, built above the maximum flood level. It was also recommended that the slope above this level should be stabilised by rock anchoring and shotcreting. An alternative was given involving placing rock fill to road level above a gabion retaining wall at the foot of the slope.

THE DESIGN AND CONSTRUCTION OF GABION WALL

11. The urgency expressed by the site investigation contractor prompted immediate action by the Regional Council who elected to adopt the alternative approach involving the construction of a protective retaining wall at the foot of the slope with rock fill. The recommendation involving stabilisation by rock anchoring and shotcreting was considered by the Regional Council to be particularly hazardous bearing in mind the expression of urgency and the possibility of danger implied by road closure. The attitude of the Regional Council in regard to the hazardous aspects of executing anchoring and shotcreting works was endorsed at a site meeting with specialist contractors in April 1978.

12. At a subsequent site meeting in May 1978, it was decided that a general survey and underwater inspection of the rock shelf forming the river bank should be carried out and also that the suitability of a gabion retaining wall should be established.

13. Discussions took place with the British Reinforced Concrete Engineering Company Limited (BRC) to ascertain the possibility of constructing a retaining wall from rock-filled gabion baskets. BRC considered this type of wall to be appropriate in the circumstances and submitted technical proposals to the Regional Council. The BRC proposals were used by the Regional Council as the basis for their own design and construction drawings prepared by the Council were forwarded to BRC for comment in June 1978. A typical cross-section of the gabion wall is shown on Fig.1.

14. Because of the importance and difficult nature of the remedial work, a pre-tender site meeting was held in June with Contractors, to ensure that each realised the difficulties involved, such as, short tender period and minimum time for completion of the works. The conclusion was reached that the

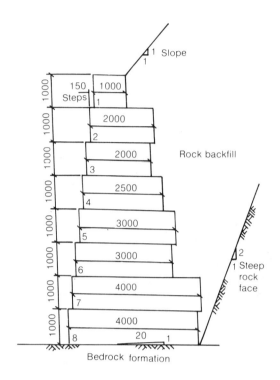

Figure 1. Typical cross section of gabion wall

Figure 2

Contract could be carried out with traffic lights and that the Trunk Road need not be closed.   Tender documents were issued in   June   1978   and   the contract was awarded   to   the   lowest tenderer in July 1978.

15.   Work   commenced on 24th July 1978 and the slip   debris and the bulk of the loose rock overburden had been removed   by the   23rd August 1978.   The question of omission of ties   to the   rock   face was raised and BRC expressed the opinion   that they were unnecessary.   The gabion wall was completed on   the 30th   September   1978 and rock fill and pitching to   the   side slope placed by the 18th October 1978.

16.   On   the   26th October a small slip was evident at   the extreme   north edge of the completed rock fill   and   pitching. On. the   27th   October 1978 further more   serious   damage   was apparent throughout the length of the pitching and substantial movement   had occurred at the toe of the gabion   wall.   This became   progressively   worse until on 30th October the   gabion wall collapsed.   No signs of distress were observed prior to 26th October   1978.   A photograph of the failure   shows   the extent of collapse, Fig.2.

INVESTIGATION OF THE GENERAL STABILITY
OF THE SITE SUBSEQUENT TO GABION WALL FAILURE
Upslope of the Trunk Road

17.   A study of aerial photographs provided by the Scottish Development Department,   and of the ordnance survey sheet for the   District,   revealed evidence of slope instability within the   steep   slopes above the   trunk   road.   Local   enquiries indicated   that   two slope failures had been   experienced   in former   times,   the   earliest major slip   probably   occurring prior to 1900.   The more recent slip occurred about 1949 and required the construction of the masonry wall which exists to the north-east of the trunk road.

18.   The available evidence indicates that,   in geological times, the River Esk progressively eroded the superficial and solid   deposits until it attained the present bed   elevation. The   erosive   power of the river probably resulted   in   steep river   banks   which   experienced   progressive   failure   and flattening to essentially stable configurations, particularly within   the superficial deposits.   There was no evidence of current   deep-seated   movements but   evidence   of   surface instability, slope degradation, and hillwash was found during the site inspections.

19.   Stability   analyses were carried out   using   various groundsurface   configurations   to permit an assessment to   be made   of the steep slopes above the road.   The analyses   in terms   of effective stresses revealed low factors of   safety and   confirmed   that critical conditions may be   expected   to have   existed at various stages of slope development.   The general   factor   of   safety   of the existing   slope   for   the groundwater regime observed from piezometers was probably   of the order of 1.25.

Downslope of the Trunk Road

20. Since the failure of the wall and the resulting state of the slope had masked effectively the original features and condition of the buried rock strata it was decided to make a general assessment of the situation using the information contained in the original geological commentary together with the results of the investigatory work after the wall failure. It would appear that the drillers who carried out the investigatory work in March 1978, experienced greater difficulty in recovering satisfactory cores than the drillers responsible for the later work. The reason for the difficulties could not be deduced from the information provided by the later investigatory work.

21. The solid deposits consisted of Carboniferous sandstones and grits with subordinate shales, siltstones and mudstones. Thin horizons of coal, fireclay and ironstone were also present. Weathering processes together with growth of tree roots and vegetation into open, often clay filled, surface joints had resulted in the degradation of steep rock faces which included loose blocks of massive sandstone separated by beds of fragmented friable siltstones and mudstones. Workable coal seams did not exist immediately below the site but a number of abandoned adits were present in the nearby Byre Burn where a thin coal was worked in the past.

22. Visual examinations of the very steep rock slopes below the trunk road together with an assessment of the condition of the rock strata from a borehole sunk from road level indicated that the general stability of the site was adequate at the present time. Local portions of the near-vertical rock faces were, however, displaying evidence of incipient surface failure and certain rock overhangs had to be removed to avoid local instability. Groundwater seepage through the rock strata was evident at several locations.

THE CHARACTERISTICS OF THE GABION BASKETS

23. The gabion baskets used for the construction of the wall were 2 metres in length, 1 metre in breadth and 1 metre in height. A central diaphragm divided each basket into 1 metre cube compartments. The sides, base, lid and central diaphragms were manufactured using BRC weldmesh fabric which consisted of 5 mm diameter bars at 75 mm centres in two mutually perpendicular directions. The mesh was galvanised after fabrication.

24. The baskets were delivered to site in a folded state with the sides fixed to the base by small diameter welded circular hoops. The four sides were rotated up into position and the vertical edges of the baskets were tied together using specially designed short helical wires together with additional lacing using 2.5 mm diameter wire which was supplied by BRC specifically for the purpose. The lids were fixed on one edge to the top of the front face of each basket by welded circular hoops and to the tops of the other three faces by an identical system of helical wire and lacing as used for the vertical joints. The stiffness of the baskets

is increased by the addition of cross tie wires. These wires are fixed at mid height of the baskets between the front and rear faces at a location 0.5 metres from either end and also between the centres of the end panels, tying them together. The cross-tie wires are 2.5 mm in diameter . The gabion baskets were carefully filled because of the angularity of the stone fill and the stone in the front baskets was placed by hand. The coarse stone fill to the gabion baskets had a grading of between 200 and 100 mm.

25. The gabion wall consisted of individual baskets tied together with 2.5 mm diameter wire. The stability of a complete wall depended on the ability of the baskets to act compositely and form a mass retaining wall. The ultimate strength of the wall depended on the compressive strength of the individual baskets and on the friction between the baskets. The efficiency of the tie wires in developing composite action between adjacent gabion baskets and in preventing slip failure between successive layers in a wall is difficult to estimate and, as no means of verifying these factors are readily available, it is impossible to establish the amount of composite action developed in a gabion wall. The life expectancy of the tying wires with regard to longterm corrosion is also a serious consideration.

26. At the outset of the investigation of the wall failure, an attempt was made to establish from the manufacturer of the baskets an indication of their likely behaviour under load. BRC indicated that it was very difficult to stipulate a maximum vertical compressive stress for a basket owing to the substantial number of variables and method of supplying and placing of stone fill (mechanical and/or hand packing). However, tests on single baskets had indicated that compressive stresses of 25 to 90 kN/m^2 could be applied without breaking the cross tie of the wire baskets. At a compressive stress of 250 kN/m^2 the baskets deformed to a substantial extent. While no indication of characteristic load/deflection relationships was given by BRC the recommendation was made that allowable compressive stresses of 100 to 150 k/m^2 could be used for the design of gabion walls.

27. In order to obtain reliable data for analysis of the wall failure a full-scale test was carried out at the University of Dundee to determine the load deflection characteristics of an individual gabion basket together with the failure stress. Load was applied to a gabion basket through a rigid plate by means of two 50 tonne capacity hydraulic jacks. The jacks were located eccentrically on top of the basket to simulate the distribution of pressure predicted for the critical gabions at the base of the actual wall. Fig.3 is a photograph of the test arrangement. The load was applied in 5 tonne increments and, at each stage, downward deflection of the plate was recorded. The initial load was approximately one tonne which was the weight of the test rig between the load cells and the basket. On attaining a load of 21 tonnes, corresponding to an average

Figure 3

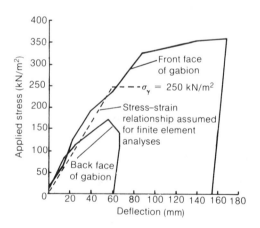

Figure 4. Results of load test on typical gabion basket

deflection of about 20 mm, the cross tie wires in the east compartment failed, allowing the front face of the gabion basket to bulge outward. At loads above 31 tonnes continuous pumping of the jacks was required to maintain the load, thus indicating creep of the basket. The remaining cross tie wire severed at a load of 41 tonnes and at approximately 50 tonnes the top south-east vertical edge of the basket burst open due to failure of the tying wires. Loading continued until a load of 51 tonnes was reached, by which time the load could only be maintained by vigorous pumping. At this ultimate load the vertical deflection of the front face increased from 90 to 140 mm and that of the back face from 55 to 65 mm. Thereafter the load could not be maintained and fell to 49 tonnes. Loading was then discontinued as the ultimate load of 51 tonnes and corresponding front face vertical stress of about 360 kN/m^2 had been reached. On unloading a vertical rebound of 15 mm at the front and 5 mm at the rear was recorded. Curves of stress v deflection for both front and rear faces of the gabion basket are given in Fig.4. In calculating the stresses on the front and rear face, an allowance had to be made for an increase in eccentricity of of the applied load due to rotation of the loading plate. Thus, although the initial eccentricity of the loading was 37 mm, the final eccentricity was 83 mm.

28. The test revealed that gabion baskets are relatively flexible under applied load. Taking a stress of 100 kN/m^2 as being representative of the conditions prior to failure of the cross tie wires,the vertical deflections of the back and front faces of the test gabion basket were 20 and 19 mm respectively.

The stiffness curve for the basket could be divided into three portions:

i.   Prior to failure of first cross tie wire
        (at 21 tonne applied load)

ii.   After failure of cross tie wire but prior to
        failure of corner of basket(at 50 tonnes)

iii.   After failure of corner of basket
        (continuous movement under no increase in load)

29. The ultimate load capacity of the basket may be expected to have been higher if the strength of the bindings at the corners had been greater. This was particularly evident since the weldmesh forming the basket, although deformed, did not appear to be in a state of imminent failure. Although the gabion basket and the stone fill was representative of that used on site, the lateral confinement of the basket could not be simulated under laboratory conditions due to the relatively large dimensions of the test specimen. While the lack of confinement of the rear and side faces could lead to an increase in vertical deflection, the

restraint conditions were representative for the front face which was the critical element under test.

30. Although only one test on a gabion basket has been carried out, it is considered that the care taken in the preparation of the specimen ensured that the characteristics measured in the test were reasonably representative of those of an average basket. The cost of each test does not readily permit mean characteristics to be determined on a statistical basis using a large number of specimens.

STRESS ANALYSES OF WALL BEHAVIOUR

31. Predictions of the forces imposed on the back of the wall prior to the failure are difficult to make and can only be approximated for the following reasons:

i.   The relative position and configuration of the steep rock face behind the wall varied considerably and, therefore the lateral pressures and total thrusts may be expected also to vary.

ii. The downward component of thrust behind the wall depends on the relative movements between the wall and the rock backfill, and the relatively large displacements of gabion baskets under load make it difficult to predict values for this component.

iii. The overlapping of gabion baskets results in variations in the "virtual back of wall" concept which is used in traditional wall designs.

iv. The transfer of horizontal thrust between baskets depends on a combination of friction and flexible restraint due to the wire ties linking the baskets and no quantitative information is available regarding this mechanism of load transfer.

v.   Gabion walls may be considered to comprise 'soft blocks' which will deform significantly in response to relatively large lateral thrusts. The deformability of these walls may result in significant forward tilting and loss of the advantageous backward tilt provided by the standard method of wall construction.

vi. The variations in the widths of the baskets and the irregularities in cross-sectional configuration of the wall as constructed may result in a variable balance in self-weight forces at each layer of baskets. An analysis of the wall configuration shown on Fig.1 has indicated the following stress distribution:

Layer No.	Vertical Stresses (kN/m^2)	
	Front	Rear
6	56	80
7	82	52
8	81	80

Forward rotation of the wall due to 'soft block' action

would modify these theoretical stresses and not necessarily in an advantageous manner.

32. In order to provide proper information for the prediction of the stresses which could have been experienced in the gabion wall at Byreburnfoot the bulk densities of the rock fill material in the gabion baskets and of the rock backfill were determined from large scale volume tests. The bulk densities of the gabion fill and the backfill were found to be 15.9 kN/m^3 and 16.7 kN/m^3 respectively.

33. Stress analyses based on traditional theory for gravity walls and assuming that the backfill moves downwards relative to the gabion wall (allowing for a downward component of thrust) indicate the following vertical stresses at the base of the eight metres high wall.

Vertical Stresses (kN/m^2)

Front	Rear
261	49

The analyses recognised the partial protection afforded to the wall by the close proximity of the steep rock face behind the wall and reductions were made in the lateral forces in an attempt to model this effect. These predictions of theoretical stresses can only be considered a rough estimate in the particular situation found at Byreburnfoot but the high order of stress, after making apparently reasonable reductions in the lateral forces, is a matter of concern if baskets deform appreciably under high stresses. Any measures which can be taken to stiffen gabion baskets would be beneficial. The use of 0.5 metres high baskets instead of one metre high baskets may provide a stiffer wall.

STRAIN ANALYSES OF WALL BEHAVIOUR

34. In order to make an approximate assessment of the deflection behaviour of a gabion wall, some finite element analyses were conducted. The mesh employed is illustrated in Fig.5. This represents a cross-section of the wall, which was assumed to deform in plane strain. The elements were 8-noded quadrilaterals, and no slippage was allowed between the base of the wall and the foundation strata. The load-deflection characteristics of the baskets were approximated as shown in Fig.4. While this is clearly a simplification of the real response in the field it does imply that in the analyses, the baskets were assigned a modulus and ultimate strength consistent with the laboratory test. The loading on the wall was assumed to be triangular, and account was taken of vertical components of load on the approximately horizontal back face steps. In reality the load was probably more concentrated towards the mid-height of the wall.

35. The elastic deformations of the gabion wall and of an equivalent mass concrete wall are shown in Figs.5 and 6, from which it can be seen that much larger displacements are developed in the former case. In addition, yield took place at the

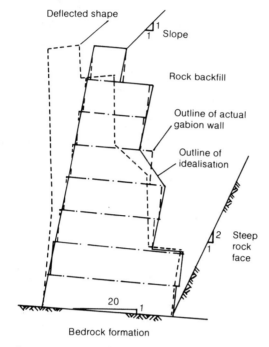

Figure 5. Wall idealisation for F.E. analysis (gabion wall)

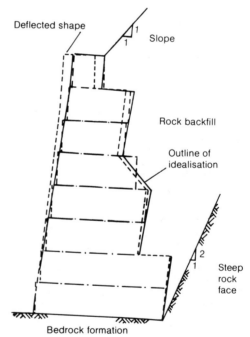

Figure 6. Wall idealisation for F.A. analysis (mass concrete wall)

290

base of the gabion wall, but the effects of this have not been shown due to the approximate nature of the physical modelling of the wall behaviour. It can be concluded that estimates of wall displacements form an important part of the design of gabion structures. The stress analyses are complicated by unknown factors such as inter-basket slip but quite simple analyses involving tying element corners together with elastic yielding ties could supply a useful insight into the manner in which gabion walls behave.

REMEDIAL WORKS

36. A study of the geological situation subsequent to the failure revealed the possibility that the rock slope was more stable than envisaged. Cost analyses of various solutions also revealed that if the remedial works to restore the road pavement and verge could be restricted to the immediate vicinity of the road itself considerable savings could be made.

37. After a careful and thorough examination of the condition of the rock strata on the steep downslope to the river and a re-assessment of the geological situation, the decision was taken to depend on the current stability of the rock slope for the design of remedial measures. Longterm degradation of the rock slope and any changes in the state of balance were to be monitored annually during normal road maintenance inspections.

38. The remedial solution involved the construction of reinforced concrete spine walls aligned transversely to the road, with a reinforced concrete slab spanning between the spine walls, and a reinforced concrete apron wall along the edge of the road adjacent to the steep rock face. The spine walls were 7.5 metres in length, 750 mm wide, and were founded on the bedrock below the road at depths of about 4 metres. The reinforced concrete slab was 450 mm thick and was designed in accordance with the requirements of the Scottish Development Department for bridges. The remedial works were completed within a contract period of 4 months and at a cost of £81,000. The debris resulting from the collapse of the gabion wall was left as a protection against scour.

CONCLUDING REMARKS

39. Although the design of earth retaining structures should be carried out in accordance with CP2 "Earth Retaining Structures" no specific guidance is given in that Code of Practice with regard to walls constructed of gabion baskets. Similarly although the use of standard materials such as reinforced concrete, mass concrete, masonry or brickwork are covered by Codes of Practice, no such guidance is available for gabions. In view of the lack of guidance, the designer must formulate his own design methods and criteria. The assessment of the soil pressures and the response of a wall consisting of gabion baskets backfilled with steeply sloping fill is extremely complex and the lack of precision with which

a designer can make predictions is such that the adoption of safety factors in excess of those quoted in CP2 should be considered. The safety factors to be applied in the design of gabion retaining walls should be greater than those adopted for traditional forms of mass retaining wall construction for the following reasons:

(i)   The lack of statistical information on the failure strength of gabions

(ii)   The secondary effects due to the high flexibility of the gabion baskets.

(iii)   The lack of information available on the friction forces and interaction generally between gabions.

40.   The test on a gabion basket representative of that used in the front layers of the wall at Byreburnfoot, indicated that failure occurred at a stress of approximately 350 kN/m^2 and at a vertical displacement of about 150 mm. The reliability of this value of limiting stress must be judged in the light of the knowledge that only one test was carried out and also no attempt was made to allow the total creep at a given loading to occur prior to increasing the applied load. Based on the results of the test it is considered probable that increased stiffness and a higher value of ultimate load could be achieved both by the use of a more robust form of lacing and tie wire and by the use of this stronger wire to bind together all edges of adjacent baskets. This would improve the wall in the following respects:

(i)   Augment the stiffness of the unrestrained front face of the wall.

(ii)   Increase the strengths of the edges of the baskets (which appear to be the weakest elements)

(iii)   Increase the efficiency of the composite action between baskets in each layer and improve the resistance to sliding between the adjacent layers.

(iv)   Improve the life expectancy of the lacing and its ties.

The requirement for life expectancy of the tie wire would appear to be most important as the wall cannot remain stable if the lacing and tying wires corrode and fail. Based on the present form of lacing and tie wire the results of the test indicate that a compressive stress of 125 kN/m^2 should not be exceeded without verification from further testing work. Even with the adoption of a maximum service stress of 125 kN/m^2 consideration of the serviceability state would be prudent for major walls as the vertical deflection of a single gabion basket subjected to a stress of 125 kN/m^2 may be expected to be of the order of 25 mm.

41. It would seem prudent to ascertain the vertical stresses in a gabion wall due to the self-weight of the baskets and obtain an indication of the balance of self-weight forces. It is considered that this initial step in the analyses is important for a wall which is not an elastic continuum exhibiting small strains at normal service stresses as in the case of concrete or brick gravity walls for which such assumptions are reasonable.

ACKNOWLEDGEMENTS

This paper is published by permission of J. A. MacKenzie Esq., F.Eng., Chief Road Engineer,. Scottish Development Department. The contributions made by W. M. Reid Esq., and the University of Dundee testing laboratory are also acknowledged.

17. The stability of excavated slopes exposing rock

G. D. MATHESON, BSc, PhD, FGS, Head (Scottish Branch)
Transport and Road Research Laboratory

SYNOPSIS. An approach to the design and excavation of highway
slopes in hard rock is outlined. This involves optimisation of
the slope in terms of the geotechnical conditions and the use
of an excavation technique which will minimise disturbance.
The approach is considered to have potential in other
contexts.

INTRODUCTION
1. Since the establishment of the basic principles of soil
mechanics one of the objectives of site investigation has been
to provide adequate geotechnical information to enable stable
slopes to be designed in cuttings. This analytical approach
has been refined to the degree that few soil slopes are now
designed by empirical methods. Such analytical techniques
have not however been so successful in the design of rock
slopes where design techniques have largely involved
superimposing standard slopes on the design drawings. These
slopes have mostly been chosen regardless of the geotechnical
conditions in the rock mass and are presumably based on
engineering experience of supposedly similar conditions on
other projects.
2. Site investigations for excavations in rock have
tended to use procedures which have been contractually easy to
control, such as diamond drilling with logging and testing of
the recovered core. However, the results often give little
relevant information for slope design and the engineer is
still left with little option but to employ an empirical
method.
3. Approaches intended to maximise the geotechnical
information by carrying out extensive drilling and testing
often meet with little better success because of the
difficulty in obtaining meaningful data on the geotechnical
conditions controlling stability and the often highly variable
nature of the rock. Such extensive, and expensive, blanket
techniques are seldom cost-effective.
4. Until relatively recently therefore, the design of
rock slopes has been largely a matter of superimposing
standard designs and the excavation technique usually the
cheapest blasting method available. The poor state of
stability in many recent rock cuttings is a direct result of

the "standard design" approach and ignorance of the damage
which indiscriminate blasting can induce.

THE DESIGN OF ROCK SLOPES
Data Collection
    5.   The stability of most slopes in hard rock is controlled
by the orientation of important discontinuities in the host
rock mass. Intact properties, as might be measured on a
single block, are of secondary importance in all but weak or
easily weathered rock types. Slope design is therefore
principally a geometric problem of matching the slope to the
discontinuities. Consequently the collection of factual data
on discontinuity orientation and importance should be the
prime objective of site investigation where rock slopes have
to be excavated.

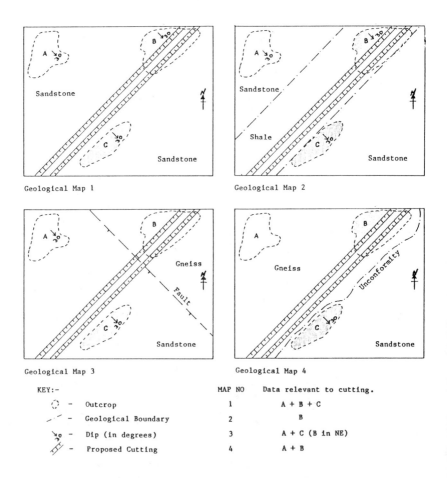

Fig. 1. Influence of local geology on the use of discontinuity
data

6. Data collection is primarily a field exercise in the recognition, measurement and recording of information on discontinuities. Suggestions for data collection and recording have been given elsewhere (Refs 1,2).

7. In highway projects it is unlikely that data on discontinuity orientations will be available from sources other than surface rock outcrop. Information from trial pits, possibly excavated to establish rock head, should however not be ignored if the rock exposed is in situ. Information from routine diamond drill cores seldom provides accurate orientation data unless a pervasive and constant reference datum, such as a mineral lineation or a schistosity, is present in the rock mass. The cost of obtaining orientation data by special diamond drilling is generally prohibitive and the data seldom reliable enough unless down-the-hole video methods are used on the drillhole walls. In the design of highway cuttings every effort should therefore be made to collect and use field data within or adjacent to the area of interest.

8. Local geology is important in determining the validity of interpolation or extrapolation of data. Geological mapping should therefore always accompany discontinuity surveys and a rock type distribution map established for the area. The influence of the local geology is demonstrated in Fig.1. Each of the diagrams depicts a different geological setting and each indicates a different use of the data obtained. Only in the case of a uniform geological setting can the data from each of the three domains be combined for the purposes of stability assessment. When collecting data in the field it is therefore important to separate the data into geological or geographical domains. Field data should only be combined if the geology permits it. Even when collecting data within a known geological domain it is advisable to subdivide the area sampled into smaller units. This will aid recognition of individual stability domains within the project.

DATA PRESENTATION AND EVALUATION

9. The field data must be presented in a form which will not only allow the discontinuity patterns to be recognised but also act as a basis for the evaluation of failure potential of any design slope. Most existing methods are graphical and based on stereographic projection (Refs 3,4). The simplest and the one recommended for manual use (Ref 2), involves plotting of the discontinuities as poles, contouring of the pole plots, and the calculation of intersections to major pole concentrations using great circles. Evaluation of the potential for plane wedge and toppling failure is then carried out using overlay techniques. A flowchart of the method is given in Fig. 2. In practice the technique is most useful for determining the optimum design slope and for estimating the sensitivity of the slope to changes in design azimuth or the bulk friction angle of the rock. A feeling for the importance

of the various parameters is thus established. Inherent in the method is of course the basic assumption that the discontinuities are representative of the rock mass being excavated.

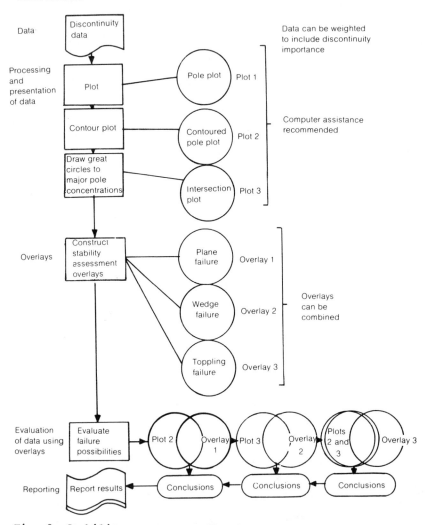

Fig. 2. Stability assessment flowchart

10. One of the major failings of the above system is that the importance of each discontinuity is ignored. This can be rectified when contour densities are being calculated by giving each pole position a value relating to its importance. Features which may not have been visible in the conventional plot are thus able to be recognised. In general any significant difference between conventional and weighted plots indicates that unusually important discontinuities are present. Every effort must then be made to establish their

precise location relative to the proposed design slope. This may involve further and more detailed field mapping, excavation, or drilling.

11. The system of data presentation described above is time consuming, repetitive, and open to error when carried out manually. Computer programs are available which considerably reduce the time taken, remove much of the tedium and make weighting of the data and calculation of intersections much easier. Programs for this purpose should either process the data to the state where stability overlays can be used (this requires both poles and intersections to be plotted) or evaluate directly the possibility of failure in terms of the proposed design.

12. The assistance given by computer programs is of considerable value and their use is recommended. The potential user should however be aware of the algorithms and programming techniques used and be satisfied regarding their validity. The programs should also be suited to the field problem and the hardware available.

STABILITY OF ROCK SLOPES
Old cuttings

13. The excavation of rock in civil engineering projects prior to approx 1900 was localised and accomplished mainly by hand wedging with manual extraction of loosened blocks. Blasting, using low explosives (black powder), was confined to areas where manual techniques had proved impractical and involved the drilling, possibly by hand, of small diameter drill holes in the larger rock blocks. The excavation work was labour intensive and time consuming but had the ability of adapting easily to local geotechnical conditions. The final slope was the steepest attainable during excavation. Rock slopes so excavated are usually remarkably steep when compared to their modern counterparts and have required very little remedial work or maintenance since their formation.

14. Following the synthesis of nitroglycerine and the commercial marketing of the dynamite-based high explosives the excavation of rock changed dramatically. It became possible to excavate quickly, safely and relatively cheaply in all rock types regardless of block size or strength. Initially the techniques used in civil engineering were those developed for surface mining where the dominant consideration was rock fragmentation. These bulk blasting techniques employed relatively large diameter drillholes and high charge weights, little attention being paid to side slope stability. Extensive post-excavation remedial and maintenance work was usually necessary.

15. The contrast in stability of slopes excavated by hand or low explosives and those formed using high explosives is marked and is most evident where old and new excavations co-exist. In many recent cuttings it is obvious that the side slope angle had been based on that visible in the early cuttings. The fact that stability was not achieved indicates

the damaging effect which modern high explosives can have on the design slope.

BLAST DAMAGE

16. The term "shattered" is in common use to describe highly fractured rock. A definition of shattering is "to break suddenly or violently in pieces" (Ref 5). The term therefore implies formation by a rapid process involving high stresses. The connection with blasting is self evident and the immediate description of the damage caused by high explosives is usually that of a "shattering" of the host rock. The description can be justified when applied to an open-cast mining environment where large diameter holes and very high charge weights are employed. In these situations the compressional shock wave can be reflected in tension at free surfaces with sufficient intensity to cause tensile failure of the rock. Many new fractures are thus formed and the rock becomes highly fragmented. However in civil engineering projects visible signs of shattering of the rock are not widespread and are common only in rock immediately adjacent to explosive charges. The common inference that instability in side slopes formed by blasting is the result of blast shattering must therefore be questioned.

FIELD STUDIES OF ROCK FACES

17. Field studies were conducted on several excavations to determine the geotechnical conditions before and after excavation. The techniques used included scan lines on exposed rock, closed circuit television in drill holes, and modified seismic refraction on rock faces. The objectives were to establish the nature and the extent of any damage induced into the rock faces by different methods of excavation.

Table 1 Fracture intensity and dilation on excavated rock faces

LOCATION	METHOD OF EXCAVATION	LENGTH SCAN (m)	NO. FRACTURES	FRACTURE INTENSITY (Fr/m)	DILATION						
					AV.(mm)	% DISTRIBUTION IN mm GROUPS					
						0	2	4	6	8	10+
A9 Calvine Snowgate Cut	Natural	37	147	4.0	0.9	62	28	8	1		1
	Presplit	21	97	4.6	1.7	52	29	5		1	3
	Bulk Blast	18	95	5.3	5.8	28	39	6		1	26
A9 Calvine Crushing Plant Cut	Natural	21	147	7.0	0.8	60	34	4			2
	Presplit (Successful)	22	148	6.7	0.8	61	30	6	2	1	
	Presplit (Unsuccessful)	25	122	4.9	1.8	45	35	10	3	2	5
	Bulk Blast	2	104	4.2	6.8	27	21	22	5	2	23

Scan lines

18. Scan line techniques were used to determine fracture
intensity and dilation before and after excavation. In most
cases poor exposure immediately over the area to be excavated
necessitated some extrapolation from nearby exposures within

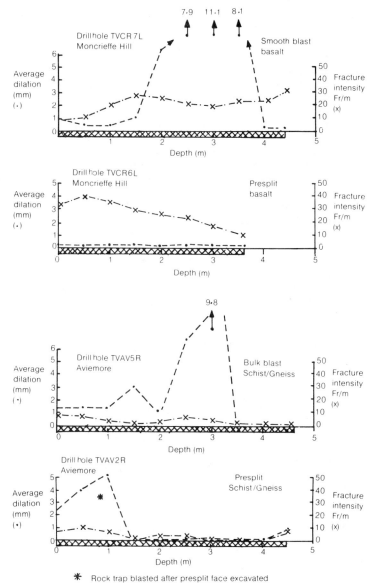

Fig. 3. Fracture intensity and dilation in excavated rock slopes

the same stability domain. Typical results are given in
Table 1. The data were collected from similarly orientated
scan lines. The results indicate that the major effect of

blasting is not a dramatic increase in fracture intensity, as might be expected from the concept of blast shattering, but one of increased fracture dilation of mainly pre-existing, natural discontinuities.

Closed circuit television (CCTV)

19.   A study of the fractures behind a number of excavated faces was made using CCTV techniques in holes drilled into the rock face. The holes were drilled at slightly inclined angle using standard 75mm diameter diamond bits. A Rees 93 TV camera, mounted on a custom made plastic housing with a skid base, was pushed into each drill hole and the depth of insertion measured using a tape attached to the camera head. Observation and video recording were performed simultaneously. Initially each hole was examined using an axial view (looking away from the rock face) and the major fractures located and the general state of the hole established. A detailed survey of each side was then made using a side viewing head attached to the camera. The depths to individual fractures were noted. A previewed scale recorded in-situ at the commencement of each traverse allowed direct measurement of fracture dilation to be made during subsequent video replay. In general sharp images and excellent resolution allowed fracture dilation to be measured to the nearest 0.5mm.

20.   Results from two locations are given in Fig.3. These illustrate the variations found in fracture intensity and dilation. The most important feature is the zone of highly dilated fractures present behind the smooth and bulk blasted faces and absent from the undamaged presplit face. At both localities there is an increase in fracture intensity towards the face but the increase in intensity is not regarded as sufficient to merit the term "shattered".

SEISMIC REFRACTION

21.   A modified technique of seismic refraction was used in an attempt to find an easy, quick and inexpensive method of

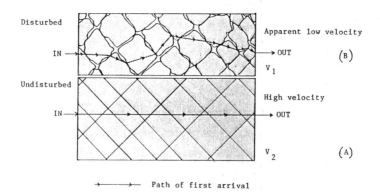

Fig. 4. Wave propagation in undisturbed and disturbed rock

detecting and measuring the extent of any damage induced into the face by blasting.

22. In a rock with perfect acoustic coupling over fractures (Fig. 4A), a compressive wave can pass directly through the rock mass, its velocity being relatively unaffected by individual fractures in its path. The seismic velocity in the rock can therefore be calculated directly from the time taken for the signal to pass from source to geophone. In a dilated rock mass acoustic coupling takes place only at points of contact. The path of a the first arrival wave will therefore not be direct but will follow the shortest path through the rock block contacts (Fig.4B). A correspondingly longer travel time will therefore be recorded and apparently lower seismic velocity calculated. It is this apparent velocity layering, with the 'lower' velocity layer occurring nearest to the rock face, which allows the seismic refraction technique to be used in the detection and measurement of blast damage.

23. The technique was employed in a study of rock faces exposed in a variety of excavations (Ref 6). A simple two channel signal enhancement seismograph (Nimbus model E-125) was used with a blow from a 14lb sledge hammer acting as the signal source. Standard geophones (Sensor model SM-4) were mounted on steel pitons driven into fractures in the rock face at a spacing up to 30m. Signal travel times from hammer impact to the first arrival of the compressive wave and the distance from the point of hammer impact to each geophone were measured. The choice of the first arrival was made mainly in the interests of simplicity and easy recognition of the measurement position on the waveform. This was greatly assisted by the enhancement facility on the seismograph. Travel times from a sequence of locations between the geophones were thus measured and the results plotted as conventional time-distance graphs.

24. Results (Table 2) indicate that a near surface layer of relatively low velocity exists behind most rock faces excavated using explosives. In this the velocity generally drops to 40 -70% of the host rock mass. The higher velocity is presumed to be the apparent velocity of the natural rock mass. Depths to the higher velocity layer, calculated using the intercept method, are given in Table.3. These show that damage was not detected in the natural slopes studied, in old faces excavated predominantly by hand, or in faces excavated by presplit blasting. Extensive damage was however detected in slopes excavated by bulk, smooth, and trim blasting, the first mentioned being associated with the greatest width. A direct comparison of the effects of different types of blasting on the same rock type is possible. The highest degree of instability (as determined visually) was associated with the greatest width of damage in all the excavations studied. The only faces excavated with high explosives in which damage was not detected seismically were those formed by successful presplit blasting.

BLAST DISTURBANCE

25. The field studies suggest that, at least in highway excavations, the damage induced in a rock face by blasting is not one of "shattering" of the rock but one of disturbance of the rock mass through dilation of mainly pre-existing fractures. The term "BLAST DISTURBANCE" is suggested for this type of damage. Some increase in fracture intensity also takes place but this is of secondary importance.

Table 2. Apparent seismic velocities in disturbed (V_1) and undisturbed (V_2) rock

LOCATION	ROCK TYPE	V_1 km/s		V_2 km/s		NUMBER OF REVERSED PROFILES
		AV.	RANGE	AV.	RANGE	
Forth Bridge Area (A90)	Dolerite	0.81	0.13-1.34	2.66	1.34-4.72	30
Craigend (M90)	Basalt	2.74	1.88-3.41	4.22	3.10-4.89	12
Aviemore (A9)	Schist/ Gneiss	1.53	0.63-1.82	4.52	3.48-5.42	7
Westfield	Dolerite	1.14	0.72-1.69	2.29	1.64-3.20	7

Table 3. Depths to higher velocity layers in rock slopes

Excavation			Location	Rock Type	Depth to Higher Velocity Layer (Metres)
Type	Method	Approx Age (Yrs)			
Natural	-	10,000	Forth Bridge Area	Dolerite	Nil
	-	"	Craigend (Perth)	Basalt	"
	-	"	Aviemore	Gneiss/ Schist	"
Rail	Hand and Black Powder	100+	Forth Bridge Area	Dolerite	"
Quarry	Bulk	0-60	"	"	6m
	"	10	Westfield, Fife	Dolerite & Sediments	8m
Highway	Bulk	20	Forth Bridge Area (A90)	Dolerite	6m
	"	5	Aviemore (A9)	Gneiss/ Schist	2m+
	Smooth Blast	8	Craigend (M90)	Basalt	4m
	Pre-split	7	Moncrieffe Hill (M90)	"	Nil
	"	5	Aviemore (A9)	Gneiss/ Schist	"

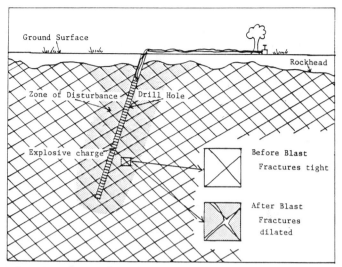

Fig. 5. Blast disturbance

26. The concept of blast disturbance is diagramatically illustrated in Fig.5. An explosive charge has been placed in a drill hole in an ideally fractured rock mass. Before detonation the fractures are predominantly tight. Immediately following detonation the rapid expansion of gases from the explosive causes dilation of adjacent fractures in a zone surrounding the hole. When the gas pressure subsides some fractures are unable to close fully. When high stresses are created in the rock adjacent to the hole, or the rock is weak, then new fractures resulting from tensile failure may be formed. In general these are close to the borehole and seldom propagate far into the surrounding rock.

27. The mechanism of disturbance is thought to be one involving rapid gas expansion following detonation of the explosive charges, wedging open of nearby fractures causing movement of the affected rock mass, and a failure of individual rock blocks to regain their former tight packing once gas pressures have dissipated. The effect is greatest when explosives with low velocity of detonation and high gas yield are used in high charge weights, and is invariably associated with significant ground heave.

OPTIMISATION OF STABILITY

28. The need to relate the design slope to the geotechnical conditions present has been emphasised. The superimposition of a standard design slope regardless of local geology is not recommended and a rigorous analytical approach is generally impractical. An alternative design approach intended to establish the geotechnical conditions controlling stability is therefore suggested. This is aimed at predicting the problems likely to be encountered during excavation and designing the rock slope to minimise them. The approach depends on the intelligent use of field discontinuity data.

305

29. It should be noted that slope design is a process of optimisation and that it may not be possible to eliminate instability entirely. In these circumstances sufficient flexibility to take care of reasonable design changes, and adequate provision for remedial work, should be included in the contract. Every opportunity to add to and update the field data should be taken as excavation proceeds and evaluation should keep pace with data acquisition. Design modifications should be made if the geotechnical conditions dictate it and the contract should allow this.

30. There is little point in optimising the design if blast disturbance is induced into the design slope. Poor stability and a need for remedial work and maintenance are likely to result. The method of excavation should therefore be carefully considered at the design stage in order to minimise blast disturbance. The only technique employing high explosives which appears to eliminate blast disturbance is presplit blasting (Ref 7).

31. Optimum stability requires careful consideration of both the slope design and the method of excavation.

ACKNOWLEDGEMENTS

32. The work described in this paper forms part of the programme of the Transport and Road Research Laboratory and the paper is published by permission of the Director. The co-operation of the Scottish Development Department is gratefully acknowledged.

REFERENCES

1. ANON (1977). The description of rock masses for engineering purposes. Q. Journ. Engng. Geol., 10, pp356 – 388.
2. MATHESON, G D (1983) Rock stability assessment in preliminary investigations – graphical methods. Department of Transport, Department of the Environment. Transport & Road Research Laboratory. Report LR 1039.
3. HOEK,E AND BRAY J W (1971). Rock slope engineering. Inst. Min. & Metal., London.
4. GOODMAN, R E (1976). Methods of engineering geology. West, New York.
5. FOWLER H W AND FOWLER F G (1964). The concise dictionary of current english. Fifth edition, Claredon Press, Oxford.
6. SWINDELLS C F (1981). The seismic inspection of excavated rock faces. PhD Thesis (Unpublished), Dundee University.
7. MATHESON, G D (1983) Presplit blasting in Highway Rock Excavation. Department of Transport, Department of the Environment. Transport & Road Research Laboratory. Report LR 1094.

Any views expressed in this paper are not necessarily those of the Department of Transport or the Scottish Development Department.

18. Quarry landslip—a design aid in highway earthworks

W. A. WALLACE, Projects Engineer, Babtie Geotechnical, and
E. J. ARROWSMITH, Geotechnical Engineer, Department of
Transport, North West Regional Office

SYNOPSIS
    A major landslip took place in an abandoned quarry on the
route of a new By-pass.   At the time of the slip the design
of the road was at an advanced stage.   Studies of this
landslip provided valuable information to aid both the
stability design of a major rock cutting and the stability
assessment of the quarry face overlooking the route.   This
paper describes the investigation of the slip, the analyses
of the findings, and their use in the designs.   It
concludes that the effect of time is an important
consideration when studying landforms and designing slopes.

INTRODUCTION
    A major landslip has provided valuable information for
stability assessments on a new highway scheme.
    The Accrington Easterly By-pass, Southern Section,
provides the remaining link in the high standard road system
joining the M66/M62 to the M65 and Calder Valley towns of
north east Lancashire.   The proposed road follows a route
over high ground, and runs parallel to and west of the
centuries old 'packhorse' road known as the Kings Highway.
    Late in 1980 the highway design was well advanced.   Then
a major landslip occurred in an abandoned quarry on the line
of the road.   The slip was investigated and was used to
refine the parameters adopted in the stability assessments.
On review, the design of the adjacent Peel Park cutting
proved to be adequate.   However, the Bluff – a part of the
quarry overlooking the road – was assessed to be unstable in
the long term.   It was therefore necessary to improve the
stability of the Bluff.
    The road is currently under construction.   The Peel Park
cutting is substantially complete with the ground conditions
being as foreseen.   The Bluff has now been regraded and its
stability assured.

THE LANDSLIP
    The landslip occurred in the disused Huncoat Quarry.
This quarry was once worked to obtain mudstone for the

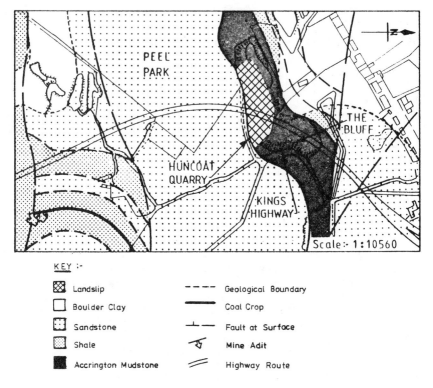

KEY :-

⊠ Landslip   – – – – Geological Boundary

☐ Boulder Clay   ▬▬ Coal Crop

▦ Sandstone   ⊥— Fault at Surface

▨ Shale   ⟆ Mine Adit

■ Accrington Mudstone   ⫽ Highway Route

Fig. 1 Geological setting

making of the renowned Accrington brick, but has now been
abandoned for some 30 years.   The quarry face extends over
350 metres and to a height exceeding 50 metres.   The
slipped mass involved about 100,000 cubic metres of the
quarry face rock.

The quarry is located on a north facing escarpment at the
northern limit of the Peel Park plateau as seen in Fig.1.
The escarpment is formed in rocks of Lower Coal Measure
age.   It is the thicker argillaceous beds of this rock
sequence that have been extensively worked for
brickmaking.   The succession of strata in the quarry face
is shown in Fig.2.   The rocks dip at a low angle (circa 3°)
generally to the south-west.   Jointing is very pronounced
throughout the rock, the major sets being parallel and
normal to the bedding planes.   In consequence the
sandstones are blocky and the siltstones/mudstones are
fissile and broken.   Regional faulting is generally
east-west but is not known to occur in the Peel Park area.

Workable coal seams are present at considerable depth
beneath the quarry site.   Total extraction in the Lower
Mountain mine took place around 1959, the workings being in
excess of 180 m beneath the escarpment.   Mining subsidence
will now be complete, but slight racking of overlying strata

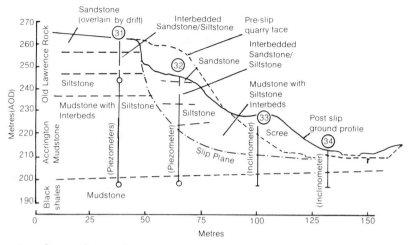

Fig. 2   Geological section through landslip

- such as opening of joints and bedding planes - will have occurred.

The ground water conditions at the quarry site are indicated on Fig.3.   These were based on piezometric information from the post slip investigation boreholes as well as boreholes of the earlier route investigation.   The piezometers recorded a perched water table within the sandstone/siltstones of the Old Lawrence Rock, separate to the main water table which stands in the underlying Accrington Mudstone.   The perched water table drains towards the south away from the quarry face.   The main water table is drawn down towards the quarry face. However, during periods of high rainfall the water table rises to become near horizontal and has been observed as a seepage line on the quarry face.   Prior to the landslip

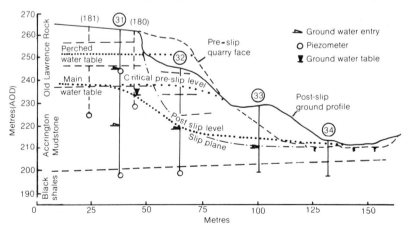

Fig. 3   Groundwater conditions at landslip

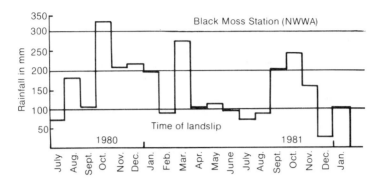

Fig. 4   Rainfall data

there was a sustained period of high rainfall - as seen in
Fig.4 - which amounted to 150% of the average monthly
rainfall for that time of year.

INVESTIGATIONS
    A number of rotary drill boreholes were sunk at the
quarry site - four in the area of the slip and two on the
Bluff.   The boreholes were drilled at H size (75 mm dia.
core) using air flush and to depths of up to 65 metres.
The core recovery was good but the presence of the slip
plane was not readily apparent as 'discing' was a common
feature in the mudstone cores.   A geological section
through the landslip has been prepared as shown in Fig.2
with the location of the slip plane being deduced from the
rock core quality.
    Piezometers and inclinometers were installed in the
boreholes.   During the monitoring period the ground water
levels remained relatively static and no further ground
movement was recorded.   Photogrammetry and conventional
crack monitoring surveys were also undertaken.   They
confirmed that no further appreciable movement was occurring
at the main quarry face.   However, surface degradation was
evident on the upper slopes of the Bluff.
    Laboratory testing included shear strength determination
using a portable 'field' shear box.   The results are given
on Fig.5.   The apparent angle of friction is represented by
the lower bound strength envelope with $\emptyset$ effective taken as
27°.   The apparent cohesion contribution (C) ranges from
50 kN/m2 to 450 kN/m2, and is taken to reflect the influence
of the nature of the fracture surfaces including weathering
effects, on the measured shear strength.
    Clay infilling is limited to a few major joints in the
mudstones and the possibility of active clay minerals being
present in the infilling has been ruled out on the basis of
index tests.   The results had indicated clays of
intermediate plasticity only.   Enhanced chemical weathering

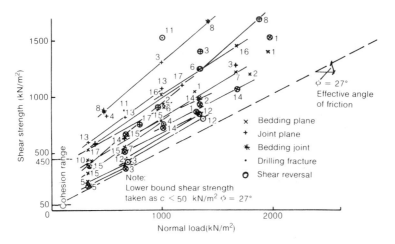

Fig. 5 Shear box test results

of the rock strata is not occurring at the quarry site as was indicated by the slightly alkaline nature of the groundwater based on the measurement of its pH value.

ANALYSES

The site evidence suggests that the landslip is a combination of rotation behind the quarry face and translation at the quarry base. In back analysing the slip three factors are considered to influence the stability.

The 'geometry of the slope' is known to have varied during the life of the quarry. The quarry was first excavated late last century with a steep working face of about 80°. When abandoned some 70 years later, the quarry face then stood at 65° but just prior to the slip the slope had reduced to 45°. Stable natural slopes are observed to stand at about 23° in the same strata along the escarpment.

The 'shear strength of the rock' on the slip plane is taken to be an all-in effective friction angle of 27° with a variable apparent cohesion contribution. No shear strength contribution is taken from the vertically jointed sandstone capping the quarry site.

The 'groundwater conditions' are a main water table at a high level within the Accrington Mudstone and a perched water table confined to the Old Lawrence Rock. After prolonged periods of rainfall the water tables can freely discharge at the quarry face.

The apparent cohesion contributing to a 'just' stable slope (F=1) has been calculated using the landslip geometry and varying the slope inclination. This approach is an attempt to model the history of the quarry face. The results of the analyses show the apparent cohesion of the mudstone to have diminished with time. The cases considered are summarised as follows:

Case	Slope History	Min. Apparent Cohesion
1.	Immediately after excavation: e.g. Huncoat Quarry working face – 80° slope.	120 kN/m2
2.	Some years after excavation: e.g. quarry face at abandonment – 65° slope.	110 kN/m2
3.	After a number of decades: e.g. abandoned quarry face prior to slip – 45° slope.	60 kN/m2
4.	Long term: e.g. Peel Park design cutting slope – 27° slope.	34 kN/m2
5.	Very long term: e.g. natural slopes of parent escarpment – 23° slope.	24 kN/m2

The analyses of the quarry history implies that the cohesion contribution to shear strength has reduced by over 50% during a period of 100 years. This is attributed to gradual stress relief – resulting from quarrying (or natural erosion) – opening up previously tight joints and bedding planes and allowing weathering of the fracture surfaces to take place. Strong iron staining on the joints in the mudstone exists down to the level of the quarry floor. The actual trigger to the slope failure was the prolonged period of rainfall resulting in the main water table rising and discharging on the quarry face.

DESIGNS
At the time of the landslip, the design of the Peel Park cutting was complete. A typical cross-section of the cutting slope is given in Fig.6. It was now necessary to check the adequacy of this design. The assumptions of the

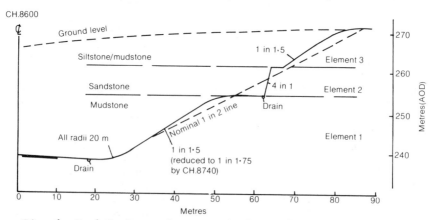

Fig. 6   Peel Park cutting – typical cross section

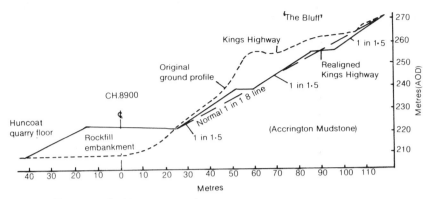

Fig. 7 The Bluff – typical cross section

design in terms of failure mode, rock strength parameters and groundwater conditions compared favourably with the findings of the quarry slip analyses. In particular the low cohesion value of 20 kN/m2 chosen for the mudstone was seen to be valid for the design life of the highway. On review the adequacy of the existing design was upheld, with increased confidence in the design parameters and assumptions.

The stability of the Bluff is seen as being at risk in the future. This spur of high ground, overlooking the route at the eastern limit of the quarry, stands with a weathered upper slope of 50° and a lower scree slope of 40° – the slope configuration of the main quarry face prior to the slip. The Bluff is predominantly mudstone and the water table within the rock is drawn down to quarry floor level. Stability analyses were carried out using the rock strength parameters from the slip back-analyses. The results highlighted a requirement to grade the Bluff back to a safe slope of 1 in 1.8 (approx. 28°). This overall slope sets the geometric limit for stability as shown in Fig.7.

The cutting design was refined to accommodate the Kings Highway by introducing a break in the slope. Consistent with the Peel Park cutting design, component slopes of 1 in 1.5 were adopted with the inclusion of a further berm to place the regraded slope within the overall 28° safe slope.

CONCLUSION

Back-analyses of the quarry landslip has proved to be a useful design aid in the assessment of stability of the adjacent earthwork features.

The case history also serves as a timely reminder that landforms, both of nature and man's construction, incorporate a fourth dimension – that being 'time' itself.

Acknowledgement. The work described in this paper was undertaken by Babtie Shaw & Morton, Preston Office, for the Department of Transport, North West Regional Office, whose permission to publish this material is acknowledged with thanks.

313

19. Kilgetty Bypass: problems related to thrust faulting

A. D. FLETCHER, MICE, Senior Engineer, W. S. Atkins & Partners,
A. B. HAWKINS, BSc, PhD, FIMM, Reader in Engineering Geology,
University of Bristol, and E. H. JONES, MSc, MICE, Senior
Engineer, W. S. Atkins & Partners

SYNOPSIS. The 6.3 km A477 Kilgetty Bypass through undulating
coastal topography across Coal Measures and Millstone Grit
Series strata required 2.6 km of cuttings, up to 32 m deep.
The cut face design involved 1:2 batters in superficial
material, 1:1 through weathered rock and 2:1 below a berm.
The southern cutting, in a zone known to geologists as the
Variscan Front, exposed clay gouge-filled thrust faults dipping
into the cut, which caused localised small-scale instabilities
in the 1:1 batter. Re-appraisal of the geological structure in
the light of a new thrust fault model led to a re-design of the
western cut batter. Although additional land take was necessary,
the extra excavation was less than 20,000 m³.

INTRODUCTION

1.   The existing A477 trunk road from St. Clear to Pembroke
Dock is severely restricted by a 3.96 m high masonary arch
railway bridge at Kilgetty. As the existing A477 in the
Stepaside-Kilgetty area is winding and it was the Welsh Office's
decision to provide a standard height route to the ferry
terminal at Pembroke Dock, the trunk road between Pen-y-bont
in the east and Begelly in the west has been upgraded and re-
routed to bypass the villages, Fig. 1.
2.   In 1978 the Welsh Office appointed W. S. Atkins & Partners
to undertake a feasibility study. Nine possible routes were
determined and assessed before a preferred route was recommended.
This has been constructed by Christiani & Nielson Ltd. and was
opened in July 1984.
3.   Although there is a short section of dual carriageway,
the 6.3 km long bypass is mainly single carriageway with over-
taking lanes on the long climbing sections. The design pre-
dates "Highway Link Design" and has maximum gradients of 5% and
minimum horizontal radii of 650 m, with appropriate verge
widening to give forward visibility of 290 m.

TOPOGRAPHIC SETTING

4.   The relief of the Kilgetty area is dominated by the east
to west strike of the strata and the river known as Ford's Lake
which flows south-eastwards into Saundersfoot Bay. The maximum
relief is about 130 m, but because of the proximity to the sea

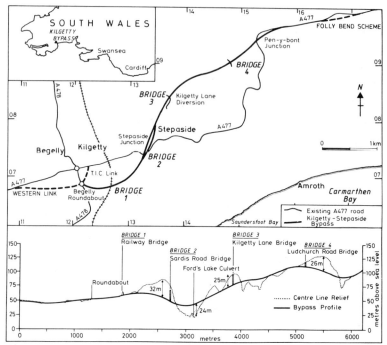

Fig. 1. Location, alignment and vertical profile of the Kilgetty Bypass.

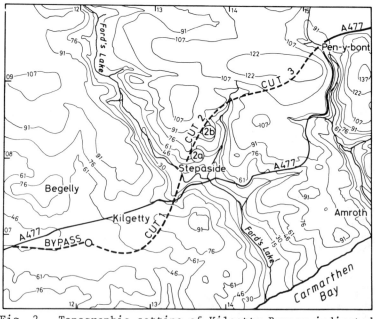

Fig. 2. Topographic setting of Kilgetty Bypass indicated by the 15 m contours.

the river valleys are deep and steep sided. Fig.2 shows the relief as depicted by the 15 m contours. At feasibility stage the topography dominated the engineering considerations, especially as the Welsh Office had specified a single track carriageway with a vertical alignment not exceeding 5%.

5.   The route selected for final design involved three main cuttings totalling over 2.6 km in length and up to 32 m deep, and embankments up to 24 m in height. The vertical road alignment and the ground profile along the centre line are given in Fig. 1.

6.   One of the design considerations was to optimise the cut/fill in such a manner that there was not too great a surplus of suitable material. The final scheme chosen involved a total excavation of 1.2 mil m³ of which 0.9 mil m³ was used for embankment fill.

GEOLOGY

7.   The client's desire to open the bypass as soon as possible meant that at initial feasibility stage the geology was considered only in so far as it affected the general route corridors, and was restricted to a desk study with limited fieldwork. The 1:50,000 Haverfordwest geological map (Sheet 228), published in 1976, was examined in detail and a check made to see what additional data was available on the 1:10,560 scale unpublished sheets. The fieldwork consisted primarily of an examination of the exposures indicated on the geological maps and those apparent from features shown on the larger scale survey drawings.

8.   The route, running obliquely over the northern limb of the Pembrokeshire Coalfield, crosses strata of Millstone Grit and Coal Measure age, Fig. 3. Overlying the solid geology are residual patches of boulder clay (till) while some alluvium is indicated in the valley of Ford's Lake. By coincidence one of the geological sections given on the 1:50,000 geological map is within 1 km of Cut 1, the main subject of this paper. The relevant part has been reproduced in Fig. 3.

9.   The geological sequence (Fig. 3), taken principally from the key to Sheet 228, shows the Millstone Grit to consist mainly of sandstones, forming the Basal Grit Quartzite at the base and the Farewell Rock at the top; the intervening strata being dominantly mudstone but containing subsidiary sandstone horizons. The overlying Lower and Middle Coal Measures are indicated to be mainly mudstones with some sandstones and a number of thin coal seams. The coal seams fit conveniently into three groups: the Kilgetty Vein, the Lower Level Vein and the Timber Vein Group. A number of marine bands occur, but these are more of stratigraphic importance than of engineering significance.

10.   Fig. 4 shows the geology along the route as given on the 1:10,560 geological maps. It can be seen that the dominant dip is southwards, but that declinations at other orientations exist, as might be expected in cross-bedded fluvial sediments. The geological maps indicate dips of 10-20° at 160-200° in Ivy Chimney Lane, within 25 m of the proposed route. Field

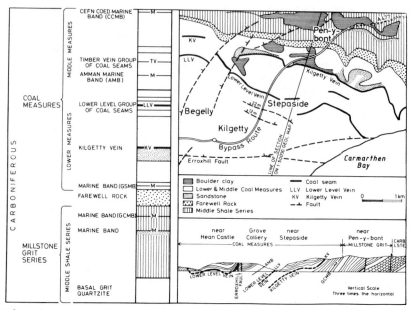

Fig. 3. Geological map, sections, and sequence; modified from the 1:50,000 geological map.

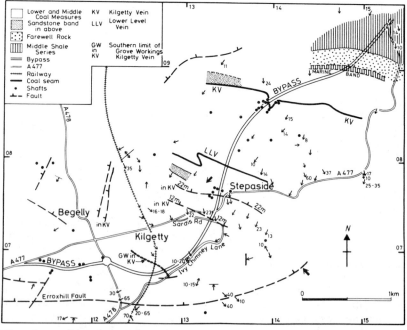

Fig. 4. Geology of the Bypass area; modified from the 1:10,560 geological maps.

examination in this area showed the dips to have a greater
variation in declination (4-18°), but in cross-bedded sediments
this was not considered unusual. In addition, being in the
area of the "Variscan Front" - the approximate line where
intense folding gives place to more minor fold disturbance
(ref. 1) a large number of small folds are likely which it
would not be possible to depict on the map.

11. Reports from the National Coal Board and the Mineral
Valuer confirmed dips of 15-20° in a south-west direction.
The reports highlight the faulting, naturally of concern in
coal mining, but do not imply the strata to be folded sufficiently for the structural complexity to inhibit mining. The
faults, shown to have apparent northern downthrows, are interpreted as being high angle reverse or thrust faults.

12. The intensity of past mining can be appreciated from the
number of shafts shown in Fig. 4, taken from the N.C.B. report.
Plans held by the N.C.B. indicate that mining in the area of
Cut 1 has been by orientated driveways (Fig. 5). They confirm
that in both the Lower Level and Kilgetty Veins there is a zone
about 50 m wide beneath Sardis Road where past mining was very
limited; probably the zone through which the east-south-east/
west-north-west fault with a 12 m downthrow passes. The Plans
show the "Average Inclination 1:4" at about 190°.

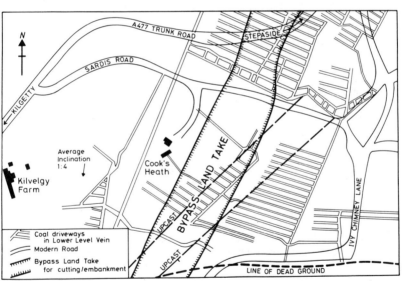

Fig.5. Records of driveway mining in the Lower Level Vein
under part of Cut 1.

13. Although it was appreciated that some remedial work
should be anticipated where the less than half-metre veins had
been worked, the fact that they were 65 and 155 m below Bridge 2
led to the conclusion that no special measures would be
necessary in Cut 1.

SITE INVESTIGATION

14.   The site investigation of the chosen route was carried out in 1980 and took the form of trial pits, cable percussive and rotary drill holes. It was carried out by Exploration Associates under the supervision of the consultants, in order to identify and confirm the lithology and depth of weathering both at the site of the main structure and within the cuttings. The trial pitting was mainly to determine the suitability of the near surface material.

15.   The cores revealed the lithologies to be similar to those anticipated and confirmed 5-20° dips, except in one borehole in the southern end of Cut 1, where vertical bedding was recorded. In several of the cores minor shearing including clay gouge zones was observed. Such signs of movement in a folded mud-stone sequence were expected and considered to be associated with bedding plane sliding. The recorded lithologies suggested the fault inferred from the mining records and shown on the geological maps would outcrop in the area of Bridge 2.

16.   As the fifty-one borehole cores, logs and core photos con-firmed the anticipated geology, it was considered that any further investigation of the structures or areas where shallow coal workings existed should be left until the main civil engin-eering contract began; at a time when the depths of drilling would be considerably reduced.

SITE GEOLOGY AND DESIGN

17.   One of the considerations in selecting the chosen route was to obtain an alignment as close as feasible to the dip direction. In Cut 1, however, in order to rejoin the A477 south of Begelly, it was necessary to increase the deviation from the average dip direction from 23° in the north to 42° in the south. It was appreciated that where lower strength, often sheared, carbonaceous horizons existed the dip angle and direc-tion would be all-important in the stability considerations. The borehole cores and the exposures in Ivy Chimney Lane indi-cated that the dip angles varied from 4-20°.

18.   In Cut 2a the alignment was within 20° of the average dip direction and with declinations of about 15° bedding plane slides were not anticipated. In Cut 2b, however, the possibil-ity of slides caused concern as the angle between the dip and the road alignment varied by up to 42°. With recorded dips of 5-15° it was considered that although some stabilisation would be necessary, notably on the western side, large scale movements were unlikely. As the main excavation in Cut 3 was in the sand-stones of the Farewell Rock, with dips of 10-15°, and the road alignment was within 20° of the average dip direction, bedding plane slides were thought unlikely.

19.   On the basis of the information obtained during the site investigation a general cutting design was selected incorpora-ting 1:2 slopes in areas of superficial material, 1:1 slopes in weathered rock and 2:1 slopes in the underlying fresher strata. The depth of weathering was determined from the core logs and found to be up to 10 m; this was subsequently proved to be

approximately the correct depth. The design included a narrow
drainage berm at the change of cutting slope, Fig. 6.

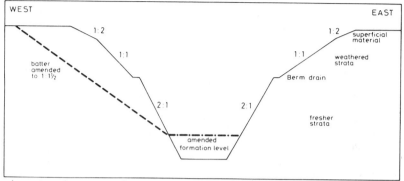

Fig. 6. Diagrammatic section through initial design slopes
and amended western batter slope in Cut 1.

20. It was considered impracticable to remove all uncertainties over the cutting slope stability, and consequently contingency sums were included to allow for reasonable remedial
works. The alternative of using a more conservative slope
design throughout the project in order to avoid the necessity
of any remedial measures would have resulted in a far higher
cost on a contract that already had a surplus of suitable fill,
would have been very overconservative in many areas, and would
have had a more adverse affect on the community.
21. At design stage it was considered that significant remedial works would be required in Cut 1 near the fault adjacent to
Bridge 2 and on the western face of Cut 2b. A contingency sum
was included in the contract to allow for such items as anchors
and drainage holes to be installed during construction. It was
also recognised that there was a risk of instability near the
southern end of Cut 1 where the bypass alignment had to swing
such that bedding planes would daylight in the west face. This
face was designed at 1:1 through intact, weathered siltstones,
sandstones and mudstones.
22. In order to provide an early warning of potential problems
it was decided to continuously monitor the bulk excavations and
map the exposed rock faces.

CONSTRUCTION
23. The £6.2 m contract commenced in spring 1982 and was virtually complete by Christmas 1983, in spite of the long wet
autumn and spring of 1982-83. Bulk excavation of the mudstones
and siltstones was by ripping and scrapers. Large backacters
and dump trunks were used in more restricted areas and for the
stronger strata whilst blasting was needed in some parts of the
Farewell Rock. Face trimming was achieved generally by backacters in the more weathered materials but a "Woodpecker"
impact hammer was needed for some of the 2:1 batters.

24.   It was known that in the Ford's Lake Valley west of
Stepaside there were shafts to the Lower Level Vein, and that
several of these lay beneath the side slopes of a 23 m high
embankment, Fig. 4.   These shafts were backfilled and reinforced concrete plugs cast at rock head.   Local areas of shallow
mining were determined by drilling and workings 0.4 m high
were proved.   Longer adits were known to run into the hillside
and a drainage gallery to pass 70 m beneath Bridge 3.   A
shallow adit, thought to be an exploratory working for ironstone, was encountered in the west batter close to Bridge 2;
this was backfilled and grouted over a distance of 20 m.
25.   In general the strata encountered at construction stage
were very similar to those predicted from the geological maps
and the site investigation.   Cuts 1 and 2 were found to be
dominantly in Coal Measure mudstones and siltstones with varying proportions of sandstone, whilst Cut 3 was mainly in the
sandstones of the Farewell Rock with mudstones at the two ends.
Due to the location of Bridge 2 the bulk earthwork programme
was scheduled to excavate Cuts 2a and 2b in advance of Cut 1,
apart from the northern section to allow the bridge construction.

Cut 2a

26.   This cutting was excavated to its full depth of 15 m with
no stability problems.   Although a small number of wedges developed on joints sub-parallel to the cutting face, these were of
minor extent and dealt with by removal and where necessary shotcreting both to add support to the remaining blocks and to
reduce weathering of the surface strata.
27.   The seepages which occurred particularly on thin clay
layers in the southern sector were used to locate effective
drain holes.

Cut 2b

28.   Three small wedge failures developed in the north face of
Cut 2b when excavation was only approximately 8 m deep (Fig. 7).
Two were associated with thin (10 mm) shear zones dipping more
steeply than the strata; the third failure being associated
with one of a number of throughgoing, planar bedding surfaces
on which a very thin layer of clay gouge was present.   The clay
gouge, identified as resulting from bedding plane movement,
often marked the natural termination of steeply inclined joints.
29.   In this location the strata had a very small component of
dip out of the north face.   The three failures occurred at a
time of heavy rainfall (50 mm in two days).   Reversal shear box
and Bromhead ring shear box tests on the gouge material indicated residual strength values in the range $\phi'_R = 10°$, $C'_R = 2$ kPa
to $\phi'_R = 14°$, $C'_R = 0$.   Following back analyses of the failures
using the method of Hoek, Bray & Boyd (ref. 2), it was decided
to use $\phi'_R = 14°$, $C'_R = 2$ kPa as the most appropriate of the
possible combinations of strength parameters for the bedding
surfaces and $\phi'_R = 25°$, $C'_R = 0$ for steeply dipping joints.
30.   Piezometric pressures measured behind the face indicated
several perched water levels and implied the saturated ground

water table may be below the bottom of the final excavation. Although a system of 10-15 m long drainholes was installed into this face and these undoubtedly reduced the ground water levels, it is clear from continued localised ground water seepage along particular discontinuities that, despite the radial drainholes, not all the water flow has been intercepted.

31.   In view of the three wedge failures in the autumn of 1982 it was considered essential to support the western face over a distance of 300 m.   This support was applied using 26.5 mm 10-20 m long Dywidag anchors.   Similar support was also applied to an area of potential toppling.

Cut 3
32.   Even when Cut 3 had been excavated to its full 26 m depth the geological structure encountered was as predicted and no discontinuity controlled failures were identified; although current bedding in the sandstones produced locally adverse dips.

Cut 1
33.   The preliminary excavation to facilitate the construction of Bridge 2 indicated the strata dipped consistently towards the south.   There was no sign of the fault shown on the geological map and inferred in the bridge area from the difference in lithology noted between BHs 151, 16a and 17a.

34.   Recent work summarised by Hancock et al (ref. 1) suggested that this fault marked the approximate northern limit of the more intensely deformed strata and hence formed the zone referred to as the Variscan Front.   Northwards the strata was expected to show minor drag folding and thrusting with intra-formational shears and an increasingly brittle style of deformation, while to the south the folding would become more intense.   During the excavations necessary to move Sardis Road southwards to facilitate the construction of Bridge 2, some very disturbed strata was encountered.   This was interpreted as being within the folded and faulted zone associated with one of the main faults, hence the structural complexity was expected to occur over a restricted lateral zone.

35.   Following the completion of Bridge 2 which carries Sardis Road across the northern end of the cut, the main excavation of Cut 1 commenced at the end of the very wet spring of 1983.   As trial pits had indicated the presence of some deformed strata and a shear plane which would dip out of the face, a prototype excavation was considered to be the only means of accurately identifying all the features relevant to slope stability.   It was decided to review the slope design once the excavation had progressed far enough to provide a reasonable overall assessment of likely conditions.   However during this early excavation the style and extent of the deformation was found to be more severe than had been anticipated and would have been impossible to predict from an even more exhaustive site investigation.

36.   In June 1983 when the cutting immediately south of Sardis Road had reached about 12 m deep, a slip occurred towards the

top of the east batter slope (Ch. 2670). This developed just
south of the mapped fault trace where the strata was locally
overturned and a carbonaceous horizon was both thickened and
intensely sheared. Whilst this was the first small slip it
was in fact the only one to occur on the eastern batter. The
hollow was cleaned out, drained, concrete filled and faced
with local stone.

Fig.7. Wedge failure in Cut 2b.   Fig.8. Lower slope 'drop away'
slip on west batter, Cut 1.

37. In mid-July a slip occurred at Ch. 2250 in thinly bedded
almost horizontal mudstones. Here the top part of the cut
appeared undisturbed yet below about 4 m a 20 m section of the
batter dropped away, Fig. 8. Examination indicated the batter
had failed on a 30-50 mm clay gouge dipping at 37-42° eastwards.
38. The next significant slip was in late July 1983 at
Ch. 2760, 60 m south of Bridge 2. Here the sliding of a wedge
on a bedding plane dipping almost parallel to the 1:1 batter
(Fig. 9) allowed the upslope portion to dilate. Later movement
involved several hundred tons of mudstone and produced a crack
up to 50 mm wide, extending to 8 m behind the top of the batters
(Fig. 10). The back of a wedge hollow within the slip was
almost vertical and showed a subhorizontally striated sheared
surface of clay gouge with a general north-south trend.
39. The slips which occurred in the 1:1 batter slopes in the
summer of 1983 led to an intensification of the mapping and a
re-assessment of the geology of the area. A quick examination
indicated about 60% of the west face between Ch. 2200 and 2700
with beds or clay gouge zones dipping into the cut, i.e. varying
60-90° from that indicated by the regional dip.

Fig.9. Shallow disturbed mass   Fig.10. Disturbed mass on west
on west batter, Cut 1.   batter, Cut 1.

STRUCTURAL INTERPRETATION

40.   As anticipated, clay gouge was encountered at some bedding planes and along thrust faults that transgressed them. What was not expected, however, was clay gouge zones of varying declinations dipping almost at right-angles to the regional dip.

41.   The Variscan structural tectonics of south-west Wales has been summarised by Hancock et al (ref. 1 and 3). The area of Cut 1 falls into their one-kilometer wide zone 1d, described as being "characterised by WNW-trending box like folds with their axial planes arranged in strongly convergent fans. The faults are accompanied by thrust on which the senses of translation are in accordance with the facing directions of the folds." However, it was not until the paper by Butler (ref. 4) that information in the published literature suggested a reinterpretation of the structures in Cut 1. Using his nomenclature of thrust tectonics and relating the model to the sedimentology of the Coal Measures it was possible to produce an explanation of the phenomena observed. The hypothesis presented here goes beyond that of Butler and suggests a logical set of geological circumstances consistent with the strata encountered.

42.   The planar nature of a contractional thrust will be broken by corrugations when a thrust plane, seeking the material with low shear strength parameters, encounters say a body of sandstone. At this point the thrust plane moves upwards or downwards to pass over/under the mass with higher shear strength. As a result part of the thrust plane will have declinations not only in the direction of movement but also at right-angles due to the over-riding of the mass of higher shear strength. Fig. 11 shows such a situation and the terminology applied. It is obvious therefore that any strata, such as fluviatile Coal Measures, in which there is a lateral variation in lithology, will be liable to develop ramps in addition to the normal flats of the planar thrust surface (Fig. 12).

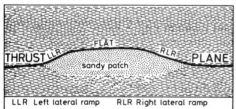

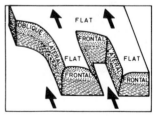

Fig. 11. A possible sedimentological explanation for a thrust to deviate from a smooth planar feature.

Fig. 12. Ramp terminol- in thrust tectonics.

43.   Observing Cut 1 with the new thrust tectonic model in mind it is clear that there are a number of gently dipping shear planes occurring mainly along carbonaceous horizons. Between these the beds are disturbed to give outward facing dips and there are also a number of "ramps" which, with clay gouge, produce obvious weakness zones. It is both the

outfacing dips and the ramps daylighting into the 1:1 cut that produced a number of slips in the western face whilst the eastern face remained stable.

CUT 1 SOLUTION

44.   As failures were occurring in the 1:1 west batters it was obvious that the 2:1 face would not be stable. It was decided therefore to reduce the western batter to 1:1½ even though this necessitated an increased land take west of the bypass. In order to minimise this the road design was changed to dual carriageway, permitting the vertical alignment to be modified. As a result it was possible to reduce the depth of Cut 1 by up to 6 m.

45.   In addition some rock anchors were installed near Bridge 2 to inhibit oblique sliding on a plane where the apparent motion surface was still steeper than the angle of shearing. In view of the change of batter near the bridge and the proximity of the fault it was decided also to place extra anchors in this area. These, taken through the shear plane, added considerable support to the bridge. To prevent deterioration of the bed resulting in the removal of support from the face above, any weak sheared horizons on the 1:1½ slope were excavated to a depth of about 0.75 m and shotcreted.

46.   Drainage holes were drilled to ensure a reduction of the ground water level. One of the problems of the clay gouge zones, related to the lateral ramps, could be the retention of high ground water levels by the low permeability bands. This would explain why the ground water level did not lower as the excavation proceeded as would have been expected.

CONCLUSIONS

47.   The road opened in July 1984. This was only three months late due to the contractors' efforts to accommodate the redesign of the Cut 1 batter as quickly as possible.

48.   Despite the works outlined above and the extra expense incurred, the scheme has still proved to be the most cost-effective solution overall. Had a much more conservative slope design been adopted, even though this may have avoided the failures which occurred and necessitated additional land acquisition, the construction costs would nonetheless have exceeded the actual incurred costs by a very significant margin.

49.   The presence of clay gouge zones at right-angles to the regional dip indicated a greater complexity of structural geology than anticipated. The model to explain the geology encountered was not published until 1982. It is clear, however, that for other works involving a similar geological setting the implications of the thrust tectonics should be considered at design stage.

ACKNOWLEDGMENTS

50.   The authors wish to thank the Welsh Office for permission to publish this paper, and Jean Bees who prepared the text figures.

REFERENCES
1.   HANCOCK, P.L.,   DUNNE, W.M. & TRINGHAM, M.E.   Variscan structures in southwest Wales. Geologie en Mijnbouw, 1981, 60, 81-88.
2.   HOEK, E., BRAY, J.W., & BOYD, J.M.   The stability of a rock slope containing a wedge resting on two intersecting discontinuities.   Quarterly Journal of Engineering Geology, 1973, 6, 1-55.
3.   HANCOCK, P.L., DUNNE, W.M., & TRINGHAM, M.E.   Variscan Deformation in southwest Wales in Hancock (Edit.) The Variscan Fold Belt in the British Isles, Adam Hilger Press, Bristol, 1983.
4.   BUTLER, R.W.H. The terminology of structures in thrust belts.   Journal of Structural Geology, 1982, 4, 239-245.

Discussion on Papers 16–19

PROFESSOR T. D. O'ROURKE, Cornell University, New York

The paper on a major gabion wall failure by Thorburn and Smith draws attention to the compressible nature of the rock-filled gabion baskets. Conventional design of retaining walls for active earth pressure is based on the assumption that the wall is incompressible relative to the retained soil or rockfill. This assumption holds for most retaining structures, but it may not apply in all cases. If portions of the wall compress and move downwards relative to the retained materials, then an upward shear will develop along the rear surface of the wall. The upward shear represents a significant departure from the equilibrium conditions assumed in conventional retaining wall design.

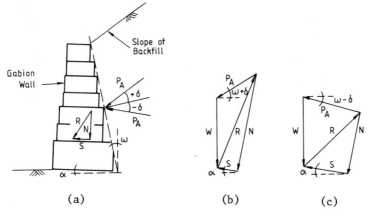

Fig. 1. Cross-section and vector equilibrium diagrams for lateral stability of a retaining wall: (a) cross-section of the gabion wall; (b) vector diagram for $+\delta$; (c) vector diagram for $-\delta$

Figure 1(a) shows a simplified cross-sectional view of a gabion wall, similar to that described in the paper. Each layer of gabion baskets is inclined with respect to the horizontal at an angle α. The effective angle of rear wall inclination is ω. An active force P_A acts on the rear surface of the wall at an angle relative to the normal of $\pm\delta$. The sign of δ will depend on the direction of relative displacement at the interface between the wall and retained rockfill. The weight of the wall is W. The factor of safety F against sliding on a plane inclined at angle α is

$$F = \frac{N \tan\phi_W}{S} \tag{1}$$

in which N and S are the normal and shear forces acting on the plane and ϕ_W is the angle of friction between the gabion baskets. For simplicity, it is assumed that the resisting shear depends on friction only.

If the retained soil or rockfill moves downwards relative to the wall, a positive value of δ results and the vector equilibrium diagram in Fig. 1(b) applies. From the diagram, it can be seen that the relative values of N and S are such that the factor of safety calculated by means of equation (1) will generally be large enough to ensure lateral stability. If the wall moves downwards relative to the retained soil or rockfill, a negative value of δ results and the vector equilibrium diagram in Fig. 1(c) applies. From the diagram, it can be seen that the relative values of N and S are such that the factor of safety may be less than unity. It should be recognized that the active earth force P_A will increase substantially for δ/ϕ close to -1 and may be as much as two to three times larger than the active force P_A which applies for conditions of positive δ.

From the vector equilibrium diagrams, it can be shown that

$$N = [W + P_A \sin(\omega\pm\delta)] \cos\alpha + P_A \cos(\omega\pm\delta) \sin\alpha \tag{2}$$
$$S = P_A \cos(\omega\pm\delta) \cos\alpha - [W + P_A \sin(\omega\pm\delta)] \sin\alpha \tag{3}$$

Combining equations (1)-(3) results in

$$F = \frac{\{W + P_A[\sin(\omega\pm\delta) + \cos(\omega\pm\delta) \tan\alpha]\} \tan\phi_W}{P_A[\cos(\omega\pm\delta) - \sin(\omega\pm\delta) \tan\alpha] - W \tan\alpha} \tag{4}$$

which is a general expression for the safety factor associated with lateral stability of the wall. Values of P_A can be determined from published compilations of earth pressure coefficients (e.g. ref. 1) or by graphical techniques.

Although truly compressible walls will exist only under exceptional circumstances in the field, it is none the less important to consider the possibility of downward wall movement for certain types of retaining structures. Many types of retaining walls are built with cells, or elements, of soil and rockfill, which can be compressible relative to the fill or in situ materials retained by the wall.

Tschebotarioff (ref. 2), for example, described the failure of a crib wall 7.3-10.4 m high in which the cribs were filled with relatively loose granular soil. His analysis showed that compression of the wall elements led to a reversal of shear at the interface between the wall and retained soil, which in turn reduced the safety factor against lateral sliding by 66%.

References
1. CAQUOT, A. and KERISAL, J. Tables for the calculation of passive pressure, active pressure and bearing capacity of foundations. Gauthier-Villars, Paris, 1948. (Translation by Bec, M. A., Ministry of Works, Chief Scientific Adviser's Division, London.)
2. TSCHEBOTARIOFF, G. P. Foundations, retaining and earth structures. McGraw-Hill, New York, 2nd edn, 1973.

DR C. J. F. P. JONES, West Yorkshire Metropolitan County Council, Wakefield

The paper by Thorburn and Smith provides detailed information

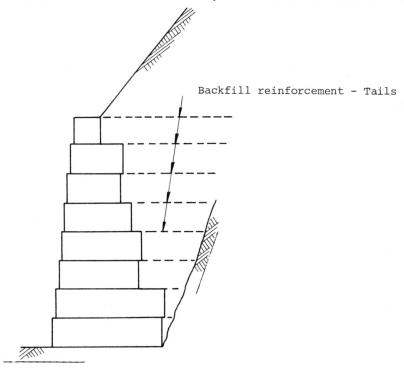

Backfill reinforcement - Tails

Fig. 1

concerning the performance and behaviour of a large gabion wall, resting on a firm foundation. The analysis showed that the gabions were compressible and Professor O'Rourke has provided an elegant explanation of how this compressibility can lead to deflexions of the gabion structure, which in the limit leads to failure.

The gabion wall was inadequate: however, it would be possible to improve the design of the structure dramatically by adding 'tails' to the gabions, as shown in Fig. 1.

The tails are formed by extending the base or top of the gabion to provide reinforcement to the backfill thereby producing an integrated structure. The advantage of this arrangement is that it would eliminate the failure mechanism described by Professor O'Rourke. Significantly, the use of tails on the gabions works best when the gabions compress, the optimum condition being when the compressions of the backfill and the gabions are the same.

The idea of the tailed gabions has been patented by Jones and Templeman (Patent No. 7941627) but the patent has been released to the public domain and is free for general use. The principle can also be used with crib walling and at least one crib walling system offers this technique.

MR L. THREADGOLD, Leonard Threadgold, Kenilworth

In relation to the design of gabion retaining walls I would like to query the need to include horizontal layers of mesh or fabric in their construction. The layer is necessary if preformed baskets are to be transported to the site but once in place it forms a plane. Interaction between the stone infilling across such a mesh and hence shear resistance can be substantially reduced.

The elements in the vertical plane extending from the face into the body of the wall can be designed to apply the necessary restraint to the wall as a whole, while narrow horizontal strips adjacent and parallel to the face can be used to restrain the face locally. This reinforcment is particularly facilitated in 'mattresses' formed using Tensar Geogrids in a zigzag pattern in plan to form triangular cells.

Such a contribution can easily be forgotten when considering the behaviour of the wall only in terms of its vertical cross-section.

MR F. HUGHES, Cementation, Piling and Foundations, Rickmansworth

Mr Thorburn, with a proper design could a crib wall solution have been successful? I assume that the foundation is not suspect and that cribs are set up correctly so that the wall compression is significantly less than that of the gabion system.

Professor O'Rourke's illustration of the shift in the force P on the back of the wall suggests that the crib wall has positive advantages over the gabions.

MR J. HARRIS, Frederick Sherrell, Tavistock

My contribution concerns the materials used and the method of filling of gabions.

I noticed·in Mr Thorburn's illustrations that the gabions were filled with randomly orientated, blocky rockfill which appears to be virtually identical with the rockfill behind the wall. In south-west England we frequently encounter slaty rocks which form flat, platy rockfill. This material makes excellent fill for gabions provided that it is placed horizontally by hand, in the manner of traditional dry stone walls. Experience shows that gabions filled in this way are more stable when the wires begin to corrode. The vertical compressibility of hand-filled units is probably less than that of gabions which are mechanically filled with randomly orientated stone.

DR A. B. HAWKINS, University of Bristol

Dr Matheson, you have correctly drawn attention to the use of pre-splitting to produce less disturbance in the permanent rock slope geology. The method of drilling a line of holes at 750-1250 mm centres to cut the rock by firing the spread explosive charge shortly before a localized loosening bulk blast explosion has been done for many years. Hoek and Bray (1974) point out that the bulk blast vibration still passes across the previously pre-split minor crack but that such minor cracks act as vent paths for the explosive gases and that this venting prevents the propagation of the radial cracks from the main blast across the pre-split line.

When we were excavating a large dry dock at Portavadie, in the Dalradian quartz-mica-schist, pre-split was used to cut a steep batter and to protect the rock behind the clean cut face. Fig. 1 shows such a situation with the bulk-blasted, jagged-edged, dilated zone adjacent to the area which had been pre-split.

Although Ritches did an excellent job at Portavadie, we had several problem areas where the pre-split line was not followed but instead the break zone was a fault plane. The significance of a fault, with a clay gouge zone, was that the rock broke back to the fault rather than taking the pre-split line. As a result, in addition to draining the rock slope, it was necessary to anchor the wedge seen in Fig. 2.

At Portavadie it became important to locate the faults which might influence the pre-split line in those areas where the slopes had to be steep to allow a crane in the dock to reach lorries on the top roadway. Although this could have been achieved by the more expensive rotary drilling, it was decided to use one of the air-percussive rigs and carefully to measure the penetration rate and colour of the risings, Fig. 3. In this way the faults, generally with a 75-150 mm gouge, were located.

Fig. 1 (left). Contrast between a disturbed bulk-blasted area and the adjacent pre-split

Fig. 2 (right). Wedge which resulted when the rock broke back to a fault rather than separating at the pre-split line

Fig. 3. Air-percussive drilling to use the penetration rate and the dust colour to locate faults which may influence the quality of the pre-split cut slope

The presence of gouge zones along faults has been known for many years. It is only recently, however, that some, such as the 25-50 mm zones at Kilgetty, are known to be related to thrusting. At a site on the A5 west of Llangollen trial pits dug in February 1985 to examine the nature of faults shown on a geologist's map have again located 25-60 mm wide gouge zones where it was anticipated that the geology would be rock against rock. The presence of these fault zones in an area of cuttings that are deeper than 15 m is likely to modify the batter angles suggested in the desk study report.

DR G. D. MATHESON

You have raised a very valid point regarding the potential
importance of major discontinuities and the success of
establishing a pre-split face. Pre-splitting should not be
considered without first of all optimizing the design to the
geotechnical conditions of the rock mass. Large important
fractures such as you described can be expected to take over
as the final face if their orientation falls within 15° of the
pre-split plane. Emphasis must therefore be placed in the
detection and location of such planes before the design is
finalized. Where final faces are not optimized to local
conditions then remedial work can be anticipated – this is
illustrated by the experiences at Portavadie. The paper
clearly suggests that optimum stability requires careful
consideration of both the slope design and the method of
excavation.

MR N. J. VADGAMA, Terresearch Ltd, Northolt

Concern has been expressed by Dr Hawkins and others that
companies in this country lack the expertise and techniques
required to take core samples which would enable relatively
weak layers within a critical soil mass to be identified. The
difficulty in determining the core orientation to find the dip
of various strata has also been discussed.

I believe that there are quality site investigation
companies in this country that are capable of producing a high
standard of work. Triple-core barrels or Mylar-lined double-
tube core barrels have been used to produce good quality cores
in weaker soils. Similarly the orientation of cores can be
determined using various techniques including the in situ
examination of rock by down-the-hole camera.

However, the site investigation companies which have the
experience and resources to undertake quality work are under
increasing financial pressure to curtail the scale of their
operations because clients are in general not prepared to pay
the extra cost.

In recent years several of the larger companies have closed
their site investigation activities for this reason. The
trend is likely to continue unless engineers change their
attitudes and seek quality rather than the lowest price.
Otherwise the standard of site investigation in this country
will degenerate.

DR E. N. BROMHEAD, Kingston Polytechnic, Kingston-upon-Thames

Mr Wallace and Mr Arrowsmith, please would you explain whether
your postulated slip surface positions were verified by
inspection during excavation, and if found to be located
elsewhere what impact would such a finding have made on the

design of the road cut? Also, where a feature in earthworks has a design life, what are engineers in the future to do when that life has expired?

Mr Fletcher commented that, although extra costs were incurred on his site due to the need for remedial measures, the total cost was less than that for the adoption of a more conservative design initially. Is this a plea for the adoption of a 'design as you go' philosophy with essentially no investigation or detailed design initially?

MR E. J. ARROWSMITH

The difficulty in locating the slip plane was accentuated by the broken nature of the mudstone. This was due partly to weathering and partly to the effect of coal mining beneath.

I confirm that the Department of Transport was satified with a designed slope in the rock cutting which was estimated to be stable for at least 100 years. All structures have a limited life and the effect of a slip in this location is unlikely to be serious.

MR W. A. WALLACE

The entire slipped mass was removed as part of the earthworks for the road construction. The practicalities and cost implications of attempting to locate the slip plane within a broken rock mass of 100 000 m^3 made the exercise unattractive – particularly so when the slip plane geometry was already reasonably well defined from site evidence and from the drillers' observations while coring through the slipped ground. A deviation on the slip plane from that adopted in the back analysis would not have altered the overall conclusions drawn from the analysis.

In times when highway structures are designed for 100 years plus and a margin for safety, it is not unreasonable to apply, when possible, a similar approach when considering highway earthworks.

MR P. HORNER, Consulting Engineering Geologist, Luton

In response to the discussions on obtaining information on weak zones in rock, which are difficult to recover in cores, we should not forget that a cored borehole provides both core and a borehole. There are numerous techniques which are now readily available, provided that the circumstances are favourable, to obtain information utilizing the hole. These include video television surveys, down-the-hole geophysics and wax impression packers. The final use of the hole for the installation of slope indicators, piezometers etc. is well established.

20. Buried services associated with transport systems including mining subsidence

N. ADDY, MICE, Assistant Area Engineer, West Yorkshire Metropolitan County Council

SYNOPSIS

Earthworks failures are usually natural occurrences or the unintentional result of some separate engineering works. In the United Kingdom the occurrence of self inflicted damage is high and growing due in part to the ageing infrastructure and the need for frequent repair and renewal. In the mining areas the results of mining subsidence may exacerbate the damage or itself be the cause of earthwork failure.

INTRODUCTION

1.   The United Kingdom has a multi-million pound investment in its infra-structure. The framework of which covers two basic elements, a transportation system and a network of utility services providing energy and communications.

2.   The transportation System. This may be subdivided into three basic elements:

>   Land – highway network and the railway network.
>   Water– inland navigational waterways.
>   Air – civil aviation.

3.   For the purpose of this paper civil aviation is precluded on the grounds that airports are few in number and do not normally have a history of earthwork problems; nor are they normally affected by mining subsidence.

4.   The Utility Services These may be subdivided into two basic areas:
>   Energy – covering electricity, gas, water, oil and chemical distribution and pumped waste.
>   Communications – telecommunication and security networks.

5.   External Influences The infra-structure must be sustained against external influences. One such intrusion

Failures in earthworks. Thomas Telford Ltd, London, 1985

is mining subsidence which has traditionally caused severe damage to both transportation systems and the utilities services. Mining in the United Kingdom embraces coal, salt, gypsum, tin, lead, clay, limestone and sands, although decline in the extraction of some minerals such as tin and lead has reduced the problem. In addition recent controls on salt extraction have now largely eradicated subsidence damage from this source, and the major problems arise from coal mining.

BURIED SERVICES ASSOCIATED WITH TRANSPORT SYSTEMS

6.   Interaction Transport systems are influenced by buried services to varying degrees. Railways and navigational waterways are frequently crossed by services but are seldom used as a route, a notable exception being the current agreement between Mercury Telecommunications and British Rail. Conversely the highway network provides the route for utilities services particularly in the conurbations. It is the maintenance and repair of these services which is a frequent cause of foundation failure in the highway.

7.   Quantitative Assessment The highway network is sometimes described as the largest service duct in the United Kingdom. The house of Commons Transport Committee in their first report on Road Maintenance 1982/83, (ref.1) identified an annual figure of 1.8 million openings in the highways. The majority of these openings occur in the urban areas and approximately 50 per cent are attributable to the Gas Industry, this may be an indication of the deterioration of Victorian gas mains and services. As an example of repair costs, West Yorkshire Metropolitan County Council annually recovers £3.9M in reinstatement charges from the Services Industries. On a national level the figure could exceed £80M.

8.   Financial Repercussions The recovery from the Utilities does not wholly reflect the damage incurred with respect to the highway. The actual damage may appear after several years when settlement of the carriageway indicates a foundation failure. Statutory provisions limit liability to the immediate damage caused when opening the highway but make no allowance for the remedial works which may ultimately be required. Consequently the recovered charges do not reflect the full cost of the failure.

9.   The Failure Mechanism Highway pavements are designed as an integrated structure in which stresses relate to the dynamic moduli of the pavement. Disturbance of the structure and the supporting earthworks threatens the integrity of the highway pavement and its foundation. Any opening in the subsoil structure disrupts the at rest

conditions and the lack of support to the sides of trenches results in the development of shear planes in the adjacent soil. Disturbance of the highway pavement increases the permeability of the foundation and the resultant ingress of water to the formation initiates failure. The problem is compounded by poor compaction which usually accompanies service repairs.

10.   The majority of openings are within the marginal quartiles on the road and occur under the inner wheel track close to the pavement edge at the point of maximum loading. Such openings, particularly in respect of large diameter mains, can result in a longitudinal shear plane through the pavement formation and earthworks. When these excavations are inadequately supported and exposed to weathering, slip conditions are common and the risk of complete carriageway failure increases.

11.   A further consideration is the equilibrium moisture content of the subsoil. Most services are laid in some form of bedding material. Large diameter pipelines are frequently laid in a granular bedding material which can act as a drain. Being of high permeability in comparison to the interstitial drainage of the adjacent subsoil, differential drainage of the earthworks can occur locally along the service trench causing trench drawdown and subsequent disruption of the pavement.

12.   The Human Element The reasons for pavement and earthworks failures are not simply explained in soil mechanics terms. Failures are often man made and result from ignorance. Close liaison between the interested parties is essential to reduce the damage to the highway foundation. It is the simple issues such as the planning of openings and control of trench loading which are likely to be contributory to reducing damage to the pavement and the supportive earthworks. Road traffic adjacent to trench workings is seldom adequately assessed or controlled, and horizontal forces resulting from vehicle braking may be aggravated by poor traffic management particularly during commuter periods and enhance axle loadings. The periphery of the excavation is weakened and the good mechanical interlock in the pavement breaks down. As a result a serviceable pavement adjacent to the opening may be reduced to a critical or failure condition.

13.   Indiscriminate opening of highway pavements and excavations into the foundations by personnel with little or no training in highway methods is a major cause of the problem. The prime objective of maintenance personnel is to maintain the utility and the reinstatement of the highway and its foundation is of secondary importance. It is unlikely that arisings will be segregated, suitable backfill material

339

may be neglected, exposed to weathering and consequently become unsuitable for use. This negligence induces highway authorities to specify expensive imported granular backfill with known low compaction fraction properties in order to limit settlement to tolerable levels. In recognition of the problem compaction plant is now increasingly apparent with services maintenance gangs, and a degree of initial compaction is being applied to reinstatements.

14.   Impact loading by heavy goods vehicles rapidly consolidates poorly compacted backfill material in transverse trenches resulting in depressions. Unfortunately, there are still many instances, more particularly in rural areas, where temporary reinstatements are left without any bituminous surfacing. This exposes both the pavement formation and the underlying earthworks to the ingress of water. An available solution is the use of thixotropic macadams, whilst the previous limitations of thermoplastic macadams have been overcome.

15.   An associated problem is the large number of surface water drainage systems damaged during the maintenance of buried services. Road gully connections are ruptured and surface water is freely admitted to the formation. Culverts are often severed or waterway restricted by services intersecting and being laid through them. The effect is to weaken the foundation which may lead to failure.

16.   Reinstatement It is almost impossible to integrate the new construction with the parent structure. Inadequate compaction of backfill to service trenches is probably the most significant contributor to reinstatement failure. It is necessarily a slow and tedious procedure in which the selection, placing and compaction of the trench backfill requires consistent competent supervision. The required supervision is rarely available, monitoring is casual and compaction verification tests are rarely carried out. In addition reinstatement of the bituminous and asphaltic surfacings which protect the subgrade and earthworks and maintain equilibrium moisture content is largely impractical. Fusion of the reinstatement to the parent surfacing is rarely attained and the mechanical interlock required to produce an integral structure is virtually impossible.

17.   Example of damage during the construction of services An example of foundation damage caused by the construction of a new service in an existing highway is illustrated in the case of the reconstruction of a town sewer along a heavily trafficked road. The new sewer consisted of a 525 mm diameter pipe laid in the Class II road with the following temporary reinstatement:

100 mm Dense Bituminous macadam
300 m Lean mix concrete
Arisings Backfill

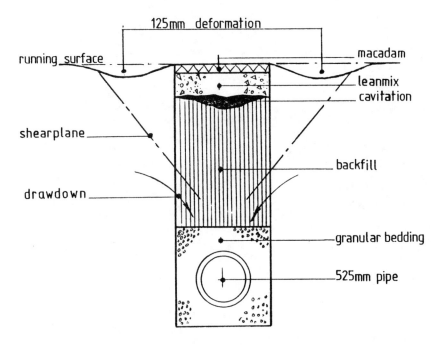

Fig.1.   Trench reinstatement failure.

The above construction failed after several months due to
trench drawdown and foundation failure as shown in fig.1.

Deflectograph readings established that the trench
reinstatement was stronger than the adjacent carriageway.
The lean mix concrete backfill reflected lower deflections
than the carrigeway adjacent to the trench, which showed
signs of foundation failure. Cores taken through the trench
reinstatement revealed voids under the lean mix concrete 125
mm in depth. fig.2. The resulting remedial works required a
carriageway repair 3500 mm wide in place of the 1400 mm which
would normally be used.

18.   Example of foundation failure caused by damage to
a Service in use The results of a foundation failure and
the resulting pavement damage are shown in fig.3. The cause
was the failure during storm conditions of a surface water
drain serving 2.0 ha of residential land. The resultant

341

surcharge on the outfall ruptured the manhole cover slab causing extensive damage, washout and failure of the road foundation. Repairs entailed reconstruction of the manhole and replacement of the limestone base of the highway.

Fig.2.   Residual cavitation following secondary compaction.

Fig.3.   Pavement failure due to washout.

19.   Similar damage can occur with water mains and
recent reports suggest that leakage accounts for some 10% of
water consumption. Much of this leakage occurs
within carriageway formations causing damage which may
eventually lead to failure.

MINING SUBSIDENCE

20.   British Railways Although British Rail are able to
reserve coal supplies beneath their track, they rarely
exercise this privilege. In general British Railways align
themselves   with the Highway Authorities and accept mining
subsidence damage recovering costs from the National Coal
Board for advanced works and terminal reinstatement. This
can be on a major scale as illustrated by the development of
the Selby Coalfield which required the diversion of the main
east coast line over a distance of 30 km at a cost of £60M.

21.   British Waterways Mining subsidence often exposes
waterways to the risk of over topping leading to damage to
the bank and flooding of adjacent land. Supportive works
need to be raised to accommodate mining subsidence and
attendant works executed to bridges in order to maintain
headroom. Locks are similarly affected and being water
retaining structures must be strengthened to accommodate
increased loading conditions.

22.   Utility Services These are particularly
susceptible to mining subsidence damage, as the ground
strains associated with mining are transmitted to the
utility pipes and cables, frequently resulting in rupture.
Tensile strain in cables may either break terminals or
breech armour leading to the ingress of water and erosion
of the conductors.

23.   As an example, fig.4. shows a high pressure gas
main laid in the verge of a trunk road which exploded as a
result of tensile strain caused during mining. The
carriageway verges were completely.blown out and the
foundation to the highway exposed. The whole of the highway
and surrounding fields were covered with toxic dust
necessitating closure of the road whilst the contamination
was removed.

24.   Highways Highways can survive subsidence to a
limited degree, with flexible constructions being better
able to accommodate the ground movements than rigid concrete
pavements, in every case the drainage system is at risk.
Severe subsidence troughs frequently disrupt the geometric
properties of the highway, an example being the condition
shown in fig.5, when a temporary speed limit had to be

Fig.4.   Exploded high pressure gas main.

imposed on a Trunk Road. This form of subsidence is
accompanied by the development of compression ridges and
tensile cracking of the pavement, which have to be repaired
quickly to preserve a satisfactory riding surface and also
to prevent uncontrolled leakage into the foundation.
Extensive remedial works are normally required when the
mining subsidence is completed. The foundation will have
been subjected to severe strains and may have failed. Even
without complete failure the residual strength of the
highway will have been impaired.

25.   Mining subsidence can produce complete failure.
The all purpose road shown in fig.6. was so extensively
damaged by fissuring that the road had to be closed. The
damage resulted from tilting of massive sandstone blocks
beneath the highway which produced fissures in the road
foundation. The repair consisted of double reinforced
concrete slabs bridging the fissured rock, fig.7.

CONCLUSION

Openings through impermeable pavements into the supporting
earth works can cause damage and failure. Ignorance makes

344

Fig.5.   Mining subsidence trough.

Fig.6.   Pavement failure from rock fissuring.

Fig.7.   Fissure slabbing.

the situation worse and the lack of appropriate protection
is a significant contributory factor in the failure
mechanism. The problems cannot be removed in an established
correlated infrastructure as in the UK. Future developments
should recognise the constraints imposed by services buried
under pavements and purpose-built ducts would eradicate many
problems.

Education of services personnel of the media in which they
work would assist in reducing needless damage factors.

Mining subsidence damage must be cost effective and a close
liaison between mineral engineers and highway engineers would
identify excessive damage and the cost effectiveness of
getting the mineral.

REFERENCES
ref.1. House of Commons Transport Committee First Report on
Highways Maintenance.

ACKNOWLEDGEMENTS
The author wishes to acknowledge the assistance of various
members of the Directorate in the preparation of this paper
which is presented with the approval of the Executive
Director of Engineering. The views expressed are those of
the Author and do not necessarily reflect the official views
of the Authority.

21. A retaining wall failure induced by compaction

T. S. INGOLD, BSc, MSc, PhD, DIC, FICE, FIHT, FASCE, FGS, MSocIS(France), Chief Engineer, Geotechnical Division, Laing Design & Development Centre

SYNOPSIS.   The distribution and magnitude of lateral earth pressure acting on a retaining structure constructed using modern techniques are generally not those predicted by classical earth pressure theories.   This situation results from the use of high energy output compaction plant which may induce high lateral pressures in the soil.   Such a phenomenon is illustrated by the performance of a long reinforced concrete cantilever retaining wall and demonstrates that the high compaction induced bending moment causing failure at the base of the wall stem could not have been predicted using classical earth pressure theories.   Bending moments at the base of the wall stem are evaluated using both a conventional approach and an analysis allowing for the effects of compaction.   Results obtained are compared with the observed performance of the wall.

INTRODUCTION

The wall, of conventional cantilever design, was built to retain an earth bund with the top of the wall constructed at the same level of 7.8m above temporary bench mark level (ATBM) throughout its entire length of 70m.   The base slab was founded at levels of either zero ATBM or 2.1m ATBM, thus giving a wall height of either 7.8m or 5.7m.   At each end the wall terminated with a 15m long battered wing wall.   Sections and foundation levels of the wall are shown in Fig.1.   The wall was founded on, and backfilled with, a very silty, slightly sandy clay and gravel soil.   As placement and compaction of the fill was completed, the top of the wall bowed giving a maximum deflection at the top of the wall of 93mm.   A full survey of the deflected form of the wall was made, Fig.2, and the wall was excavated for inspection when samples of the wall fill were taken for testing.

In the initial stages of the investigation trial pits were excavated at the front and back of the wall.   Undisturbed U100mm diameter tube samples were taken at one metre vertical centres by pressing an open-drive sampler into the fill.   In situ density of the fill was determined at one metre vertical centres by the sand replacement method   and

Failures in earthworks. Thomas Telford Ltd, London, 1985

to determine the degree of compaction four determinations were made of the maximum dry density using the B.S. Heavy (4.5 kg rammer) compaction test. Effective shear-strength parameters of the fill were determined from six sets of consolidated-undrained multistage triaxial compression tests, with pore-water pressure measurement, carried out on undisturbed samples with effective wall friction characteristics being assessed using drained shear box tests with concrete blocks cut from the back of the wall in the lower half of the shear box and fill compacted at the appropriate density and moisture content in the upper half of the box. Typical geotechnical proper-ties are given in Table 1.

Table 1.   Typical Geotechnical Properties

Unit Weight	:	2.16 Mg/m³
Moisture Content	:	11.5%
Plastic Limit	:	16.0%
Liquid Limit	:	25.0%
Angle of Shearing Resistance	:	35°
Effective Cohesion	:	1.7 kN/m²
Angle of Wall Friction	:	32°
Effective Wall Adhesion	:	1.6 kN/m²
Degree of Compaction	:	87.2% BS Heavy

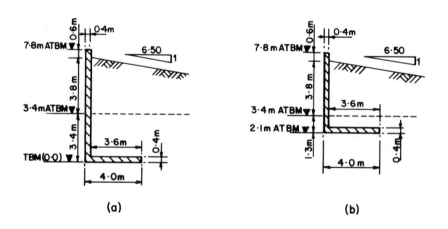

(a)   (b)

Fig.1.   Typical Wall Cross Sections.

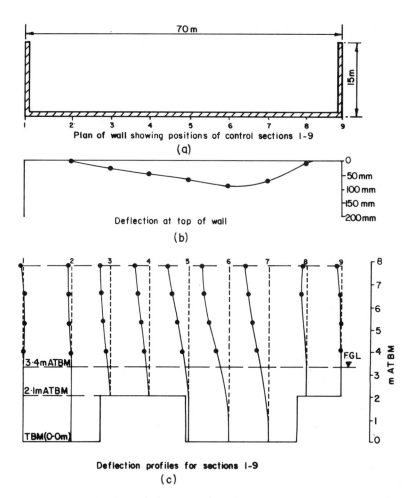

Fig.2.   Deflected form of wall.

FAILURE OF WALL

Fill was placed in front of and behind the wall keeping the same fill level on both sides of the wall, thus minimising bending moments and risk of sliding due to unequal filling. This process was continued until the fill was up to the finished ground level in front of the wall. At this stage, the fill behind the wall was continued up to full height in the normal manner. Compaction was by four passes of a Stothert and Pitt 54T vibrating smooth-wheeled towed roller with the fill placed in 400mm thick layers. The backfilling to sections 3 to 9, Fig.2(a), was completed at dusk and at this stage no signs of distress were noticed. The following morning inspection showed that the wall had tilted and at section 6 was 93mm out of plumb. A survey was carried out and the deflected form of the wall was found to be as shown in Figs. 2(b) and 2(c). Several vertical cracks were found in the front face of the wall between sections 5 and 8 where the maximum bowing occurred. These cracks, typically 2mm wide, were fitted with tell-tales, which indicated no further widening of the cracks after the discovery of the failure. Excavation behind the wall at the point of maximum deflection showed a horizontal tension crack 2mm wide running along the base of the wall stem near its junction with the base slab, Fig.3. Groundwater level was approximately at finished ground level. When all the fill behind the wall was removed to base-slab level, the wall slid back 11mm at the base and 22mm at the top of the wall indicating a translation of 11mm and a reduction in tilt of 11mm. The backward translational movement was due to the fill in front of the wall causing a conventional sliding failure, since with no fill behind the wall the active thrust generated by the remaining fill in front of the wall exceeded the base sliding resistance. A crack 3mm wide and running parallel to the wall was observed at finished ground level 1.5m in front of the wall. The base slab showed no signs of cracking or distortion. Levels taken on the toe and heel of the base slab after the failure of the wall were compared with the levels at the time of construction of the base. This comparison indicated that there had been

Fig. 3. Detail of Crack at Base of Wall Stem.

no rotational movement of the base.   On refilling after remedial measures, the wall movement did not increase once light
compaction was carefully completed.   From this it seems likely
that the original failure of the wall occurred at the time of
compaction but passed undetected in the poor light.

CONVENTIONAL ANALYSIS

At the design stage, the effective shear-strength parameters for the fill material were derived by inspection of the
fill and reference to the current British Code of Practice,
(ref.1).   This resulted in the selection of an angle of shearing resistance of 35° with zero effective cohesion.   In operation, the wall was to be subjected to vibration transmitted
from large-diameter subterranean oil mains passing through the
wall from a nearby pumping station.   Under such conditions
the Code specifies that the angle of wall friction should be
taken as zero.   Using these parameters, the resulting
coefficient of active earth pressure, K_a, was calculated as
0.25 using Coulomb theory with a value of 3.69 for the coefficient of passive earth pressure, K_p, calculated using Rankine
theory.   In evaluating the bending moments at the base of the
stem, a factor of safety of three was applied to the calculated
passive pressures in order to limit wall deflection.   In the
design, the groundwater table was assumed to be at a level of
3.4m ATBM.   The resulting calculated design moment at the
base of the stem was 206 kN.m anticlockwise.   The distribution
of effective earth pressure and thrusts for these conditions
are shown in Fig.4(a).   When the wall was analysed using the
geotechnical data given previously it was found, using Coulomb
theory, that K_a decreased to 0.22 with the value of the normal
component K_{an} being 0.19.   The value of K_p was determined as
9.0 from a graphical solution employing the "ϕ-circle" method.
This value was confirmed by extrapolation from charts published
by Packshaw, (ref.2).   For the active case wall adhesion and
effective cohesion were taken into account using the expression
proposed by Bell, (ref.3).

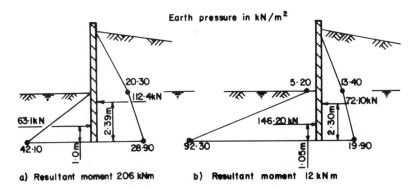

a) Resultant moment 206 kNm   b) Resultant moment 12 kN m

Fig. 4.   Distribution of effective earth pressures
from conventional analysis.

For the passive case K_{pc}, the coefficient of passive earth pressure with respect to effective cohesion,was determined as 10.5 by graphical solution, this value being verified by extrapolation from Packshaw's chart. The resulting effective earth-pressure distribution and thrusts are shown in Fig.4(b). As before, a factor of safety of three was used in evaluating passive pressure. From inspection of Fig.4(a) it can be seen that the calculated bending moment at the base of the stem, based on the originally assumed soil parameters, was 206 kN.m. Using the measured soil parameters, the calculated bending moment reduces to a mere 12 kN.m, Fig.4(b). The wall was reinforced with 32mm diameter mild steel bars at 200mm centres in the back face and nominal 12mm diameter bars, also at 200mm centres, in the front face to give an ultimate moment of resistance at the base of the stem of approximately 315 kN.m. It was obvious from inspection of the wall that the ultimate moment of resistance had been closely approached or even achieved, yet using classical earth-pressure theory and the measured soil strength parameters, a bending moment of less than 5% of the ultimate moment of resistance was predicted.

ANALYSIS ALLOWING FOR COMPACTION EFFECTS

In classical earth-pressure theory it is primarily the self weight of the soil that is taken to induce the active pressure. It is of course implicity assumed that there is sufficient lateral yield to mobilize the active condition. An adjunct to this theory allows the effects of compaction to be taken into account (ref.4). For the wall in question the lateral earth pressure increases from zero at the surface of the fill to a maximum value at a depth z_c, equation 1, below the surface.

$$z_c = Ka \sqrt{\frac{2p}{\pi\gamma}} \qquad \dots \ 1$$

Below this critical depth, z_c, the lateral earth pressure, in this case, maintains a constant value given by equation 2.

$$\sigma'_{hm} = \sqrt{\frac{2p\gamma}{\pi}} \qquad \dots \ 2$$

The line load p in the above equations is the quotient of the equivalent roller weight, being the sum of dead weight and centrifugal force, and the roll width. A Stothert and Pitt 54T Vibroll with an equivalent weight of 66 kN and a roll width of 1.4m was used to compact the fill, thus giving a line load p of 47.8 kN/m. Since the value of effective cohesion was small it was considered permissible to use the foregoing theory. Although the soil contained some clay, it was considered appropriate to use effective stress analysis assuming the initial pore-water pressure to be negligible. The subsequent increases in pore-water pressure due to applied loading were also assumed to be small since the low degree of saturation of 79% was taken to be indicative of low B values.

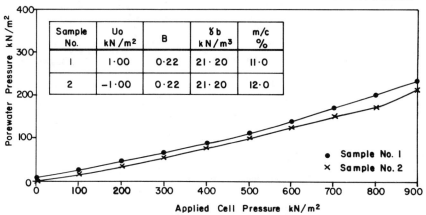

Sample No.	Uo kN/m²	B	૪b kN/m³	m/c %
1	1·00	0·22	21·20	11·0
2	−1·00	0·22	21·20	12·0

● Sample No. 1
× Sample No. 2

Fig. 5.   Pore Pressure Parameters.

These assumptions were substantiated by measuring the B values for two 100mm diameter samples.   Both samples were compacted to the mean unit weight of the in situ fill, one sample having a moisture content of 0.5% less than the mean measured value for the fill, the other having a moisture content of 0.5% greater than the mean value.   As can be seen from the test results shown in Fig.5 the average value of B was 0.22 for a range of cell pressure from zero to 900 kN/m². The average value of initial pore-water pressure, u_0, was zero.   The pore-water pressure parameter A at failure was found from undrained triaxial tests to be in the range -0.2 to 0.1, which correspond to an $\bar{A}$ value in the range -0.4 to 0.02.   In determining z_c from equation 1, a value of K_a = 0.235 was used. This was the mean of 0.22 and 0.25 which were, respectively, the K_a values with and without wall friction.   The calculated value of z_c was 0.28m.   The value of maximum horizontal pressure, determined from equation 2 was 25.6 kN/m².   Fig.6 shows the resulting pressure diagram which leads to a calculated K_a moment at the base of the stem of 568 kN.m.   It is not certain that the rotation of the wall was sufficient to generate full passive pressure in front of the wall.   The actual horizontal movement at finished ground level was 35mm which is equivalent to a rotation of 0.01 rad.   Full passive pressure for such a rotation was observed by Broms and Ingleson, (ref.5) for dense fill and is recommended by the South African Code of Practice for very stiff clays (ref.6).   However, the fill in front of the retaining wall was of firm consistency for which the South African Code states that a rotation of up to 0.05 rad is reruired to fully mobilize passive thrust. Using the graphical construction prescribed by the Code, a mobilized earth pressure coefficient of 5 corresponds to the observed rotation of 0.01 rad.   The pressure distribution relating to this condition is shown in Fig.6 from which it can be deduced that the mobilized resultant bending moment is 314 kN.m, which is almost exactly the calculated moment of resistance of the stem.   To

353

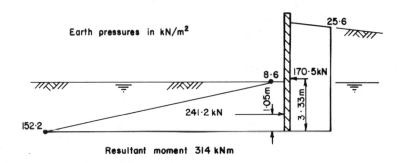

Fig.6.   Distribution of effective earth pressures
allowing for compaction.

further check the magnitude of the actual bending moment induced at the base of the stem a bending moment was backfigured
using the measured deflection at the top of the wall.   This
calculation was based on the assumption that the wall deflected
as a propped cantilever as shown in section 6, Fig.2(c).   The
elastic modulus for the concrete was derived from the empirical relationship taken from the British Code of Practice for
Reinforced Concrete (ref.7).   The design strength of the
concrete was 21 N/mm² and the actual mean cube strength was
25 N/mm².   This mean crushing strength relates to an elastic
modulus of 26.3 x 10³ N/mm² for the concrete.   The second
moment of area of the cross section was evaluated on the
assumption that the concrete was cracked to the neutral axis
and that the steel area could be expressed as an equivalent
area of concrete using a modular ratio of 7.6.   On this basis,
the second moment of area of the section was 2.37 x 10^{-3} m⁴.
Taking the measured deflection of 93mm and a cantilever length
of 7.4m, the theoretical bending moment was calculated using
$M = 3 EI\delta/L^2$.   The resulting backfigured bending moment of
318 kN.m agreed very closely with the calculated ultimate
moment of resistance of the stem section of 315 kN.m which
was derived using standard reinforced concrete theory.   Using
measured soil parameters and classical earth-pressure theory
the calculated effective active moment at the base of the wall
stem was 166 kN.m and the full effective passive moment was
462 kN.m.   No combination of these moments could be shown to
equal or exceed the ultimate moment of resistance of the stem,
which was approximately 315 kN.m.   This conclusion was arrived
at by the fact that the stem was badly cracked indicating that
the ultimate moment of resistance of the section, 315 kN.m,
had been either closely approached or achieved.   Also a
moment of 318 kN.m had been backfigured from consideration of
the measured deflection at the top of the wall.   Application
of earth-pressure theory taking account of stresses induced by
compaction shows that an effective moment of 568 kN.m could

be induced in the same sense as the active moment. Using data from the South African Code of Practice it was deduced that a passive bending moment of 254 kN.m was mobilized. The resultant of these two opposing bending moments was 314 kN.m, which is almost exactly the calculated ultimate moment of resistance at the base of the stem. The calculated deflection at the top of the wall, using this bending moment, is 91mm compared to the measured value of 93mm. Since the proposed method of analysis is not rigorous, it is obvious that the close agreement between observed and predicted performance is purely fortuitous. Conversely, the predictions made using classical theory are clearly lacking.

ACKNOWLEDGEMENT

The author wishes to thank the Directors of John Laing Construction Limited for their kind permission to publish this paper.

REFERENCES

1. "Earth Retaining Structures", Code of Practice No. 2, Institution of Structural Engineers, London, 1951.
2. PACKSHAW, S. "Earth Pressure and Earth Resistance". Proceedings, Institution of Civil Engineers, London, Vol. 25, 1946, pp.233-256.
3. BELL, A.L. "The Lateral Pressure and Resistance of Clay and the Supporting Power of Clay Foundations". Proceedings, Institution of Civil Engineers, London, Vol. 199, Part 1, 1915, pp.233-272.
4. INGOLD, T.S. "The Effects of Compaction on Retaining Walls". Geotechnique, London, Vol. 29, No. 3, 1979, pp.265-283.
5. BROMS, B., and INGLESON, I. "Earth Pressure against the Abutments of a Rigid Frame Bridge". Geotechnique, London, Vol. 21, No. 1, 1971, pp. 15-28.
6. "Lateral Support in Subsurface Excavations". Code of Practice, South African Institution of Civil Engineers, Johannesburg, South Africa, 1972.
7. British Code of Practice for Reinforced Concrete - CP110, British Standards Institute, London, 1972.

22. Failure of a large diameter pipe

C. J. F. P. JONES, BSc, MSc, PhD, FICE, Assistant Director, West
Yorkshire Metropolitan County Council

SYNOPSIS

The failure of a large diameter sewer under a motorway
embankment required extensive remedial works. The cause of
the failure and the subsequent cracking of the replacement
pipe was closely related to the geotechnical conditions.
Field measurements of the conditions near to the replacement
pipe showed close agreement with the results of finite
element analyses.

INTRODUCTION

The construction of the M62 Motorway required the replacement
of a 1.8m foul sewer at Clifton Interchange where it passed
under the motorway embankment, fig.1.   It was decided to
lay the replacement sewer (Line AA) parallel to and 14m to
the south of the existing sewer using 1.8m diameter
reinforced concrete Class 5+ pipes laid on a plain concrete
bedding. The design of the new pipe was based upon the
assumptions of the induced trench theory and the analysis for
the new sewer followed the recommendations of Special Report
No.37 ASCE Manual for the Design and Construction of Sanitary
and Storm Sewers. The design was checked using the theories
of Spangler and Marston. The analysis indicated that the
Class 5 pipe with a concrete bedding would be satisfactory
for the 21m of fill which would be placed over the pipe. No
special precautions in the design of the pipe were taken to
accommodate an adjacent bridge abutment which was to be
built with a footing level 12m above the pipe invert. This
latter decision was justified on the grounds that the load
from the bridge was less than that of the complete
embankment. The line of the original and subsequent sewers at
the interchange are shown in fig 1.

A soil survey indicated made ground of a cohesive nature to
a depth of 3.5m overlying a dense gravel. The invert level
of the new pipe varied from 2.4 – 5.5m below existing ground
level.

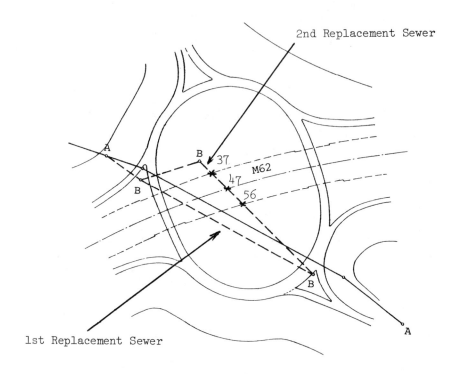

Fig.1.   Clifton Interchange - Foul Sewer

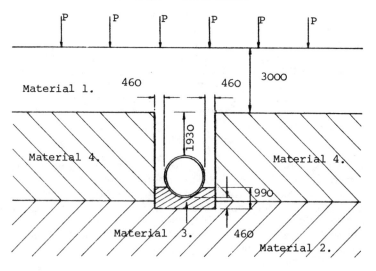

Fig.2.   Initial pipe laying conditions (dimensions in mm).

The bridge abutment was founded on a well compacted good quality cohesionless (Class A) bearing pad, with a minimum thickness of 2m. Prior to the construction of the pad, but after the pipe had been laid, it was decided that the made ground was an inadequate material upon which to construct the bridge pad. The made ground from beneath the pad was removed and replaced with a dense granular free draining material (Class D). The cohesive material was left in place immediately adjacent to the pipe. The conditions prior to and after the removal of the made ground under the bridge pad are shown in figs. 2-3.

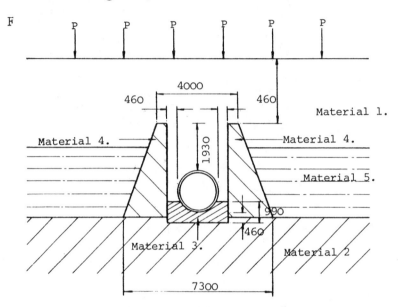

Fig.3.   Pipe laying conditions after replacement of made ground (dimensions in mm).

FAILURE OF BURIED PIPE AA

Inspection of the pipe after approximately 12m of fill had been placed revealed serious cracking. The cracks quickly developed until they represented structural failure. The failure was evident in most of the pipes under the embankment, although the major cracking occurred in the pipes under the motorway carriageway adjacent to the bridge pad. Tests on representative samples of the 1.8m pipe at the Manufacturer's works and at Bradford University established that the pipe used to form this new sewer did not fully conform with the requirements of the contract. The contract called for a class 5+ pipe, which does not feature in BS556 (1956).

Two different causes for the failure were suggested. In the first it was argued that the substandard pipes were responsible; in the second it was proposed that failure of the pipe was caused by failure in the induced trench in which the pipe had been laid, brought about by the manner in which the Class A bridge bearing pad had been constructed and by the removal of the cohesive material close to the pipe, fig 3. To determine the validity of the two conflicting failure hypotheses a finite element analysis was undertaken.

Analysis of Pipe AA The stresses in the pipe induced under the two conditions shown in figs. 2-3 were compared using an elastic isoparametric twelve node element, Ergatudis 1972. The use of a linear elastic model was justified on the grounds that the pipe under consideration was contained within the centre of the embankment. Other studies have indicated that under this particular condition linear models produce similar stress conditions to non-linear elastic and fully plastic models, Jones and Edwards (1976). The material properties, together with the loading intensities used in the analysis, are shown in Table 1.

TABLE 1. Pipe Dimensions and Properties

  i)   Internal diameter 1.8m
 ii)   Wall thickness 135mm
iii)   Concrete strength 58 N/mm^2
 iv)   Reinforcement,- 2 layers HT indented wire
        4.5×10^5 kN/m^2
  v)   Diameter of strand 8.2mm.

Material Properties

Material	Type	Density Mg/m^3	Elastic Modulus kN/M^2	Poisons' Ratio
1	Class A fill	2.02	69×10^3	0.45
2	Gravel base	2.04	69×10^4	0.2
3	Concrete bedding	2.36	24×10^6	0.17
4	Soft fill	1.88	34.5×10^3	0.45 & 0.2
5	Class D fill	1.88	13.3×10^4	0.2

Loading Intensities P

        Loading Case 1   324 kN/m^2
        Loading Case 2   417 kN/m^2

The values for the densities used were taken from site
records. The elasticity of the pipe was taken from the pipe
tests at Bradford University. Relative stiffnesses were
assumed for the gravel base, the made ground, and the Class D
granular fill. The elastic properties of the bridge bearing
pad (Class A material) were based upon the results of a
seismic investigation undertaken jointly by West Riding
County Council and Leeds University on structural back fills
and filling materials, Thompson (1971). Two values for
Poisson's ratio for the stiffness of the materials were
considered, a value of 0.45 was taken to represent the fully
saturated condition and a value of 0.2 assumed for the
drained cohesive material.

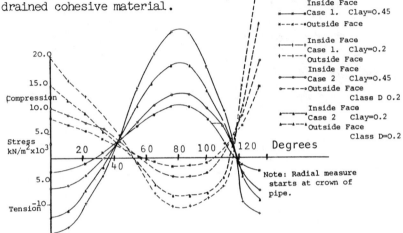

Fig.4.   Results of analysis of Pipe AA

Figure 4 shows the results of the analysis. From this it can
be seen that the stresses on the inner and outer faces of the
pipe were reduced when the conditions indicated in fig.3.
prevailed, provided that the cohesive soil (material 4) is
drained. If the made ground is saturated and the assumption
of a Poisson's ratio of 0.45 is valid, then the stresses
induced in the pipe would be greater in the conditions shown
in fig.3. than if the soft cohesive material had been left in
place. As the water table is below the level of the cohesive
soil (material 4) it is reasonable to assume that dissipation
of pore water would occur over a period of time. If it is
assumed that this dissipation results in a reduction
of Poisson's ratio then it is possible to conclude that
stresses in the pipe would increase with time and be at their
highest if the fill had been left intact, fig.2.

Pipe BB Following the failure of the first replacement sewer,
Pipe AA, it was decided to relay the damaged section of sewer
(a length of approximately 120m), on a new line using

redesigned pipes. The basic design concept of an "induced trench" was retained; the line of the second replacement sewer is shown as BB in fig.1. The second replacement entailed a major excavation of the motorway embankment and the removal and replacement of approximately 160,000 cubic metres of fill material, fig.5. The opportunity was taken to instal field instruments to study the adequacy of the specified pipes, to confirm the assumptions involved in the design of the 1.8m sewer, and to monitor its performance. Three pipes on the new sewer line, Line BB, located under the hard shoulders and under the central reservation of the motorway, were chosen as centres for the instrumentation; pipes 37, 47 and 56 fig.1. Oil filled total earth pressure cells were laid both above and to the sides of pipes 37, 47 and 56 in the configuration shown in fig.5. In addition six pneumatic piezometers were placed in the pipe trench each side of the pipes near the earth pressure cells. A supplementary soil survey was undertaken to investigate the cohesive material which formed part of the new trench. fig.5

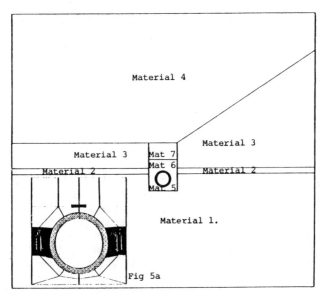

Fig.5.   Finite element analysis - Foul Sewer

Instrumentation Results

1. Pore Pressures Some initial low pore water readings were recorded immediately after installation of the piezometers. These resulted from the method adopted in placing the instruments. Once these had dissipated no subsequent pore pressures were recorded in the fill adjacent to the new sewer.

2. Earth Pressure Measurements The results of the earth
pressure measurements are shown graphically in figs 6, 7,
8.

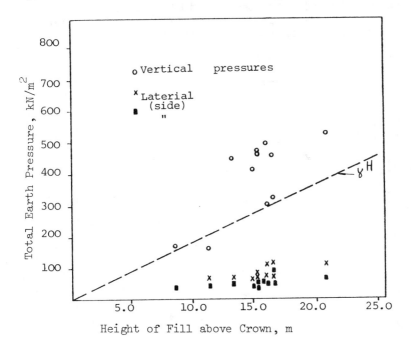

Fig.6.   Earth pressure measurements - Pipe 37 line BB

The results are consistent and close correlation was observed
between the gauges recording pressures adjacent to pipes 37
and 56 where the embankment geometry is the same for each
pipe. The measurements indicated that the earth pressures on
the side of the pipes were low and also that the earth
pressures acting vertically on the pipes were in excess of
the overburden pressures. Only when the filling was within
the confines of the new trench was the vertical pressure
less than the theoretical gravitational load (ɣ H), fig 8.
The low pressures recorded at the side of the pipe suggest
that the backfilling adjacent to the pipe was not well
compacted and, therefore, provides little support for the
pipe. This is understandable as the material forming the
sides of the trench was of a loose friable nature. In
addition the trench required sheeting to ensure the safety of
the works and many large voids were observed behind the sheet
piles. The Contractor elected to withdraw the sheeting once
the trench had been filled to original ground level. The
conclusion from these observations was that the analytical
conditions assumed in the "induced trench" theory had not
been fully obtained.

CRACKING OF PIPE BB

As filling above the pipe BB proceeded cracks were seen to develop in individual pipe sections in two areas centered on pipes 17 and 31 at a position where the line of the second replacement sewer, Line BB, crossed the line of the original sewer, fig.1.

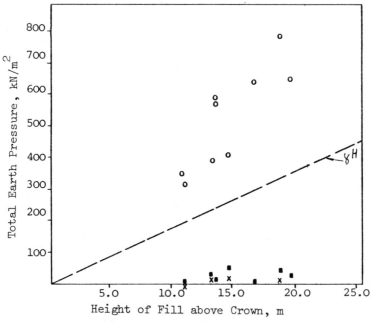

Fig.7. Earth pressure measurements - Pipe 47 line BB

Analysis of BB Whilst laying the replacement sewer line BB, a second finite element analysis was undertaken to determine the stresses within the pipe. The analysis took the form of a parametric study using the same twelve node isoparametric element used with the analysis of pipe AA. The geometry of the analysis is shown in fig.5. and the material properties used are given in Table 2. The results of the analysis were compared with the readings from the field instrumentation.

Three analyses were undertaken. In the first, Analysis A, it was assumed that the material immediately surrounding the concrete sewer (material 6, fig.5.) was well compacted. This analysis represents the case where ideal trench conditions were achieved during back filling. The results of the analysis are shown in figs.9,10. These indicate a marked increase in the vertical load immediately above the

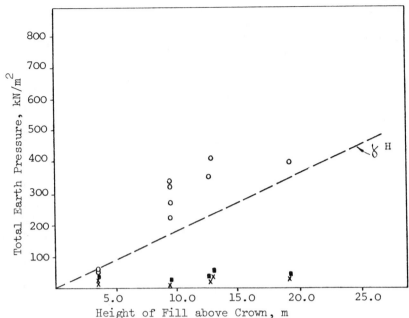

Fig.8.   Earth pressure measurements - Pipe 56 line BB

TABLE 2.

Analysis of Line BB - Material Properties

Material	Type	Density Mg/m^3	Elastic Modulus kN/m^2	Poisson's Ratio
1	Foundation Material	2.09	69×10^4	0.2
2	Soft Material	1.88	34.5×10^3	0.2
3	Previous fill	2.06	69×10^3	0.2
4	Fill	2.09	69×10^3	0.4
5	Concrete bedding	2.36	24×10^6	0.17
6	Fill adjacent to pipe	2.04	$6.9-69 \times 10^3$	0.4
7	Fill at top of trench	2.0	$0.1-69 \times 10^3$	0.4
8	Pipe	2.36	24×10^6	0.17

pipe similar to those recorded by the earth pressure cells. The maximum vertical load indicated by the analysis is 690 kN/m^2 and this falls between the minimum and maximum values of 550–890 kN/m^2 recorded in the fill once the embankment was completed. The lateral pressures adjacent to the pipe also indicate correlation with the measured values although the boundary between the backfill and the concrete base distorts the results. The analysis suggests a maximum internal tensile stress at the soffit of the pipe of 4500kN/m^2. Large compressive stresses on the inside of the pipe adjacent to the quarter points are indicated.

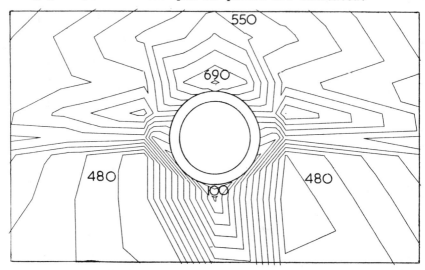

Fig.9. Well Compacted Backfill - vertical stresses (kN/m^2)
Maximum soil pressure at crown 690 kN/m^2

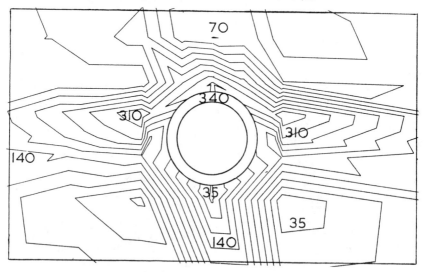

Fig.10. Well Compacted Backfill - horizontal stresses (kN/m^2)

The second and third analyses (B and C) were undertaken to
study the effects of extreme differences in backfilling
conditions. In Analysis B, material 6A was assumed
to be soft and compressible representing a condition of poor
compaction and the presence of voids, fig.5a. The results
of the analysis are shown in figs.11,12. These show that the
maximum vertical pressure acting on the pipe in these
conditions would be in excess of 890 kN/m² whilst the
lateral pressures on the side of the pipe would be very
small, less than 10 kN/m². At the same time the maximum
internal tensile stress in the pipe would be in excess of
6400 kN/m², a stress which would also be repeated on the
outside face of the pipes near the quarter points. These
results compare well with the field measurements adjacent to
pipe 47.

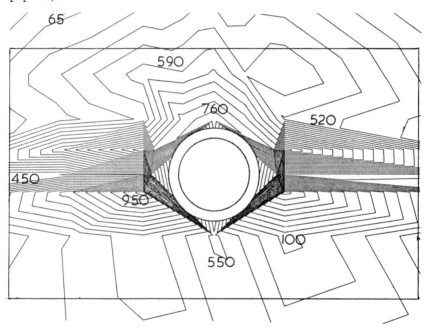

Fig.11. Poor side compaction – vertical stresses (kN/m²)
Maximum soil pressure at crown is greater than
890 kN/m².

Difficulties were experienced with the backfilling in the
areas of the old sewers and it is probably that little or no
effective drainage existed in this area. If this is the
case then the analytical solution represented by Analysis B
is relevant.

The third analysis C, represented the other extreme in the
backfill conditions in which a soft material (material 6B,
Fig 5a) was positioned immediately above the pipe, whilst

367

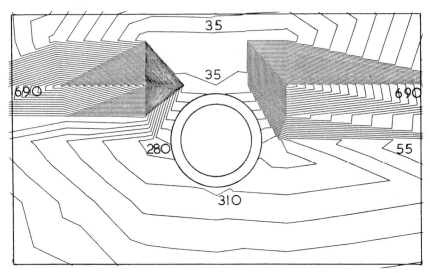

Fig.12. Poor side compaction - horizontal stresses (kN/m^2)

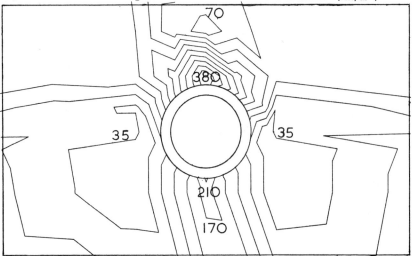

Fig.13. Soft material above pipe - vertical stresses kN/m^2)

the material to the side of the pipe was assumed to be well compacted. This analysis represents the text book "bale of straw" configuration. The results are shown in figs.13,14.

CONCLUSION

The designs for the replacement sewer at Clifton Interchange rely upon support from the adjacent soil to reduce stresses within the pipe sections. In the case of the first replacement sewer it is probable that failure occurred due to lack of strength of the pipe rather than through loss or lack

368

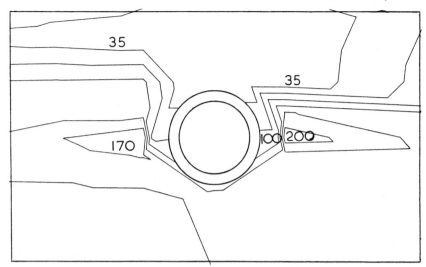

Fig.14, Soft material above pipe - horizontal stresses
kN/m^2)

of support from the surrounding soil. In the case of the
second replacement pipe there is clear evidence that the
induced trench was not obtained where the pipe crosses the
line of the two abandoned sewers. In this position the
relief assumed from the adjacent supporting soil was lacking
and is the probable cause of the cracked sections of the full
strength pipe.

REFERENCES
1. JONES C.J.F.P. & Edwards L.W. 1876 "Finite element
analysis of the Trent Embankments" Report to NERCU, WYMCC,
Wakefield
2. THOMPSON J.F. 1971 "A seizmic investigation
of backfill materials" MSc Dissertation. Department of Earth
Leeds University.

ACKNOWLEDGEMENTS
The Author wishes to acknowledge the assistance of the
members of the Structural Engineering Unit of West Yorkshire
Metropolitan County Council in preparing the paper which is
presented by permission of Mr. J. Jefferson, Director
(Transport) Harrogate and Mr. J.A. Gaffney, Director of
Engineering Services West Yorkshire Metropolitan County
Council. The views expressed in the paper do not necessarily
represent the official views of the Department and
West Yorkshire Metropolitan County Council.

23. Ground movements from a deep excavation in sands and interbedded clay

T. D. O'ROURKE, Associate Professor, School of Civil and Environmental Engineering, Cornell University

SYNOPSIS. The paper presents and discusses the ground movements caused by an 18-m-deep braced excavation in sands and interbedded stiff clay. The excavation was part of the construction of a metro station in Washington, D.C. The maximum horizontal and vertical ground movements were relatively small, being on the order of 18 mm. Nevertheless, significant damage was observed in adjacent buildings, one of which was a four-story masonry structure which was underpinned before excavation was started. The characteristics of the ground movements and their effects on adjacent structures are evaluated.

INTRODUCTION

1. This paper presents a case history of ground movement caused by a deep braced excavation and its influence on adjacent buildings. The open cutting was performed during construction of the Gallery Place Station on the Washington, D.C. Metro System. The 18-m-deep excavation led to 18 mm of settlement in a nearby masonry structure. The nature of the ground movements and the measures taken before construction to safeguard the buildings raise important questions about soil-structure interaction, methods of protection, and building response in terms of serviceability.

2. The case history includes a description of the location and surroundings of the excavation and a discussion of the soil profile and construction methods. Measured soil displacements are shown at various stages of construction. The ground movement effects on the adjacent buildings are evaluated, and recommendations are made for protecting structures during future excavations.

CONSTRUCTION SITE AND SOIL CONDITIONS

3. Fig. 1 shows a plan view of the excavation and surrounding buildings. The excavation was approximately 21 m wide with a local width of 37 m at the intersection of 7th and G Sts., where a cross-over for two rapid transit lines was planned. The most prominent structure in the vicinity of the excavation was the Fine Arts Building, located as close as 1.5 m from the southern edge of the cut. The Fine Arts

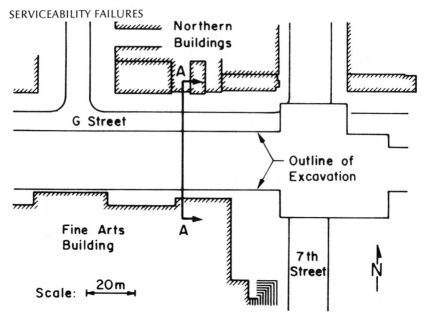

Fig. 1. Plan view of excavation and surroundings

Building houses the U.S. national portrait collection. It is a four-story masonry structure which was built principally in the years from 1836 to 1856. The building is composed of brick walls and vaulting with wrought iron beams for floor support. The building contains a 4-m-deep basement and derives its bearing from rubble stone foundations. Approximately 15 m north of the excavation were two and three-story brick buildings, with 200 to 300-mm-thick bearing walls typically separated by 6 m on center. The structures were built during the 1890's with timber joists to support the floors and 2.5-m-deep basements.

4. Before construction, the Fine Arts Building was underpinned with 300-mm-diameter, concrete-filled pipe piles which were jacked in 1.2-m sections to tip depths of 20 and 22 m below the ground surface. The design capacity of each pile was 712 kN, and each was tested to 150% design capacity during installation. No special protective measures were taken for the buildings north of the excavation.

5. The excavation was undertaken in terrace deposits of Pleistocene age. At the site, the Pleistocene soils lie uncomfortably on hard clays and dense sands of Cretaceous age at a depth of approximately 24 m below the ground surface. The depth to bedrock is approximately 39 m. The soils encountered at the excavation are illustrated in Fig. 2, which represents the soil profile at cross-section A-A in Fig. 1. Five layers of soil are shown in sequence from the street surface. They are the Upper Brown Sand, Middle Gray Clay, Gray Sand and Interbedded Stiff Clay, Lower Orange Sand, and Hard Cretaceous Clay. Typical values of the standard penetration resistance are given for each stratum. A 4.5-m-thick

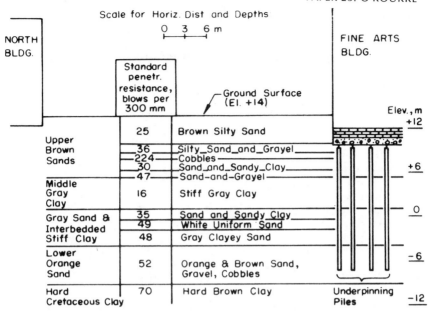

Fig. 2. Profile view of soils at cross-section A-A

layer of gray clay was encountered below the Upper Brown Sand. The clay had an unconfined compressive strength of approximately 145 kN/m^2.

CONSTRUCTION SEQUENCE

6. The water table in the Pleistocene deposits was initially at a depth of 11 m below the street surface. Before excavation, 300-mm-diameter wells containing submersible pumps were established along the line of intended construction, and the water table was drawn down by pumping to a depth of 21 m.

7. Soldier piles (H-piles, 610-mm-deep × 1.75 kN/m) were installed in pre-drilled holes on 2-m centers and set to a tip depth 4.5 m below the final bottom of the excavation. The sides of the braced cut were lagged with oak boards installed behind the flanges of the soldier piles. Five levels of cross-lot braces were installed on 4-m horizontal centers. With the exception of the first level braces, which were not preloaded, each strut was jacked to one-half its design load and wedged. Excavation was taken to a maximum depth of 18 m, after which a 2.5-m-thick concrete invert was constructed. The fifth level braces were then removed, and the north and south walls of the subway were built. Each reinforced concrete wall was approximately 4 m high and varied in thickness from 3 m at its juction with the invert to 1.5 m at its upper portion. Upon construction of the subway walls, the fourth and third level braces were removed and the concrete arch was poured. Finally, the second and first level braces were removed as granular fill was placed and compacted to the base course elevation of the street.

MEASURED GROUND DISPLACEMENTS

8. Ground movements were measured within and adjacent to the excavation by means of inclinometers, heave point extensometers, optical leveling, and survey taping. A description of the field instrumentation and observational methods is given by O'Rourke (ref. 1) and O'Rourke and Cording (ref. 2). Measured ground displacements and associated stages of construction pertaining to cross-section A-A in Fig. 1 are summarized under the following headings.

Displacements for Excavation Depth of 10.5 m

9. Fig. 3 summarizes the observed movements just before installing the third level braces. Two sets of inclinometer measurements at the edges of the cut are shown: 1) horizontal displacements, designated by A, which pertain to an excavation depth of 6 m, and 2) horizontal displacements, designated by B, which pertain to an excavation depth of 10.5 m. The placement and compaction of a 3.5-m-deep layer of fill on the south side of the excavation contributed to a northward shift throughout the upper portions of the excavation walls. The horizontal displacement profiles are approximately triangular in shape, and are indicative of the cantilever displacements of the walls that occurred before the second level braces were installed. The first level braces were not preloaded and some displacement adjacent to them occurred as loads developed in response to the open cutting. Heave point extensometers in the cut indicated an upward movement in the clay stratum of 13 mm, whereas those north of the excavation showed a maximum settlement of 9 mm.

Displacements for Excavation Depth of 18 m

10. Fig. 4 summarizes the measured movements shortly after the excavation had been extended to subgrade and the fifth level braces had been preloaded. Two sets of inclinometer measurements at the edges of the cut are shown: 1) horizontal displacements, designated by C, which pertain to an excavation depth of 15 m, and 2) horizontal displacements, designated by D, which pertain to an excavation depth of 18 m. Inward bulging of the soldier piles increased to over 16 mm beneath the fifth level braces. Significant lateral displacement occurred to the base of the soldier piles. Surface settlement immediately adjacent to the north side of the cut increased to 13 mm. During this period, the underpinned north wall of the Fine Arts Building settled 13 mm. A 6-mm separation between the sidewalk and building, 15 m north of the cut, was observed. Separations between sidewalk slabs, 13 mm wide, were observed 12 m north of the cut.

Displacements after the Bottom Braces Removed

11. Fig. 5 summarizes the measured displacements after the fifth, fourth, and third level braces were removed. Two sets of inclinometer measurements at the edges of the cut are shown: 1) horizontal displacements, designated by E, which

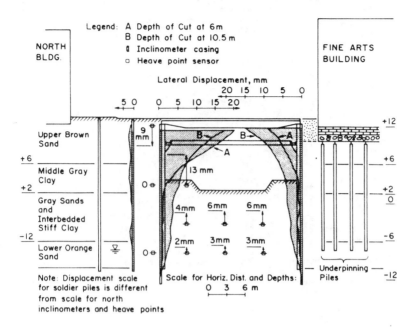

Fig. 3. Displacements for excavation depth of 10.5 m

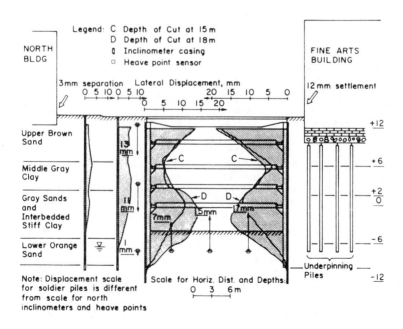

Fig. 4. Displacements for excavation depth of 18 m

375

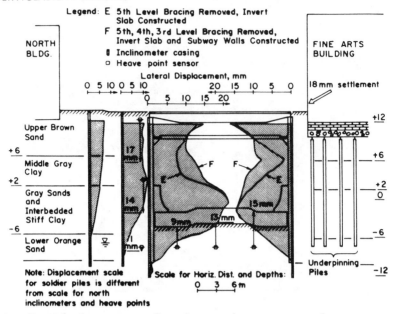

Fig. 5. Displacements after bottom braces removed

pertain to the construction of the concrete invert and subsequent removal of the fifth level braces, and 2) horizontal displacements, designated by F, which pertain to the construction of the north and south station walls and subsequent removal of the fourth and third level braces. The removal of the braces showed a consistent pattern with respect to displacements of the excavation walls. Following the removal of a bracing level, maximum inward movements were approximately 5 mm. The movements were localized between the brace level immediately above the removed level of support and a distance of 3 m below the removed braces. The invert and subway walls restrained lateral movements in the lower portions of the cut. In response to the weight of the subway structure, the heave point sensors beneath the slab showed a maximum settlement of 3 mm. The settlement of the Fine Arts Building increased to 18 mm, and 3 to 6-mm-wide cracks were observed in a brick building, 15 m north of the excavation.

DAMAGE IN ADJACENT BUILDINGS
12. Problems in the Fine Arts Building developed shortly after the excavation had been extended to its maximum depth, when ornamental tiles fell from the walls in a second floor room adjacent to the cut. The tiles had been attached to the walls without anchors, using gypsum cement to establish a bond with the brick wall surfaces. Accordingly, they were susceptible to cracking and detachment along their cement contacts. Cracks, varying from 1 to 6 mm, were observed in floors, walls, and arches. The appearance of new cracks and the growth of old ones continued as the fifth, fourth, and third level braces were removed in the adjacent cut. Several

rooms in the northeast section of the building were closed to the public during this time.

13.   Cracks and separations of about 3 mm in thickness were observed in a three-story brick building north of the cut shortly after the excavation had reached its maximum depth. The cracks and separations were located near the front facade wall of the structure.   One conspicuous line of cracking appeared in the floor of the building's display window.   Since the building was used as a commercial art gallery, the damage drew considerable attention.   The crack opened from 3 to 6 mm as the fifth, fourth, and third level braces were removed in the adjacent cut.

GENERAL PATTERNS AND MAGNITUDES OF MOVEMENT

14.   The most important movements were those which developed during excavation to subgrade, particularly those associated with inward bulging of the walls as the excavation was taken from a depth of 12 to 18 m.   These deep displacements resulted in surface settlements as far as 28 m from the cut and caused distortion in the buildings both south and north of the excavation.

15.   The displacements which occurred during brace removal were a significant portion of the total movement, representing roughly 40 percent of the total volume of inward wall displacement.   Fig. 6 shows the horizontal displacement profiles at various stages of the excavation and station construction.   The darkened areas represent the volume of lost ground caused by brace removal after the maximum depth of the cut had been achieved.   Although these displacements were associated with continued deformation of the neighboring buildings, their area of influence diminished steadily as successively higher levels of braces were removed.

16.   The settlements related to the excavation adjoining the Fine Arts Building as well as those associated with other portions of the metro construction in similar soil profiles are summarized in Fig. 7.   The settlements and distances are expressed in dimensionless form as fractions of the maximum excavation depth.   In comparison to the settlements summarized by Peck (ref. 3), the settlements are small.   The relatively small movements and the corresponding damage to adjoining structures, despite protective measures, raises important questions with respect to the response of the underpinned Fine Arts Building and the specific nature of the building distortion north of the cut.

BUILDING RESPONSE TO GROUND MOVEMENT

17.   Fig. 8 compares the settlement profile of the pile-supported building with the settlement profile of the ground surface north of the excavation.   The settlement corresponds to excavation at the west end of the Fine Arts Building, but typifies behavior along the entire north side of the structure.   The zero datum for this settlement was established when the excavation at the west end was approximately 9 m

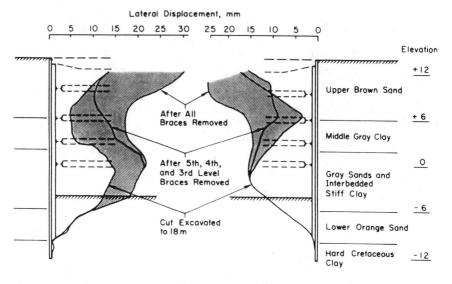

Fig. 6. Displacement profiles at various stages of construction

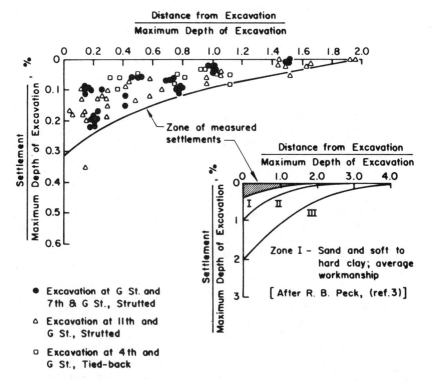

Fig. 7. Settlements expressed in dimensionless form

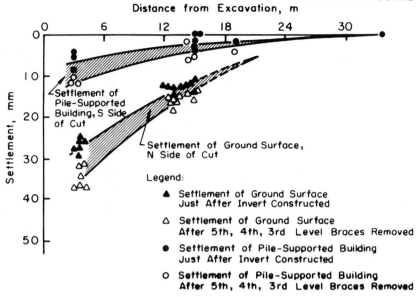

Fig. 8.   Settlements of Fine Arts Building and ground surface

deep.   Consequently, these movements occurred in response to excavation and construction in the bottom 9 m of the cut.   As the bottom three levels of braces were removed, maximum ground settlement increased from 27 to 34 mm, and maximum settlement of the pile-supported building increased from 6 to 12 mm.   Measured settlement of the building was approximately 40% of the ground surface settlement north of the excavation.

18.   The measurements suggest that down-drag of the soil as well as ground movement at and slightly beneath the pile tips resulted in settlement of the pile-supported building.   Although the underpinning did reduce the magnitude of potential settlement, it did not prevent movements which impaired the use of the building.   Field measurements showed horizontal soil movements which were comparable in magnitude to settlements.   These movements, in conjunction with building cracks and separations showing net extension, indicate that underpinning was not effective in protecting the building from horizontal soil strains.

19.   Fig. 9 shows a profile view of the damaged brick building north of the excavation.   As is common for many commercial buildings, an underground vault was connected to the structure as a means of making freight deliveries from the street.   The vault was constructed with concrete walls and a concrete floor slab.   The front wall of the vault was located 7.5 m from the edge of the excavation.   The measured vertical and horizontal movements at the ground surface adjacent to the vault are shown.   Although maximum settlement of the building was only 3 mm, maximum settlement of the vault was as high was 12 mm.   No cracks or signs of distortion were apparent in the vault.   This observation and the linear dis-

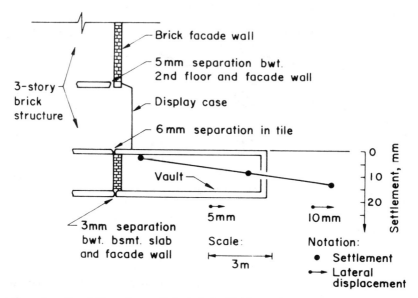

Fig. 9. Profile view of brick building displacements

tribution of settlement suggest that the vault rotated and pulled away from the building as a rigid body, thereby concentrating deformation at its junction with the front facade wall. An inspection of the building showed separations between the facade wall and all floors of the structure at exactly the same location as the 6-mm-wide crack in the floor of the display window.

CONCLUDING REMARKS

20. The term "failure" in reference to building damage should be used cautiously. Judgments about damage will depend on the use of a given structure and will generally involve consequences which are not catastrophic or even unsafe. In this case history, the building response did not involve structural damage, but did involve architectural damage where a pleasing display was necessary for proper use of the structure. Moreover, deformation of the Fine Arts Building caused ornamentation to become unstable, thereby posing a threat to the art and the people who used the facility.

21. A monumental masonry building represents a special class of structures. This type of structure should be reviewed on an individual basis, with special attention to potentially unstable areas of the architectural cladding and ornamentation.

22. It is clear from the case history that underpinning was not successful in preventing damage. Whereas underpinning is effective in reducing settlement, it carries some noteworthy disadvantages. Small-diameter piles are not sufficiently stiff to resist horizontal soil movement, and down-

drag from vertical soil movement may add considerably to their axial loads. The underpinning procedure, itself, may be a source of local deformation as underpinning pits are excavated and piles jacked into place.

23. The most effective means of protection are those which prevent or limit movements at the source. The use of concrete diaphragm walls in conjunction with preloading braces and controlling the excavation depth beneath the lowest previously installed level of support are measures that can reduce ground displacements to levels acceptable for masonry buildings.

24. The brick building north of the excavation was damaged even though its maximum settlement was only 3 mm. The damage was related to the settlement and horizontal displacement of an underground vault. This somewhat unlikely source of deformation emphasizes the need to consider structural details and to properly characterize structures as part of a comprehensive site exploration in urban areas.

REFERENCES

1. O'Rourke, T. D., Ground movements caused by braced excavations, Journal of the Geotechnical Engineering Division, ASCE, Vol. 107, No. GT9, 1981, Sept., 1159-1178.

2. O'Rourke, T. D. and Cording, E. J., Measurement of strut loads by means of vibrating wire strain gages. Performance Monitoring for Geotechnical Construction, ASTM STP 584, 1975, 58-77.

3. Peck, R. B., Deep excavations and tunneling in soft ground, Proceedings Seventh International Conference on Soil Mechanics and Foundation Engineering, State-of-the-Art Vol., 1969, 225-290.

Discussion on Papers 20–23

DR R. C. OWEN, British Gas Corporation, Newcastle-upon-Tyne

Mr Addy's paper raises three matters on which I should like to comment.

Firstly very briefly the British Gas Corporation (BGC) mains network separates into high pressure transmission mains of steel and low pressure distribution mains, typically of cast iron and polyethylene. The high pressure system is subject to very rigorous inspection both during and after construction with regular surveys along the route. The photograph in Paper 20 suggests a cast iron main for the low pressure distribution system not the transmission system.

Secondly in Fig. 1 of Mr Addy's paper there is no reference to trench depth. Certainly BGC are aware of the large and damaging ground movements associated with some deep excavations made for sewer renewal. It was for this reason that an extensive series of joint field trials was carried out by the BGC's Engineering Research Station (ERS) and the Water Research Centre. This culminated in the adoption by BGC and the Water Authorities Association of the Model Consultative Procedure for Deep Excavations, which sets out safe proximities and renewal lengths for cast iron mains adjacent to or crossing deep excavations.

Finally I should like to draw attention to the programme of research and development being carried out by BGC into the reinstatement of trenches, some of which is detailed in a forthcoming paper (ref. 1). For two years now the ERS has been running training courses in highway engineering for all levels of distribution staff. BGS is developing tools and equipment to enable more efficient excavation causing less damage to the adjacent pavement structure. This has already resulted in greater use of pavement saws and excavators fitted with narrow buckets. In the future road planes and chain trenchers might be used more frequently. For compaction, vibrotampers have been developed with narrow feet and extended legs for working in confined spaces. Quality control of graded materials and supervision of compaction in layers is a problem and the ERS is developing ways of stabilizing and improving backfills to render them less sensitive to

Fig. 1

Fig. 2

Fig. 3

compaction. To assess the condition of the existing urban road network and to monitor the performance of different reinstatement options the ERS has developed a road profile gauge (Figs 1 and 2), to measure surface settlement, and a portable stiffness beam fitted with a simple electrolytic transducer (Fig. 3) for use with any suitably ballasted vehicle.

Reference
1. OWEN, R. C. and HOWE, M. Research into trench reinstatement materials. Highw. Transpn, 1985, vol. 32, no. 4, 9-14.

MR N. ADDY

The terminology 'high pressure' gas main was tendered by North Eastern Gas Board officers attending the incident. I have continued this reference and apologize if there has been any misrepresentation.

The trench condition featured in Fig. 1 of the paper refers to a trench typically 2-2.5 metres in depth through the failed section.

It is encouraging that British Gas Corporation are conducting research into the effect of their apparatus on the parent highway pavements and formation.

I reiterate that the main laying operation must be a total design concept and liaison between the statutory undertakers and the highway authority is a prerequisite. The fact remains that damage to pavement structures is rife in metropolitan areas and the informal openings are the prime source of irritation and concern to the highway authorities. I welcome the concept of formal training of British Gas Corporation manual workers in highway engineering but regret that the end product remains to be proved.

Evolving techniques such as diamond sawing are being evaluated in West Yorkshire but reservations remain concerning the effectiveness of sharp arris inlays against the loss of interparticle friction and bond previously associated with other methods of trench cutting. The main benefit arising from saw cutting is the arrest of lateral distribution of impact energy in breaking out the bituminous surfacings.

Continuous shear faces through the depth of the trench is a retrograde step. Paving course joints are traditionally offset, distributing shear stress and limiting strain in high modulus materials such as asphalt. It is accepted that the narrow trench condition considerably improves stability at the trench invert; however, slip plane formation along the trench wall is a function of dynamic loading from the adjacent carriageway and is possibly initiated during the excavation rather than after.

Low compaction fraction backfill materials are frequently. accompanied by high permeability properties and the risk of

subsoil degradation with particle migration in granular subsoils. This could be resisted by geotextile separators.

High energy compactors rarely produce better than 8-10% air
void content in trench conditions. The results from low
energy vibrotampers must be significantly inferior.

MR M. P. O'REILLY, Transport and Road Research Laboratory,
Crowthorne

A recent estimate by Mr Clare of Southern Electricity Board
put the replacement value of statutory undertakers'
distribution plant at £117 billion, and the greater part of
this would be located in highways. The value of the road
system in this country must be of a similar order of magnitude
and it is not surprising that when these two elements rub
together there is friction. There is evidence that the
standard of reinstatement of sewer trenches is poor (ref. 1)
and that ground relaxation is inevitable in the vicinity of
trenches (refs 2 and 3): trenching in roads must reduce the
life of the road surface by 10-20%.

With regard to reinstatements themselves my recent
experience in the street where I live is that the highway
authority is not much better than the statutory undertakers.
However, I do not think the position is helped by your remark,
Dr Owen, that a road with perhaps 2-3 in surfacing and no more
hardly complied with Road Note No. 29. Can you tell us what
proportion of the cast iron mains in the British Gas
distribution system complies with the current British
Standard? The important point for both of us to remember is
that perhaps half of our infrastructure was constructed before
the turn of the century and our task as engineers is to make
the best of this invaluable legacy from our Victorian
forebears.

References
1. FARRAR, D. M. Some measurements of differential
    settlement above reinstated sewer trenches. Highw. Publ.
    Wks, 1980, vol. 49, 30-32.
2. SYMONS, I. F., CHARD, B. and CARDER, D. R. Ground
    movements caused by deep trench construction. Proc. Conf.
    Repair of Sewerage Systems 1981, Institution of Civil
    Engineers, London, 1981.
3. TOOMBS, A. F., McCAUL, C. and SYMONS, I. F. Ground
    movements caused by deep trench construction in an urban
    area. Transport and Road Research Laboratory Report 1040,
    Departments of the Environment and Transport, 1982.

MR N. ADDY

It is considered that the deterioration in life expectancy is
significantly greater than 20% in metropolitan areas.

Strengthening overlays often do not survive more than 6 months before they are disrupted by statutory undertakers.

It is not suggested that highway authorities are beyond reproach. They should be more aware of the damage factor in opening the highway. The number of openings will be insignificant compared with the aggregate arising from statutory undertakers' activities.

MR D. F. LAWS, Cambridgeshire County Council, Huntingdon

I should like to endorse the points made by Mr Addy regarding the deleterious effects of trenching work in the highway.

Trench cutting, coupled with subsequent consolidation of poorly compacted backfill, leads to movement, settlement and cracking and a general loss of integrity of the adjacent pavement structure.

Even saw cutting, as adopted by the British Gas Corporation can allow lateral movements with the development of adjacent cracking (Fig. 1).

Fig. 1

These movements, although minor in geotechnical terms, cause a substantial amount of damage countrywide and the local authorities bear the cost.

Improvements in trench backfill compaction and in pavement materials appropriate to trenching works will be welcome, but

I must take issue with Dr Owen of the British Gas Corporation regarding the use of wet-mix macadam as trenching roadbase. The necessarily tight specification limits for moisture content are unlikely to be consistently met on typical public utility works resulting in a poorly compacted unbound layer unable to carry traffic loading or to give support to the adjacent pavement layers.

MR N. ADDY

Perhaps of more significance than lateral movement is the danger of longitudinal creep, particularly on gradients, which may occur under the braking effort movements of heavy goods vehicles. The low friction face presented by diamond sawing provides little resistance to creep.

I agree with your comments, particularly with respect to wet-mix macadam. There is a real danger of cavity bridging.

MR B. L. PARKER, Surrey County Council, Guildford

Although I support other local authority engineers present who have expressed concern about statutory undertakers' works, I commend the research being undertaken by the British Gas Corporation under Dr Owen's direction at Newcastle and I wish that other undertakers, notably British Telecom and the cable workers would apply the results to their own daily operations. With respect to the 'narrow' trenches proposed I do not agree with Dr Owen's ideas which implied a continuous vertical face between the base of the trench and the carriageway surface: such a configuration required that all the strain between the trench backfill and the adjacent existing ground be resisted by the bond at the interface (Fig. 1(a)). This is not possible - the only satisfactory way to reinstate the carriageway is by means of a 'stepped' configuration as shown (Fig. 1(b)).

After the Clarkston Toll disaster the Department of the

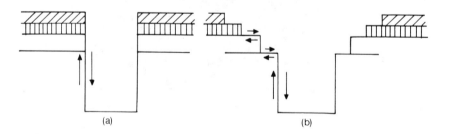

(a)   (b)

Fig. 1. (a) Strain resisted by the vertical interface bond alone and (b) strain resisted by the vertical and interlayer bonds

Environment issued Circular No. 70/73, but 11 years later many
engineers are unaware of the dangers and warnings. Fig. 2
shows a typical operation in a built-up area having basement-
type properties adjacent to the works. In this example the
remedial works required the removal of the bound layers and
the construction traffic is running on a weak sub-
base/subgrade. In such instances, in cold weather, brittle
cast iron pipes might fracture allowing gas to escape into
adjacent basements. Geotechnical engineers present are
probably best qualified to assess the risk in any situation
and should therefore be on their guard.

Fig. 2

MR N. ADDY

I have reservations on the practicalities of achieving
satisfactory densities in narrow trench conditions
particularly at depths exceeding 750 mm. The stepped
configuration prescribed by many highway authorities ensures
both strain relief in surfacings and a greater surface area
for sealing against the ingress of water into the unbound
pavement courses.

DR R. T. MURRAY, Transport and Road Research Laboratory,
Crowthorne

Dr Ingold, you have described the need for taking account of
compaction plant in the design calculations: do you not
perhaps consider that restricting the weight of plant close to
the wall might be a better solution?

DR T. S. INGOLD

The magnitude of lateral earth pressures induced by compaction will, among other things, be a function of the effective weight of the compaction plant and the proximity of this plant to the back of the retaining wall or structure. A more general expression for lateral earth pressure induced by compaction is (ref. 1)

$$\sigma'_{hm} = \frac{w}{w+d} \left(\frac{2p\gamma}{\pi}\right)^{1/2}$$

where w and d are the rolling width and the distance of the roller from the back of the wall respectively. The other variables are defined in the paper. As can be seen from this expression compaction-induced lateral pressure decreases as the roller is distanced from the back of the wall. A more dramatic reduction in pressure can be obtained, where possible, by turning off the vibrating mechanism when a vibratory roller is to be used close to the back of a wall.

It would be uneconomic to design a wall to sustain unnecessarily high lateral pressures induced by using a 'heavy' roller close to the wall. The objective of the paper was merely to present the engineer with a method of assessing the effects of compaction whereby a sensible choice of compaction plant and mode of use can be made. The purpose of the case history was to demonstrate the type of failure that can occur if compaction effects are ignored.

Reference
1. INGOLD, T. S. (1980). Discussion. J. Geotech. Engng Div. Am. Soc. Civ. Engrs, vol. 106, GT9, 1062-1068.

MR I. ELLIS, Fondedile Foundations Ltd, Yiewsley

I wish to comment on a statement you made in your paper, Professor O'Rourke.

In your concluding remarks and description of the settlement of the Fine Arts Building in Washington, DC, you stated that small diameter piles (mini-piles) are not sufficiently stiff to resist horizontal soil movement.

I suggest that had the piles adjacent to the proposed excavation been in the form of a reticulated Pali Radice structure both horizontal and vertical movement would have been considerably reduced, if not eliminated altogether.

Pali Radice piles are essentially small diameter grouted piles, typically 100-200 mm in diameter, normally constructed by rotary drilling methods, and reinforced with either a single bar or a small cage.

A reticulated Pali Radice structure is a three-dimensional lattice soil/pile structure built directly in the soil in situ according to a predetermined plan, with many Pali Radice forming a special resisting network. In the resistant complex, the piles are lines of force, while the soil encompassed supplies the weight, rather like a gravity wall. The whole is intended to resist compressive as well as tensile and shear forces.

Figure 1 is an example of a reticulated Pali Radice structure to an old building. If necessary the piles can be constructed directly through the building's foundation.

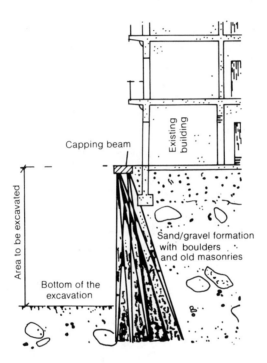

Fig. 1. Typical scheme of reticulated Pali Radice

Reticulated Pali Radice structures have been used successfully for more than 20 years to protect many old buildings when excavation, either in trench or tunnel, is to be carried out in close proximity. In all cases where the walls have been monitored it has been difficult to measure the very small movements and no damage to the 'protected' structures have occurred.

Reticulated walls have usually been used where excavation depths are up to 12 m and I recognize that in the 18 m deep excavation adjacent to the Fine Arts Building a reticulated Pali Radice structure would have been expensive and your suggestion to use a diaphragm wall in conjunction with preloading braces and controlled excavation would have been adequate.

PROFESSOR T. D. O'ROURKE

Piles in the form of a reticulated Pali Radice structure could
have been sufficient to reduce horizontal as well as vertical
soil movement. With reticulated pile structures, the ground
is reinforced both horizontally and vertically so that, by the
nature of the system, restraint against settlement and lateral
displacement would be expected. There are other construction
and reinforcing schemes that would have the same effect, such
as the concrete diaphragm wall approach referred to in the
paper. A prime consideration in choosing the support system,
however, is cost. Ground engineering always is based on a
choice of alternative measures that are appropriate for the
ground conditions and structures at a given site. A support
scheme may work in concept, but be disadvantageous in cost
when compared with other schemes. As you have pointed out,
reticulated Pali Radice structures generally have been used
for excavation up to 12 m deep; their cost competitiveness for
deeper cuts would have to be carefully examined.

MR F. HUGHES, Cementation Piling and Foundations,
Rickmansworth

Professor O'Rourke, you have not indicated what, if any,
remedial measures were undertaken at the metro station in
Washington, DC. We know that compaction grouting whereby a
toothpaste-like mix is injected with strict control of
pressure and level instrumentation to restore or maintain
structures in position is quite commonly used in the USA. Was
it considered for this problem? We have used the process in
the UK but there is much more experience in the USA and it
would be interesting to learn more.

PROFESSOR T. D. O'ROURKE

The remedial measures undertaken were to close portions of the
Fine Arts Building until a thorough inspection had been made
and to initiate a monitoring programme with special
instrumentation to detect further building response. In
addition guide-lines which had existed for the depth of
excavation and the installation of the support were strictly
enforced.
    Because of the small magnitude of the soil movements, damage
to the building was in the form that we would typically regard
as architectural. The building was not affected by
distortions which were adverse with respect to overall
structural stability. Consequently, a careful 'watch-and-
wait' approach made sense, given the extent to which the
adjacent excavation and construction had progressed. What was
especially important about the building response was the
detachment and falling of decorative tiles. Although this

type of response might conceivably belong within the category
of architectural damage, it represented a threat to the people
and art objects within. Once this source of difficulty had
been recognized and rectified, the galleries were opened even
though the construction continued outside with additional
small soil movements and cracks in the building.

The problem associated with locally weak ornamental fixtures
is one better treated by preventive than by remedial medicine.
Large monumental masonry buildings should be regarded as a
special class of structures, which need to be thoroughly
inspected and fortified before construction starts. In
general, it will be extremely difficult to eliminate
architectural damage in such structures. Some cracks and
separations should be anticipated, and locally weak fixtures,
if present, should be stabilized.

MR A. R. DAWSON, University of Nottingham

Mr Adestam's technical note refers to a settlement profile
gauge. The gauge operates by electronically measuring the
water pressure in a probe which is pulled along a hose.

The measuring unit (Fig. 1) consists, in principle, of two
plastic tubes with different diameters. The small diameter
tube which contains air and an electric cable is inserted into
the large tube. The annular space between the two tubes is
filled with a fluid (generally water). The lower ends of the
tubes are connected to a measuring head which contains an
electric fluid pressure transducer. The upper ends of the
plastic tubes are connected to an open stand-pipe. Thus, the
difference in level H between the measuring head, which can be
inserted in the hose under the embankment, and the fluid level
at the read-out unit can be measured directly and compared
with previous measurements.

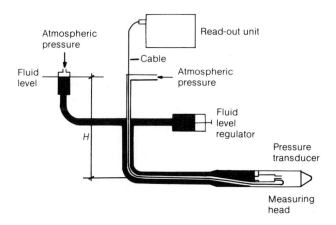

Fig. 1

The present equipment has been tested in a stable plastic hose and the accuracy was then found to be +3 mm.

The equipment can also be used to measure the elevations of buried pipes, e.g. sewers. A special adapter has been constructed to measure the lowest point of a circular section.

DR A. D. M. PENMAN, Sladeleye, Chamberlaines, Harpenden

However well a road pavement may be designed to withstand the adverse effect of a stiff clay formation, cyclic traffic loading etc., it can be ruined by being cut through in all sorts of structurally undesirable positions for the installation of underground services, as has been shown in this session. Pipe jacking offered a way of pushing a pipe under a pavement and nowadays developments of hammer moles of various sizes and powers gives a method of producing ducts of various diameters under pavements without a need for trenching. Hertfordshire County Council is currently calling for 300 mm dia. ducts to be passed under some trunk roads to take a variety of cables and while this may seem to be rather large for cables which presumably could use several smaller, and therefore more easily placed, ducts the fact that this size is being installed by impact boring shows that piped services for water, gas and sewerage could use these ducts.

Longitudinal services laid under footpaths in urban areas could be taken to the other side of the road by these methods and leave the perfectly designed pavement to survive for its very long design life, thereby drastically reducing maintenance costs and avoiding the traffic disruption caused by road (destroying) works.

MR A. D. LEADBEATER, Derbyshire County Council, Matlock

Arising from the content of Paper 3, Dr Clark's contribution and further comments made by Mr Garrett, Mr Addy and particularly Professor O'Rourke it has become increasingly clear from the majority of papers submitted that

(i)   engineers have been very frank in describing their
        failures
(ii)   engineers have given excellent advice from the various
        case histories written up to give good guidance for
        future engineering decisions.

However, the financial, political and risk factors have not been discussed and this must be the subject for future discussion and papers. A study of the papers highlights the following points

(i)   Paper 3 recommends that the stability is satisfactory if
        an increase in the factor of safety of 0.15 is obtained

from unity
(ii)   Paper 9 recommends that the stability is satisfactory if
         an increase in the factor of safety of 0.30 is obtained
         from unity
(iii) Professor Skempton showed reasonable grounds for
         assuming a factor of safety between 1.7 and less than
         unity
(iv)   Paper 4 shows extreme difficulty in predicting the
         behaviour of slopes
(v)   Papers 5 and 6 show unpredictable patterns of behaviour
         of clay slopes.

An analysis of virtually all the papers reveals that in every
case something unpredictable happened. The value of this
symposium in demonstrating these points is justification
enough for it.
    Thus several questions must be identified.

(i)   What degree of risk is acceptable in the design of
         remedial works based purely on the factor of safety?
(ii)   What life should the remedial work be designed for?
(iii) A method of qualitatively comparing schemes that brings
         in all factors must be produced. Such a comparison must
         be capable of comparing all features in a budget (i.e.
         comparisons of land instability, retaining walls and
         bridges or whatever else).
(iv)   The cost of increasing the factor of safety must be
         assessed (in Paper 9 pumping boreholes increased the
         factor of safety from 1.2 to 1.3: why was 1.2
         unacceptable and 1.3 acceptable?).

The basis of this problem and the answer to questions (i)-(iv)
are what engineers do subconsciously.
    Risk analysis as applied to schemes is something which is
already done in many areas such as

(i)   concrete and steel designs - load factors and
         serviceability and ultimate load cases are a form of
         risk analysis
(ii)   the design of land drainage and surface water drainage
         is based totally on a 1 in 10 year storm, i.e. can we
         risk the system surcharging once a year or once in 100
         years etc.?

    It may be that these questions cannot be answered when
considering the risk of failures in earthworks but they must
be asked and by identifying the correct questions the correct
answers may be obtained.

TN1. Instabilities with large horizontal deformations

A. JONKER, Senior Geotechnical Engineer, and R. J. TERMAAT,
Senior Geotechnical Engineer, Rijkswaterstaat

SYNOPSIS. A construction method by which instabilities were
accepted was not an economical solution because of the great
horizontal deformations in this case. For the afterdictions
of those instabilities the slip circle analysis had been
used more in detail. It was assumed that in a zone of the
subsoil covered by the set of slip circles with low
stability large shear strains or large plastic deformations
occured.

INTRODUCTION

In 1976 the Netherlands government decided to close the
Oosterschelde estuary with a storm surge barrier in
combination with a number of secondary dams. One of these is
the Markiezaatsdam, some 4300 m. in length (Fig.1).

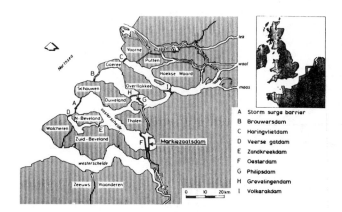

A   Storm surge barrier
B   Brouwersdam
C   Haringvlietdam
D   Veerse gatdam
E   Zandkreekdam
F   Oesterdam
G   Philipsdam
H   Grevelingendam
I   Volkarakdam

Fig. 1. Location of the Markiezaatsdam

An existing part of the dam had to be reconstructed. To make
a save reconstruction of the dam a berm is necessary as the
subsoil consist of very soft clay and peat (Fig.2).

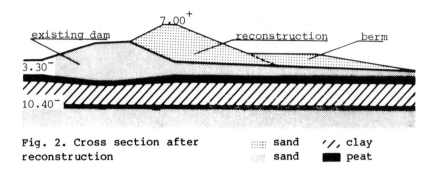

Fig. 2. Cross section after
reconstruction

::::: sand   //, clay
::::: sand   ■ peat

Nearly 30 - 40% of the quantity of sand will be used for the
berm. A solution without a berm seems economical, also when
instabilities occured. When the dam is constructed on such a
way, the safety is 1 immediately after construction. To avoid
instabilities in the final situation the extreme loading,
1:100 years, had to be simulated in the construction period
(equivalent load).

DETERMINATION OF THE EQUIVALENT LOAD
The equivalent load is chosen in such a way that the load
simulates the failure mechanisms in the extreme situation
i.e. lower or the same stability factor in the same part of
the subsoil. The following mechanisms has been considered
(Fig.3):
(a) circular surfaces;
(b) straight sliding surfaces;
(c) squeezing.

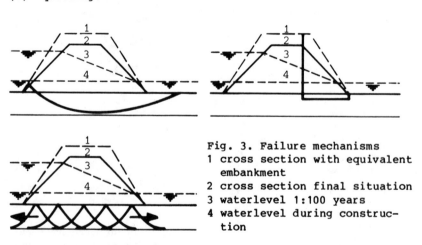

Fig. 3. Failure mechanisms
1 cross section with equivalent
  embankment
2 cross section final situation
3 waterlevel 1:100 years
4 waterlevel during construction

EXPERIENCE AND CONCLUSIONS

During construction of the test section unexpected large horizontal displacements occured. Only with use of great quantities of sand it was possible to make the required level. The profit of less sand, compared with a solution with a berm decayed entirely. To make a good extrapolation to the reconstruction of the existing dam, a reliable afterdiction of the observed horizontal displacements was necessary.

For lack of time we only considered more in detail the slip circle analysis. In the area covered by this set of circles with a low stabilityfactor (less then 1.1), we assumed that large shear strains or plastic deformations occured. For the test section it is a great part of the subsoil. This may be an indication for large horizontal deformations. These assumption has slightly confirmed by a theoretical study for an other part of the Markiezaatsdam. In this study this assumption had been compared with an finit element method (Ref. 1). For the existing dam also a set of circles had been considered. Also in this case a great part of the subsoil has a low stability (Fig.4).

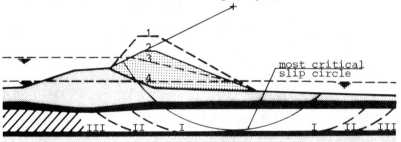

Fig. 4. Areas with low stability,   ⋯ sand   '/, clay
based on slip circle analysis   ▓ sand   ▬ peat

I   SF<1   ⌐ for   1 cross section with equivalent
II   1<SF<1.1 │ equivalent   embankment
III 1.1<SF   ⌐ load   2 cross section final situation
                              3 waterlevel 1:100 years
                              4 waterlevel during construc-
                                tion

Both on these calculations and engineering judgement it was decided to reconstruct the exixting dam with a berm.

It is well known that for constructions on soft soils the stability will be overruled by the large plastic deformations. Such situations can be recognized of course with an finit element method, but a cheaper and more simple method to get an indication will be important for the geotechnical engineer.

REFERENCE

1. Termaat R.J., Vermeer P.A. and Vergeer G.J.H. Failure by large plastic deformations, will be published, XI ICSMFE, San Francisco 1985.

TN2. Failures during river dike construction

R. CARPENTIER, Chief Engineer, Belgian Geotechnical Institute,
P. KERSTENS, Senior Engineer and W. GRARE, Engineer, Ministry
of Public Works, Sea Scheldt Service, and G. VAN ALBOOM,
Engineer, Belgian Geotechnical Institute

1.   In Belgium, the river Scheldt and its affluents (Fig.1)
are subjected to a semi-diurnal tidal regime over a total
length of 300 km. At several occasions in the past extreme high
tides due to north west storms on the North Sea caused inundations by overflow or dike failures. Since 1977 the Belgian Ministry of Public Works started heightening and reinforcing
            the existing river dikes. As in many locations
these works have to be executed on soft alluvial layers, and
although special precautions are taken, local failures sometimes occur during construction. In these cases and also at
places where the existing dikes, river banks or other defence
structures had already undergone some damage, special repairing or stabilizing techniques had to be applied.
2.   In this technical note only general information about the
works can be given.

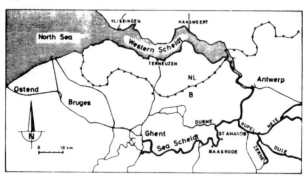

Fig1: The Scheldt and its affluents

3.   Most of the existing dikes have been constructed in the
course of time on the soft alluvial layers in the winter beds
of the rivers. In many cases safety factor of those dikes was
very near to unity and slopes crept continuously, necessitating
frequent levelling of the dike platform. Heightening and widening of the dikes therefore necessitated flattening of the
slopes, sometimes in combination with toe loading with quarry
stone, blast furnace slag or other stone material. At many places toe loading is realized near the low water level with one
or more rows of gabions.

4. In the following some special chosen solutions will be illustrated.

5. On the right bank of the river Scheldt upstream Baasrode, part of the existing dike turned away from the river, leaving a wide haugh between the dike and the river (Fig.2). In first instance it was decided to construct the new dike along the

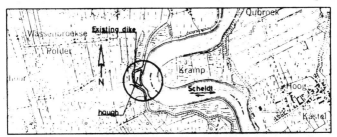

Fig 2: Dike construction site upstream Baasrode

river,crossing over the haugh. However, already in the beginning of the works, a slide occured. Medium heavy CPT tests on the haugh revealed soft alluvial layers over a thickness of about 6 to 7 m (Fig. 3). The original project was therefore adapted with preservation of the old dike site and strengthe-

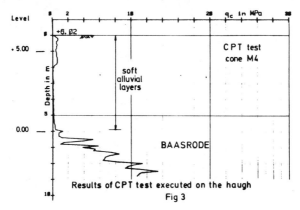

Results of CPT test executed on the haugh

Fig 3

ning of the haugh at the riverside against erosion by the strong water currents and the propeller action of ships. A scheme of the adopted dike profile is shown in Fig. 4.

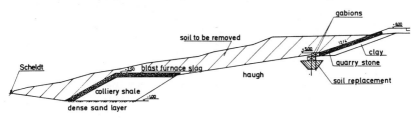

Fig 4: Dike profile on the right bank of the Scheldt upstream Baasrode

6. Further downstream the river shows a narrowing at the historical church of Baasrode (Fig. 5). To protect the village and the church against inundations a sheet pile wall was constructed in the early 70's at the riverside. By the increased currents the toe of the sheet pile wall was continuously eroded despite of frequent dumping of quarry stone, causing inacceptable deformations of the sheet pile wall over a length of about 150 m. The position of the deformed sheet pile wall is shown in Fig. 6. After an extensive geotechnical investigation, and taking into account the necessity to preserve the church and its surroundings, the solution of Fig. 6 was retained. In this solution the new river bank is formed by a slope up to the level + 6.50, combined with a reinforced concrete cantilever wall and partly with a sheetpile wall reaching the required level + 8.0. This solution however necessitated the pulling down of a series of houses on the old river bank.

Fig 5: The Scheldt at Baasrode

7. When constructing an enlargement at the land-side of the dike at St. Amands, a slide started. Borings and CPT tests revealed the existence of soft alluvial layers over a thickness of 5 to 6 m. In order to assure continuity of the dike construction works the solution was chosen to place the fill in layers of 1 m thick at a slow loading rate over the whole length of 1200 m of the dike construction site. To accelerate consolidation a drainage layer protected against contamination by a geotextile, was placed on the soft layers.

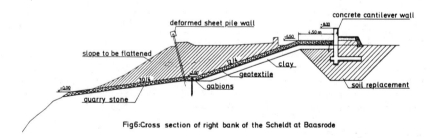

Fig6:Cross section of right bank of the Scheldt at Baasrode

TN3. Wedge analysis of embankment instability during construction of the M25 motorway

J. R. GREENWOOD, BSc, MEng, MICE, MIHT, Regional
Geotechnical Engineer, Department of Transport

SYNOPSIS: This note considers the mode of failure of embankments associated with overconsolidated clay soils. A simple stability analysis is used to demonstrate that wedge failure can be more critical than circular failure.

INTRODUCTION

1. The failure during construction in 1981 of the M25 embankment at Hill Hall was fully reported by Finlayson et al (ref 1). Discussion on this failure by Moore (ref 2) brought out other cases of instability during construction on adjacent sections of the M25 in Essex.

2. These problems of instability all involved slippage of a wedge of embankment material along a near horizontal planar surface. A review of the literature shows that most highway embankment construction failures and failures of embankment dams (such as the recent Carsington dam failure (ref 3)) occur by wedge type failure often with some rotational component of the material behind the wedge. Investigation of the failures usually reveals a plane of weakness in the foundation or basal layers of the fill which was either present before construction or which developed during construction.

3. The embankment designer is faced with a difficult task. He can carry out numerous calculations and demonstrate convincing factors of safety but they all become meaningless if the strength parameters appropriate to the weakest layer are not correctly predicted. Alternatively he can assume the worst parameters throughout but this may lead to construction of the scheme being ruled out on economic grounds. A balance must be struck and this is usually achieved by careful field inspections and monitoring to ensure safe construction where design assumptions are in doubt.

WEDGE ANALYSIS

4. In the author's experience wedge type analyses are not normally carried out for highway embankment design. This is possibly explained by the fact that design engineers are presented with exploratory hole logs showing strata of finite

Failures in earthworks. Thomas Telford Ltd, London, 1985

403

thickness and design parameters are assigned to that total thickness. If a narrow weaker band is identified it is often removed and replaced before embankment construction. Furthermore comparison of conventional wedge and circular analyses (ref 2) invariably gives a lower factor of safety for the circle than the wedge and designers adopt the results of the circle believing this to be more conservative. (Case A, Fig 1).

5. However application of the simple stability equation (1), which takes account of the horizontal stress acting on the failure surface realistically predicts that for overconsolidated soils where high Ko values are present near the ground surface the wedge can become more critical than the circle (Case B, Fig 1).

Stability equation (ref 4):

$$F = \frac{1}{\sum W \sin \alpha} \sum \left[c'b \sec \alpha + W(1 - r_u)(1 + K \tan^2 \alpha) \cos \alpha \tan \phi' \right] \quad (1)$$

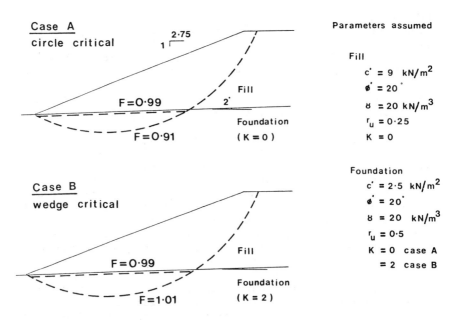

Case A
circle critical

F=0·99
F=0·91

Fill
2
Foundation
(K = 0)

Case B
wedge critical

F=0·99
F=1·01

Fill
Foundation
(K = 2)

Parameters assumed

Fill
$c' = 9$ kN/m^2
$\phi' = 20°$
$\delta = 20$ kN/m^3
$r_u = 0·25$
$K = 0$

Foundation
$c' = 2·5$ kN/m^2
$\phi' = 20°$
$\delta = 20$ kN/m^3
$r_u = 0·5$
$K = 0$ case A
$= 2$ case B

Fig 1. Influence of a high Ko value on the shape of the critical failure surface.
(Based on M25 Hill Hall failure (ref 1)).

6. The simple equation (1) particularly lends itself to 'hand' (computer assisted) calculations by the method of slices where the designer can check the sensitivity of the factor of safety to the parameters assumed and thereby develop a 'feel' for the problem. The equation, based on conventional

shear strength theory, gives sensible factors of safety for both circular and non cicular slip surfaces giving comparable results to other recognised methods (ref 4). As there is often considerable doubt over the appropriate parameters to apply in an analysis the simple equation is considered sufficiently accurate for many routine applications where more rigorous techniques are not justified. Although useful in parametric studies, in most practical situations the ratio of horizontal to vertical effective stress, K, appropriate to the slip surface analysed cannot be readily estimated or measured. The effect of changes in the K value on the calculated factor of safety are small and for routine calculations it is usual to make the slightly conservative assumption K = O.

CONCLUSIONS

7.   Most embankments associated with overconsolidated clay soils which fail during construction do so by a wedge type failure. Many wedge failures occur due to the presence of weaker soil layers but even if these planes of weakness are not present it is demonstrated by application of the simple stability equation that for overconsolidated strata with a high Ko value a wedge shaped failure surface tends to be more critical than the circle.

ACKNOWLEDGEMENT

The co-operation and assistance of Consulting Engineers and colleagues involved in this work is appreciated. Views expressed are those of the Author and should not be attributed to his employer.

REFERENCES

1.   FINLAYSON D M et al   Lessons to be learnt from an embankment failure.   Proc Inst Civ Engrs, Part 1 1984, 76, 207-220.

2.   MOORE P, GREENWOOD J R etc.   Discussion on 'lessons to be learnt from an embankment failure'.   Proc Inst Civ Engrs, Part 1, (To be published 1985).

3.   SKEMPTON A W, COATS D J   Failure during construction of Carsington embankment dam.   Proc of ICE Symp on Failures in Earthworks, London, March 1985.

4.   GREENWOOD J R   A simple approach to slope stability. Ground Engineering 1983, 16, No 4, 45-48.

TN4. Monitoring pore water pressures in an embankment slope

I. CRABB, BSc, AMICE, and G. WEST, BA, MPhil, DIC, FGS, Ground Engineering Division, Transport and Road Research Laboratory

SYNOPSIS.   A section of 7 m high embankment on the Cambridge Northern Bypass (A45) has been instrumented with the aim of elucidating the mechanism of shallow slope failures that have affected adjacent lengths of the embankment. A comprehensive array of automatically recording hydraulic piezometers was installed to monitor changes in pore water pressure, an automatic rain gauge was set up to record the rainfall and automatic sterocameras were positioned to photograph the slope daily so that the onset and development of any slip can be traced.

INTRODUCTION
   1. More than thirty embankment slope failures have occurred on the A45 trunk road and the M11 motorway to the north of Cambridge in the period since the construction of the embankments in 1977–1978. The embankments involved were made from the overconsolidated Gault Clay that occurs in this region. The slope failures were typically shallow translational slides affecting the upper 1 – 1.5 m of the embankment slope and tend to be initiated in the winter or spring, often after high rainfall. North facing slopes were particularly likely to fail. This paper describes a field experiment made to investigate the role that changes in pore water pressure play in shallow embankment slope failures of this kind.

EXPERIMENTAL SITE
   2. The site chosen was the north facing slope of a 7 m high embankment made of Gault Clay on the Cambridge Northern Bypass (A45). A section of intact embankment slope having a slope of 1 : 2 was selected adjacent to a section that failed over a length of 60 m in the winter of 1982–83. Typical values of soil properties, measured during a site investigation made of the site between September and November 1983, are given in Table 1.
   3. The instrumentation of the site consisted of (a) a comprehensive array of automatically recording hydraulic piezometers to measure the pore water pressure distribution within the embankment and any changes of it with time, (b) an automatic rain gauge to record the rainfall, and (c) automatic

Table 1. Typical properties of the Gault Clay on site at 0 - 1.5 m depth

Liquid limit, per cent	66
Plastic limit, per cent	26
Plasticity index, per cent	40
Casagrande classification	CH
Moisture content, per cent	28
Vane shear strength, kN/m^2	80

stereocameras positioned to photograph the slope daily so that any slope movement can be monitored. The leads from the hydraulic piezometers are brought down the slope to a small instrument house at the foot of the embankment shown in Fig 1; this contains the hydraulic scanning valves and the associated

Fig.1 Piezometer scanner and logger

control and data logging system that utilises replaceable integrated-circuit memory cartridges. The site is visited once a month to change the cartridge, any piezometers which need de-airing being attended to at the same time. Back at the Laboratory the data stored in the cartridge is transferred via a cartridge reader to a microcomputer system where it is recorded on magnetic disc. The data is processed and output is provided in the form of graphs, shown in Fig 2a, of the pore water pressure at each piezometer tip plotted against time. A cartridge from the rain gauge is dealt with in a similar way to give a histogram of rainfall against time shown in Fig 2b.

4. The objective of the work is to monitor the site over several years, in particular to study the changes in pore water pressure near the surface of the embankment slope in the winter and spring so that a better understanding is gained of how the pore water pressure in this upper layer affects slope stability. If a failure should occur, then the piezometric and rainfall data immediately before failure will be invaluable, particularly as the time-lapse photographs should enable the time and rate of slipping to be pinpointed.

PRELIMINARY RESULTS AND DISCUSSION

5. Figures 2a and 2b show the pore water pressure recorded by three piezometers and the rainfall for the site for the period 23rd February to 3rd April 1984. The diurnal cyclic pressure variation is caused by temperature changes affecting the system and can be disregarded. It can be seen that the rainfall peak occurring on 23rd to 26th March had little effect on the pore water pressure at 1.5 m depth in this instance. Figure 2c shows the pore water pressure distribution over the instrumented cross-section during this period. It can be seen that the distribution is consistent with the initially dry clay having become wetter by moisture movements, both up from the water table and down from rainfall on the slope, so that high negative pore water pressures remain only in the core of the embankment.

6. The pore water pressure distribution is required for stability analysis in terms of effective stress, but for the study of seepage the results are required in the form of a total head distribution, and this is shown in Fig 2d. It can be seen that at this time of year the down-slope component of the hydraulic gradient is small compared with that normal to the slope surface, so that seepage is mainly into the slope and not down it.

CONCLUSION

7. The experimental installation briefly described in this paper was commissioned in December 1983. Results for only the succeeding winter and spring period have therefore been obtained to date. In the event, that particular winter and spring were exceptionally dry so that the likelihood of an embankment slope failure would be low, and in fact none has

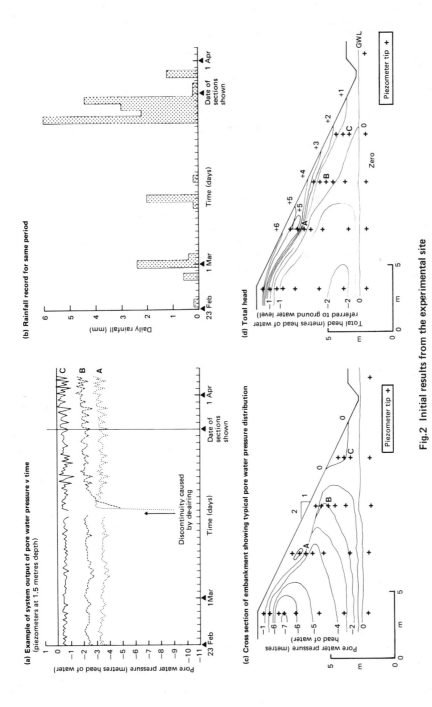

(a) Example of system output of pore water pressure v time
(piezometers at 1.5 metres depth)

(b) Rainfall record for same period

(c) Cross section of embankment showing typical pore water pressure distribution

(d) Total head

Fig.2 Initial results from the experimental site

occurred at the site or nearby. However, monitoring of
piezometric conditions will continue in the expectation of
a more normal winter and spring to come.

ACKNOWLEDGEMENTS
    8. The work described in this paper forms part of the
programme of the Transport and Road Research Laboratory and
the paper is published by permission of the Director. The
authors would like to thank the Eastern Regional Office of
the Department of Transport and Cambridgeshire County Council
for permission to make the experimental installation described.
Northamptonshire County Council Engineering Services Laboratory
is also thanked for installing the piezometers. The authors'
colleague, Mr M W Todd, is thanked for installing and attending to the stereocameras on site.

TN5. Field observations of embankment failures on soft clay

R. S. PUGH, BSc, MSc, PhD, DIC, MICE, Senior Engineer, Sir
William Halcrow and Partners

SYNOPSIS. Observations from three instrumented trial loadings
of very soft Thames alluvium are presented. Horizontal
displacements and strains, divided by fill height, are plotted
versus time and depict the onset of local failure of the
alluvium beneath three embankments. Local failure commenced
beneath the embankment crests and progressed outwards to beyond
the embankment toes in those cases where failure occurred. It
is suggested that such plots may be used for construction
control purposes by detecting the onset of local failure beneath
the embankment crest.

FIELD OBSERVATIONS

1. The embankments described were part of extensive pre-
construction trials for the heightening of flood protection
embankments which confine the River Thames in Essex. Three
embankments were constructed of sand and induced plane strain
conditions in the 7.5m thick, very soft clay foundation. The
clay, which was underlain by sands and gravels, was lightly
overconsolidated. Because of desiccation, the upper 1m was both
more heavily overconsolidated and of significantly higher
undrained strength. This upper material was removed from
beneath and beyond the toe of the first embankment (B1/1). The
failure of B1/1 at 4.0m fill height was thus confined to the
very soft, lightly overconsolidated clay. The second embankment
(B1/2) was constructed on virgin ground and failed at 4.5m fill
height. The third embankment (B2) was constructed on virgin
ground to a height of 2.8m, with a safety factor of 1.5 using
limit equilibrium analyses.

2. Under initial loading there were elastic pore-pressure
changes and significant pore-pressure dissipation in the foundation clay. At fill heights of about 1.25m, elastic behaviour
gave way to yielding plastic behaviour. Yielding was associated
with a sharp increase in pore-pressure response, virtually zero
pore-pressure dissipation and increased displacements and
strains. Subsequent local failure of the foundation produced a
further increase in displacement, strain and pore-pressure
responses. Local failure was observed to progress from beneath
the embankment crest towards and beyond the embankment toe.
Undrained creep was observed at post-yield stress levels.

CONSTRUCTION CONTROL

3. The data from the trials were analysed to determine which measurements could be used for construction control. When plotted versus fill height, horizontal displacements of the foundation showed non-linear behaviour throughout and gave no clear indication of yielding or local failure. However, if horizontal displacement was divided by fill height, and plotted versus time, the onset of local failure could be detected. This is illustrated in Fig. 1 for points approximately 1m below original ground level; the data were obtained from inclinometers through the embankment side slopes.

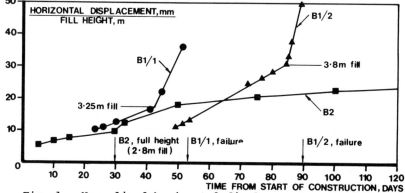

Fig. 1.   Normalised horizontal displacement versus time.

4.   Distinct changes in behaviour are evident at 3.25m fill height for B1/1, 3.8m for B1/2 and 2.8m for B2. Based on these observations construction would have been stopped at 3.5m, 4.25m and, say, 3.25m fill heights for the respective embankments. Continuing observations would then be required to monitor the rate of displacement, and to determine if it was accelerating towards failure or diminishing to a stable condition. Local failure must be detected by monitoring beneath the embankment crest in order to permit continuing observations and remedial action if required. Once local failure has spread beyond the embankment toe overall failure may be inevitable.

5.   Horizontal displacements recorded from extensometers beneath the embankments enabled average horizontal linear strains to be monitored over the entire embankment width. Fig. 2 shows these strains, divided by fill height, plotted against time for extensometers 1m below original ground level at positions beneath the embankment crests.

6.   The strains shown in Fig. 2 are tensile and suggest that local failure developed beneath the crests of B1/1 at 2.75m, B1/2 at 3.8m and B2 at 2.7m fill height. Similar plots beneath and beyond the embankment showed a change from tension beneath the crest to compression beyond the toe. Local failure at the embankment toe was observed beneath B1/1 at 3.5m and B1/2 at 4.25m fill height. At the toe of the ensuing failure surface local failure coincided with overall embankment failure.

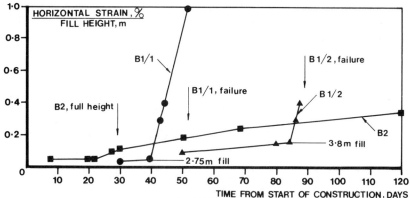

Fig. 2 Horizontal strain/fill height versus time.

7. Vertical displacements were not considered for construction control because of complications caused by consolidation settlements, although, similarly to the pore pressures, these data clearly illustrated plastic yielding of the foundation. The measured pore pressures from 48 piezometers appeared extremely reliable, and their use in limit equilibrium analyses with effective stress shear strength parameters 'predicted' both the failure heights (peak strength) and post-failure 'heights' (residual strengths) of the embankments. However, definite indications of local failure were only observed beyond the embankment toe. Increased pore-pressure response at this location accompanied the first indications of significant heave and closely preceded failure.

CONCLUSIONS
8. It is suggested that initial elastic behaviour of the soft clay foundation was superseded by plastic yield and subsequently by 'progressive' local failure commencing beneath the embankment crests and finishing at the toes of the failure 'surfaces'. Construction control requires the identification of local failure beneath the embankment crest which appears to coincide with a limit equilibrium safety factor of about 1.5. This may be achieved by plotting average horizontal linear strains (from extensometers) and maximum displacements (from inclinometers), both divided by the appropriate fill height, versus time across the entire foundation width likely to be affected by the embankment construction. Outward spread of local failure may also be detected by observations of pore pressure and heave beyond the embankment toe.

ACKNOWLEDGEMENTS
The trials were performed by the Anglian Water (formerly Essex River) Authority and designed and supervised by their consulting engineers, Binnie and Partners. The author thanks both parties for permission to publish the data presented herein, and also Professor A.W. Bishop who supervised the associated research which was funded by the Science Research Council.

TN6. The failure of Açu Dam

A. D. M. PENMAN, MSc, DSc, Consulting Geotechnical Engineer,
Sladeleye, Chamberlaines

SYNOPSIS. The Açu embankment dam which was to be 40m high,
failed on 15 December 1981 during construction, when it was
still 5.2m below crest level. Changes made to the original
design introduced an impervious blanket of clay under the
upstream shoulder, through which the slip surface passed.
Failure occurred rapidly - about $\frac{1}{2}$ hour - causing the
construction surface to fall about 15m and the upstream toe
to move horizontally 25m.

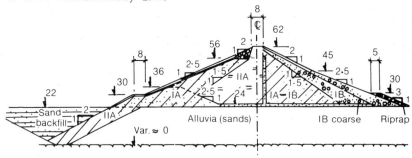

Fig 1. Section showing original design

DESIGN AMENDMENTS

1. The designed section of the dam is shown by Fig 1. Two
main materials available for construction were IA) a dark-grey
to black flood-plain silty clay and IIA) an unsaturated red
terrace clayey sand and gravel.

2. The water table in the sand foundation was close to
ground surface. To construct the cut-off the sand was
de-watered to allow a 22m deep excavation to be made in the
dry to expose bedrock. Side slopes of 1 in 2 were used and a
layer of the black silty clay placed against the bedrock and
the downstream slope as the excavation was backfilled. To
save time, this excavation was made upstream of the dam so
that dam construction could be concurrent.

3. An international specialist was called in from North
America to advise on instrumentation and rapid drawdown
conditions. He suggested, from visual inspection and study of
three grading curves, that the red clayey sand and gravel would

not be sufficiently impervious to provide the link between below ground cut-off and the dam core. He therefore proposed a modified design, to which the dam was built, with a 7m thick horizontal blanket of the black clay over the sand.

4. In accordance with good practice, the specification for the black clay core called for placement wet of optimum. This was used for the first layers of the blanket over formation level. The core shape had been modified but part of the original upstream shape had been retained as a cofferdam. In anticipation of the wet season the cofferdam had to be built ahead of other fill, so that, in little more than a month, a black clay bank with a 1 in 1.5 upstream slope, was raised 14m.

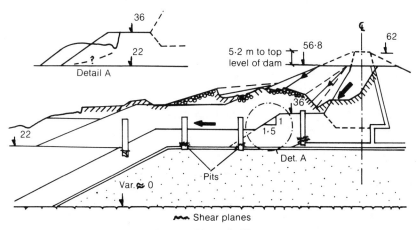

Fig 2. Section after failure

FIRST SLIPS

5. Suddenly two slips occurred, as indicated by detail A in Fig 2, each about 150m long and separated by some distance, involving about $16 \times 10^4 \text{m}^3$ of the clay fill. In the absence of pore pressure measurements, it was not considered possible to use effective stresses and back-analysis by total stress showed that the slips had mobilised a shear strength $c_u = 49\text{kN/m}^2$. The cofferdam was repaired with a slope of 1 in 2.5.

6. It was apparently agreed that for the black silty clay $\varphi' = 20°$, but discussion on values to be assigned to c_u, probable pore pressure values and use of circular arc slip surface for stability calculations was still continuing as construction reached a height of 34.8m. The specification for placement water content had been changed in March 1981 to optimum - 1.5% to + 0.5% but unfortunately no piezometers had been installed in the black clay.

THE FAILURE

7. Major failure of the upstream slope occurred on 15 December 1981, almost exactly a year after the cofferdam failures. Eye witnesses described tension cracks forming in the level construction surface along the downstream edge of the core. These were followed by sinking of the core and upstream

shoulder over a length of about 600m and a massive pushing outwards of the flatter lower section and berm by about 25m. The whole event occurred in only about 30 minutes. $1.2 \times 10^6 m^3$ of material was involved in the slide.

8. The appearance of the failed dam was very similar to that of Carsington, which failed early in June 1984. The heights of the two dams at the times of the failures and the lengths affected were similar, as were the movements that occurred. A difference was that the Carsington failure took several days.

ANALYSIS

9. Trial pits, shown by Fig 2, revealed slip surfaces in the lower layers of the black clay blanket and cofferdam that had been placed wet under the original specification. The back scarp and other trial pits indicated that a curved slip surface through the black clay core connected to this more nearly horizontal surface in the blanket. A back analysis using total stresses has given $c_u = 48kN/m^2$ showing remarkable agreement with the value obtained from back-analysis of the cofferdam failures.

10. Conventional unconfined compression tests gave values of $c_u = 80$ to $90kN/m^2$. Undrained tests on specimens cut from block samples and tested slowly gave $c_u = 65kN/m^2$ and $\varphi' = 6^o$. During excavation of the trial pits, many layers were found to be intensely laminated by overcompaction. It has been argued that powerful modern earthmoving and compaction equipment working on wet soil soon brings it to a state of near saturation and then simply remoulds, shears and laminates a silty clay. In trial pits in the upper part of the slide mass, considerable tension cracking was evident. Consideration is being given to the effects such cracks and lamination may have had on the overall strength of the black clay. Recent work has shown the clay to have a remarkably high salt content and strength parameters $c' = 10kN/m^2$ and $\varphi' = 18^o$.

RECONSTRUCTION

11. The failed length of the dam has been reconstructed from the red clayey sand and gravel IIA, using the well-known Brazilian section with a central filter in place of a core (Vargas 1970).

12. The failure of Açu dam has been described by de Mello (1982), de Carvalho (1982) and Pessoa (1982).

REFERENCES
1. DE CARVALHO L H (1982) Contribution to discussion on Q55. Trans.14th Int.Congr.Large Dams, Rio de Janeiro,vol.5, 551-554.
2. DE MELLO VFB (1982) A case history of a major construction period dam failure. Amici et Alumini Em.Prof.Ir.e.e.de Beer. Comité d'hommage au Prof.E de Beer, 1040 Bruxelles, Belgie, 63-78.
3. PESSOA J C (1982) Contribution to discussion on Q 55. Trans.14th Int.Congr.Large Dams, Rio de Janeiro,vol.5,680-682.
4. VARGAS M (1970) The use of vertical core drains in Brazilian earth dams. Trans. 10th Int. Congress Large Dams, Montreal, vol.1, 599-608.

TN7. Reinforced soil techniques for the reinstatement of failed slopes using Geogrids

T. L. H. OLIVER, BSc, MICE, MIHT, Area Civil Engineer, Netlon Ltd

SYNOPSIS. The most common method of reinstatement has been excavation and substitution with granular fill. Development of Geogrid reinforcement techniques has made reinstatement using the foundered soil an economic alternative. These techniques are illustrated by two recent case studies.

INTRODUCTION

1. The technique of using Geogrid reinforcement for the reinstatement of failed slopes has been shown to offer economies over alternative methods (ref. 1). A number of reinstatements to embankments and cut slopes have now been completed in the UK and USA (refs 2-4) using this technique.

EMBANKMENT SLIP REINSTATEMENT, M11, ESSEX

2. The major benefits of Geogrid reinforcement over conventional granular substitution methods are well illustrated by examining a recent embankment reinstatement by Essex C C.

3. A shallow slip, 20m wide, had occurred on a section of embankment supporting the M11 Motorway near Bishops Stortford. The embankment at this point is 12m high, with a slope of 1:3, and consists of London Clay fill.

4. The conventional method of repair would have been to excavate the failed material and substitute an imported stone. Excavation is usually carried out by dragline from the hard shoulder, loading directly into dump trucks; this necessitates closure of the slow traffic lane. The estimated cost of conventional reinstatement was £30,000.

5. For the first time in Essex, Geogrid reinforcement was adopted as an alternative. The terraced excavation was completed in two and a half days using a D6 dozer, stockpiling the clay adjacent to the excavation. This soil was then replaced and compacted in layers, with the inclusion of 'Tensar' SR2 Geogrid reinforcement at 1.5m vertical centres. The spacing of reinforcement had been determined using the method proposed by Murray (ref. 5). 'Tensar' SS1 was installed as secondary intermediate reinforcement, to prevent localised failures occurring at the surface, Fig. 1. Working in such a restricted area, compaction was found to be adequately achieved using the D6 alone.

6.   No drainage layers were installed, as in this case it had been decided to provide a cut off drain 300m long at the top of the embankment, with the intention of preventing additional slip failures.

7.   The reinstatement was completed in 8 days, with final costs as follows

plant and labour (2 operatives, 1 lorry, D6 dozer)	£ 2,400
'Tensar' reinforcement	£ 4,000
300m length of interceptor drain	£ 4,500
	£10,900

excavated volume 1,200 cubic metres = £6.60/cubic metre

8.   The above rate represents a saving of 65% when compared with conventional methods of reinstatement.   In addition to this financial benefit, as the method of working did not require closure of the slow lane, the safety of road users was not impaired.

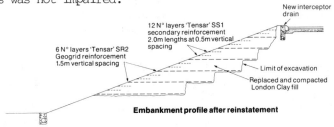

Embankment profile after reinstatement

Fig. 1 Slope reinstatement – M11 Motorway, Essex

STEEP SLOPE REINSTATEMENT, DUNDRY, AVON

9.   Where steeper slopes are to be formed the reinforcement may be continued up the slope face to enclose the layers of fill.   Temporary formwork may be required to support the face during construction.   This method has been used to good effect by the Highways and Engineering Department of Avon C C.   A steep slope supporting an unclassified road had failed immediately behind a 200-year-old cottage, near the village of Dundry.   The slope was reinstated as a 4.5m high terraced wall with a turf face approximately 12m wide at the base.   The foundered soil, consisting of hillwash material overlying stiff Lias Clay, was excavated and recompacted with 'Tensar' SR2 Geogrid reinforcement. Secondary reinforcement was included to reduce bulging of the face.   The steep   face was supported by temporary formwork during construction.   Turfs placed behind the reinforcement prior to backfilling have produced a wall with an extremely attractive appearance, Fig. 2.

10.   The estimated cost of the reinstatement, excluding roadworks and ancillary works, was as follows

drainage layer	£ 1,030
earthworks	£10,900
prelims, site clearance etc	£ 1,900
reinforcement	£ 4,180
	£18,010

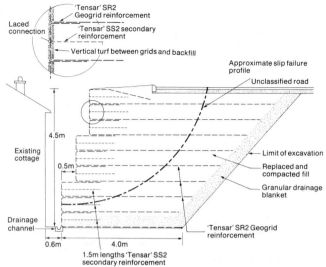

Fig. 2 **Steep slope reinstatement – Dundry Lane, Avon**

11.   The relatively high cost of the earthworks in this case can be attributed to the severely restricted area of the site and difficult access to the excavation.   The benefits of adopting this form of reinstatement were seen to be the simplicity of construction, requiring no specialist skills, and the avoidance of disruption to the adjacent building foundation that would possibly have occurred with either a sheet piled or traditional gravity wall.

ACKNOWLEDGEMENTS
The County Surveyor, Essex County Council
Department of Transport, Eastern Regional Office
The County Engineer and Surveyor, Avon County Council

REFERENCES
1.   GREENWOOD J.R., HOLT D.A. AND HERRICK G.W.   Shallow slips in highway embankments constructed over consolidated clay.   Proc, Symp on Failures in Earthworks, London, March 1985.
2.   SZYMONIAK T., BELL J.R., THOMMEN G.R. and JOHNSEN E.L. A Geogrid reinforced soil wall for landslide correction on the Oregon coast.   Proc, 63rd annual meeting of the Transportation Research Board, Washington D.C., January 1984.
3.   FORSYTH R.A. and BIEBER D.A.   La Honda slope repair with Geogrid reinforcement.   Proc, Symp on Polymer Grid Reinf, I.C.E., London, March 1984.
4.   Tensar Corporation.   Reinforced repair of a slip failure in an embankment on highway U.S. 69, a 'Tensar' case study.
5.   MURRAY R.T.   Reinforcement techniques in repairing slope failures.   Proc, Symp on Polymer Grid Reinf, I.C.E., London, March 1984.

TN8. Slip repairs using reinforced soils

R. JEWELL, Binnie and Partners

INTRODUCTION
1. Three types of slip failures in embankments, cuttings and slopes formed in overconsolidated clays have been described in papers to the conference. The application of reinforced soil techniques to such problems is briefly described with reference to these cases.

SLIP FAILURES
2. The types of slip failure are summarised in Table 1. The table draws attention to the geometry of the slip, the time elapsed before the event, and the vital role of water and relic weakness (in a cut or embankment foundation) in causing the failures.

TABLE 1   Types of slip in overconsolidated clay earthworks

Geometry of slip	Shallow typically < 2.5m	Deep seated	Deep seated
Time after construction	Long term	Long term	Short term
Water pressures	Water ingress at crest and ground water	Groundwater (sometimes perched water tables)	Construction induced pore water pressures
Relic weakness	-	Weak planes	Weak plane(s) in the foundation
Type of Structure	Embankments and cuttings	Slopes steepened by toe or surface erosion	Embankments
Paper to Conference	5, 6, 7	3	7

    Failures in earthworks. Thomas Telford Ltd, London, 1985

SHALLOW SLIPS

3. The study reported by Parsons and Perry (ref.1) shows
that shallow slips (the depth below the surface of the
slope rarely exceeding 2m) have occurred over significant
lengths of motorway embankments and cuttings since
construction. The incidence of events in embankments as
well as cuttings, suggests that water and pore water
pressures are perhaps more important in the cause of
shallow slips than local relic weakness in the soil. The
interaction of cracking of the plastic clays and rainfall,
of groundwater rising close to the slope surface, and, in
the case of highway embankments, run-off and seepage from
the pavement and sub-base drainage, are all significant
factors (refs 2 and 3). Water helps speed the softening of
the clays at low effective stresses, and the pore water
pressures reduce the effective stresses available to
mobilise shear resistance.

4. Reinforced soil may be used in the repair of highway
slips. Instead of replacing the slipped and softened soil
with granular materials, the soil may be recompacted with
the inclusion of reinforcement layers, Fig.1.

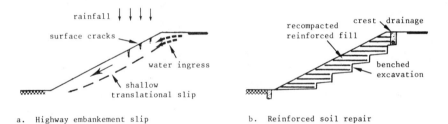

a. Highway embankement slip   b. Reinforced soil repair

Fig.1. Shallow slips (a) typical causes and
(b) reinforced soil repair.

5. Benching at the back of the replacement zone (in this
case reinforced) would eliminate safely the possibility of
slumping when the clay surface behind the strengthened soil
softens, Fig.1. The inclusion of drainage materials behind
the replacement zone to control the rise of groundwater,
particularly in a cutting, would improve stability further.

6. There must still be discussion on whether drainage behind the replacement zone (described above) and simple excavation and recompaction of the softened soil would be a sufficient repair, and clearly many factors need to be considered. In the case of embankments, any supply of water to the top of the slope, and the combination of cracking in the soil and subsequent rainfall, could arguably re-establish the conditions for shallow slipping, Fig.la. The addition of reinforcement may help reduce the size and depth of surface cracks, as well as providing strengthening of the soil. Alternatively, the mixing of lime before recompacting the soil may reduce cracking by reducing the plasticity of the clay, and provide a sufficient increase in soil shear strength; data on the long term low effective stress behaviour of saturated lime stabilised clays would be needed to answer the latter question.

DEEP SEATED SLIPS

7. <u>Long term stability</u>. The case reported by Leadbetter comprises a slope being gradually steepened by toe erosion, and relic weakness and groundwater having a decisive influence on stability. Where flattening of the slope is precluded, reinforced soil may be a convenient technique for improving stability, Fig.2. The key factors are (1) to halt further erosion by toe protection, (2) to increase the toe weight with a steep reinforced zone and, if required, (3) to reduce the driving soil forces by steepening and reinforcing the slope near the crest (using insitu reinforcement where temporary excavation is not possible). The sequence of construction is crucial, building up from the toe first.

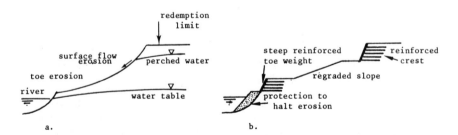

Fig.2. Deep seated slips (a) typical causes and (b) reinforced soil repair.

8. <u>Short term stability</u>. Where there are relic weaknesses in the overconsolidated clay foundation the overall embankment geometry may be governed by the excess pore water pressures generated at the end of construction, rather than long term conditions (ref.3). Reinforcement may be included across the base of an embankment to improve short term stability, until the excess pore water pressures have dissipated, Fig.3. Relatively stiff reinforcement to contain lateral displacements would be required (a) to reduce the severity of loading on the foundation by supporting the outward thrust from the embankment fill and (b) to restrain at the ground surface lateral displacements in the foundation of soil blocks sliding over the relic weak surface, Fig 3. A design that has adequate stability in the long term without need for reinforcement forces would probably be most desirable, and would eliminate the need to consider the long term behaviour of the reinforcement.

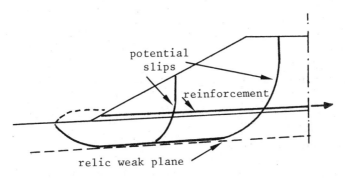

Fig. 3. Short term (end of construction) stability on overconsolidated clay with a relic weakness.

REFERENCES

1. PARSONS A.W. and PERRY J. Slope stability problems in ageing highway earthworks. Proceedings of Conference on Failures in Earthworks, London, 1985.

2. GREENWOOD J.R., HOLT D.A and HERRICK G.W. Shallow slips in highway embankments constructed of overconsolidated clay. Proceedings of Conference on Failures in Earthworks, London, 1985.

3. GARRETT C. and WALE J.H. Performance of embankments and cuttings in Gault clay in Kent. Proceedings of Conference on Failures in Earthworks, London, 1985.

The reference numbers in Table 1 are those given to the papers to the conference.

TN9. Stabilization of slips in cohesive soil by grouting

D. J. AYRES, FSE, FGS, Soil Mechanics Engineer, British Railways Board

SYNOPSIS. Injection as a treatment for embankment slips has been known for over 40 years. Controlled hydrofracture systems are used. A simple but specific site investigation is usually required. It may be applied with minimal disturbance to the running line in the case of railways at a much lower cost than other conventional methods.

1. Many embankments for railways and motorways have been constructed using cohesive fill, rarely with a very safe slope angle. With the passage of time, under the effects of weather, traffic and changing vegetation slips occur with loss of support to the subgrade. For railways, especially, it has been necessary to develop techniques which maintain safety while traffic can continue to pass, if necessary under speed restriction, while remedial work is carried out. Such work should restore the integrity of the track to a fine quality permitting speeds of 125 mph of modern trains. The rail itself is an excellent gauge to note ground movement and forms an integral part in noting the local morphology of the slip.

2. Rail level is maintained by adding more ballast or other granular fill, including ash, under the track to restore level and, as movement continues, a "ballast pocket" is formed under the rail immediately over the tension crack. This pocket collects precipitation water which maintains a high water table in the slope. Permeation grouting to displace this water has had only a limited measure of success. The usual measures of providing support by retaining walls, based on driven or bored piles, give a safe slope but still permit some minor loss of level as the ballast pocket compacts.

3. The system developed by BR is to locate the slip and apply controlled hydrofracture grouting, using plant developed to overcome difficult access. Boring equipment is avoided with all tubes being driven. A steel pipe of 19mm nominal bore is driven into the ground using a hand-held petrol hammer, adding pipe in 1m sections. The expandible point at the base is driven forward and polyethylene tube of 12mm bore and 17 mm external diámeter is fed down the tube. A sintered poly-

ethylene end piece 150mm long and 12mm OD and capped at one end, made specially for BR since 1957, with 100µm pores is fitted at the bottom of the plastics tube as a push fit without adhesive. The steel tube is then withdrawn, leaving the polyethylene tube loose in the hole. A mandrel is lowered on a string to prove clear passage and the tubes on site are subsequently monitored for deformation. When the mandrel will not pass a particular point it represents a ground movement of 10mm to 20mm.

METHOD OF INJECTION

4. A grid of injection points is drawn with the base of the points below the slip plane and a quantity of grout is specified for injection into the various points based upon experience, the local conditions, the rails affected and depth of injection. More than one level of injection may be required. Points near the track are driven at an angle to ensure the proper gauge clearance for traffic.

5. Grouting proceeds in sequence up the slope, using a constant flow pump delivering at 30 litres per minute. For economy a 30% sand cement grout is used which is aerated by 15% to 25% to increase the viscosity and flow properties. The air bubbles also maintain pressure against the sheared surfaces in the hydrofracture plane up to the time of set. The ground is lifted during hydrofracture and so the grout proportions are modified where it flows near and into the ballast pocket to prevent track movement. The grouted pocket of ash and ballast provides a solid track foundation.

6. The usual embankment height treated is from 5m to 9m at a cost of one third of alternative conventional measures. Most slips are of circular form with a few of slab form. On larger sites there can be more difficulty and expense in driving tubes but the cost savings are greater.

TWO CASE HISTORIES

7. To the south of Durham the double track main line crosses the R.Browney on a viaduct and then passes on to an embankment 18m high. The natural ground beneath comprises a 1.5m band of alluvium overlying shale. A slab slide about 30m long occurred in the down (east) side of the 35° slope at 60m from the viaduct. There was a history of previous movement of the slope, as evidenced by ash pockets 3m to 6m deep near the top, and water was impounded further down the bank where the 3m layer of ash and clinker gave way 4m above the toe to a cohesive mass. The body of the embankment below the ash was a mixture of clays, sandy clays and gravels. The driving of steel tubes for the installation of slip indicators was difficult but was found sufficient to pass the slip surface. This, in turn, indicated that treatment based on a driven tube system was possible. The nature of the site precluded the driving of piles and there was no room within the railway boundary to alter slope profile. The west side of the

embankment already extended on the site of a previous railway junction and so the tracks were slued off the slip to a temporary alignment with an appropriate restriction of speed. Removal and replacement of the moving mass with rockfill was considered, with the possibility of treating all the slope as far as the viaduct.

Grouting for the whole 90m of bank was estimated to cost less than 15% of the alternative work and was started in July, 1979. Large masses of water were expelled as the first 4 rows were injected and, after the top row was reached, extra ballast was added to make up the shoulder of the slope. An extra 23m^3 of grout bonded this to the slope and the track was returned to its alignment. Track is now stable and inclinometer access tubes show no slope movement. A total of 1444m^3 of sand-cement grout was injected and work was completed in 16 weeks.

8. To the north of the Firth of Forth, between Kinghorn and Kirkcaldy, the railway runs along the coastal slope about 20m above beach level and there are various areas of instability. At one such, Wee Abden, between two cuttings, the railway has been constructed on ash and clinker fill placed at the top of a natural slope which extends at 30° reducing after 38m along the slope to 25° for a further 24m before a steep toe ending at an outcrop of basalt lava. Soft strata lay parallel to the slope and comprised 2.6m of Boulder Clay overlying 1.5m of micaceous sandstone with silt laminae underlain by basalt lava. Past slipping in a planar form has resulted in much of the Boulder Clay being replaced by ash fill for the first 18m down the slope from the track.

9. Three inclinometers installed at 9m, 18m and 30m from the track indicated that movement was occurring at depths of 3m, 3m and 2.5m depth respectively, i.e. virtually at the interface of the overburden and the bedrock in the form of a slab. Further down the slope, at 46m from the track, the Boulder Clay lay directly on the lava at 6m depth, which coincided with the inclinometer movement. This may have been indicative of a circular movement in the larger mass of soil in this lower section of the slope. Above the railway there is a steep face of weathered lava with ground swiftly rising 50m above track level inland.

10. Slope movement in 1979 undermined the up track on the seaward or east side and track level was corrected by regular filling. A standard pattern of injection points was produced with single level grouting in the planar movement as far as 29m from rail. The soil system made it difficult to drive tubes to depths much lower than the slip plane. Provision was made to regrout in rows where a low average amount was injected. Grouting work took six weeks to stabilize the 29m of track affected, using 209m^3 of grout.

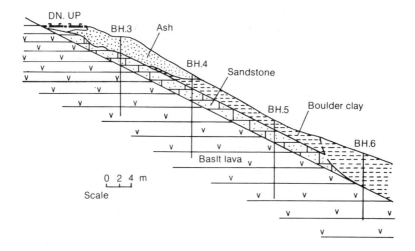

Fig. 1

11. A drain was installed at trackside at the base of the slope on the landward side to channel water coming from the slope above rail level. After treatment the inclinometers in the slab showed no further movement and the track has since remained stable. The lowest inclinometer, in the deeper movement at the base which was ungrouted, continued to show downhill movement. This lower part was then grouted and no subsequent movement has been noted in this inclinometer.

ACKNOWLEDGEMENT
The author is grateful to M.C.Purbrick, Director of Civil Engineering, British Railways Board, for permission to publish this paper.

TN10. Example of a submarine slope failure during installation of piles

E. H. STEGER, Consulting Engineer

One interesting example of a submarine slope failure during the installation of piles for a number of jetties may be mentioned.

The strata at this locality in the Middle East consisted of very recent deltaic and estuarine deposits, some 30 to 35m deep and overlying decomposed bed rock. The alluvial deposits consisted of alternating bands, lenses or layers of silty clays, silts and sands, laminated and with much horizontal interbedding.

Table 1. Berth Extension Site Investigation Report (1)

SOIL LAYER	units	a	b	c	d	e	f	g
Water Content	%	29	33	24	27	28	22	21
Unit Weight	t/m^3	2.0	1.9	2.0	2.0	2.0	2.05	2.1
Clay Content	%	18	32	-	20	36	-	45
Liquid Limit	%	30	35	-	30	30	-	38
Plastic Limit	%	20	22	-	18	18	-	20
Plasticity Index	%	10	13	-	12	12	-	18
Undrained shear strength S_u	t/m^2	2-8	2-3	-	4-20	8-40	-	30
Sensitivity	-	2-4	2-3	-	2-3	2-3	-	-
Effective internal angle of friction ϕ'	degrees	-	31	-	27	33	-	-
Effective cohesion c'	t/m^2	-	1.0	-	0	0	-	-

NB: Estimated values are not shown
- a - Soft to medium silty clay and clayey silt
- b - Very soft to soft silty clay
- c - Fine silty sand and coarse silt
- d - Fine silty sand and coarse silt
- e - Silty clay
- f - Sand and silt
- g - Clay and silt

Prior to construction, ground level was about 2.5m. This was raised to +3.8m along the line of the relief piles, which were installed first, whilst the sea-bed was dredged to an angle of

Failures in earthworks. Thomas Telford Ltd, London, 1985

Table 2. Berth Extension Site Investigation Report (2)

SOIL LAYER	units	a	b	c	d	e	f	g
Water Content	%	34	31	23	23	29	20	28
Unit Weight	t/m³	1.9	1.9	2.05	1.9	2.0	-	-
Clay Content	%	29	31	-	-	27	-	32
Liquid Limited	%	39	35	-	30	32	-	32
Plastic Limit	%	21	20	-	19	19	-	18
Plasticity Index	%	18	15	-	11	13	-	14
Undrained shear strength S_u	t/m²	1-10	2-5	-	3-30	5- 30	-	30
Sensitivity	-		2-4	2-3	-	2-3	2-3	-	-
Effective internal angle of friction ϕ'	degrees	-	31	-	-	34	-	-
Effective cohesion c'	t/m²	-	1.4	-	-	1.2	-	-

NB: Estimated values are not shown

    a - Soft to medium silty clay and clayey silt
    b - Very soft to soft silty clay
    c - Fine silty sand and coarse silt
    d - Fine silty sand and coarse silt
    e - Silty clay
    f - Sand and silt
    g - Clay and silt

Table 3. Summary of drained parameters adopted for stability analyses

	After Failure	Design
Effective internal angle of friction ϕ'	34° - 38°	33°
Effective cohesion c'	0.4 - 0.8 t/m²	0.0t/m²

about 1:3. This had not quite been reached at the time of the first pile failures, which occurred in the pre-cast relief piles, after a great number had been driven in quick succession and at right angles to the shore. Subsequent investigations showed that a considerable number of piles were broken along a reasonably circular line, resulting in heave and distortions of the dredged sea-bed slope. In addition, the tidal range was increasing at the time of failure and an exceptional amount of rain had fallen in the preceding days.

After dredging was completed to 1:3 the main, hollow, spun marine piles were installed and soon seaward dislocations of up to 50cm were noted. Parts of the sea-wall also moved outwards.

In investigating the reasons for these events, it appeared that the original design concentrated only on the final condition of the submarine slope. It further seemed to me, as the result of local experience and despite the parameters obtained from three site investigations, that these parameters were much too high.

The original stability analysis was based on partial factors of safety; that carried out by me on the methods of Bishop, Price and Morgenstern using total factors.

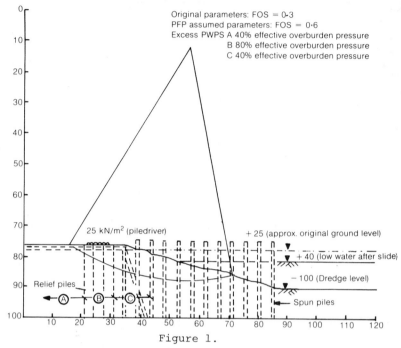

Figure 1.

Table 4. Stability Analyses
(Incomplete Dredging Minimum Tide Level = -1.4m)

| | TYPE OF ANALYSIS | SOIL PARAMETERS | | | SOURCE | POREWATER PRESSURE CONDITION | MIN. FACTOR OF SAFETY |
		Su kN/m²	c' kN/m²	∅'			
A	Total stress (undrained)	10 to 26	-	-	Original Report	-	1.3
B	"	9 to 23	-	-	EHS assumed 10% reduction	-	1.2
C	Effective stress (drained)	-	0.0	33°	Original Report	Normal ground water lvl	2.1
D	"	-	0.0	33°	Original Report	Excess PWPs due to relief pile driving	0.8
E	"	-	0.0	25°	EHS assumed	normal grd water lvl	1.5
F	"	-	0.0	25°	EHS assumed	Excess PWPs due to relief pile driving	0.6

(rows A–F grouped under side label: BEFORE)

430

	TYPE OF ANALYSIS	Su	c'	ø'	SOURCE	POREWATER PRESSURE CONDITION	MIN. FACTOR OF SAFETY
G	Total stress (undrained)	10 to 26	-	-	Original Report	-	1.7
H	"	10 to 26	-	-	Original Report	-	3.1 (for failure plane analysed in A&B above)

(AFTER)

Su - this is a range of values of undrained shear strength
c' - effective cohesion ø' - angle of effective internal friction

Table 5.   Proposed 1:3 Profile Stability Analyses
Relief piles in place (completed dredging-Min. Tide lvl=2.4m)

	TYPE OF ANALYSIS	SOIL PARAMETERS Su	c'	ø'	SOURCE	POREWATER PRESSURE CONDITION	MIN. FACTOR OF SAFETY
J	Total stress (undrained)	10 to 26	-	-	Original Report	-	1.0
K	"	9 to 23	-	-	EHS assumed 10% reduction	-	0.9
L	Effective stress (drained)	-	0.0	33°	Original Report	normal ground water lvl	1.7
M	"	-	0.0	33°	Original Report	Excess PWPs due to Row M piling and part relief piling	1.2
N	"	-	0.0	25°	EHS assumed	normal grd water lvl	1.3
P	"	-	0.0	25°	EHS assumed	Excess PWPs due to Row M piling and part relief piling	0.9

The analyses indicate that the stability of the dredged slope
was assured only in the long-term. However, it appeared highly
unlikely that this condition could ever be achieved because of
the steepness of the slope and of the high porewater pressures
generated by the pile driving.

In the context of this very brief summary, I have naturally
only been able to concentrate on the bare essentials. I do want
to stress however the importance, in design, of using a 10%
reduction to the undrained shear strengths to allow for the
effects of piling, sample disturbance and other unknowns. This
applies particularly to sensitive, recent, deposits, where it
is equally essential to use parameters which are known to work
rather than data obtained or tested under difficult or
questionable conditions. Subsequent investigations incidentally confirmed the lower parameters. Construction proceeded
satisfactorily by pre-boring pile positions and by using
special wicks and pumps to reduce porewater pressures to a
minimum.

431

TN11. The use of reticulated Pali Radice structures to solve slope stability problems

I. W. ELLIS, FICE, Managing Director, Fondedile Foundations Ltd

SYNOPSIS. The in-situ construction of earth reinforced structures by the installation of small diameter grouted piles can offer a practical and economical solution to slope stability problems. Two examples of such solutions are briefly set out in this technical note:-

A ROAD BRIDGE

1. A tightening of the expansion joints of a bridge deck over a newly construction motorway was the first indication that the Northern abutment had undergone both sliding and rotation.

2. Fondedile Foundations Limited were approached to determine if their specialist techniques could provide a viable solution.

3. Analysis suggested that a circular slip plane approximately 2.5 metres below the underside of the abutment foundation had a calculated factor of safety of approximately 0.9.

4. This mode of failure was supported by the recorded horizontal and vertical displacements of the abutment.

5. The solution adopted, to arrest the movement and stabilize the abutment, consisted of direct underpinning of the abutment using 170/133 mm diameter Pali Radice piles having a safe working capacity of 150 KN, penetrating approximately 12.5 m below the assumed slip plane at an intensity of 4 No. per metre run of abutment. In addition a Reticulated Pali Radice structure was constructed in front of the abutment. See Fig. 1. This consisted of 10 No. 133 mm diameter piles per metre, installed in a network pattern designed, to act as an in-situ reinforced earth retaining wall, penetrating through the slip plane to provide a shear key to resist the motive slip forces. These forces were calculated to have a residual value of approximately 600 KN per metre.

6. The Reticulated Pali Radice Structure was connected at the surface by a continuous R.C. capping beam, constructed by The Main Contractor to contact directly with the toe of the abutment foundation to produce a degree of structural continuity between the abutment and the reticulated structure.

Failures in earthworks. Thomas Telford Ltd, London, 1985

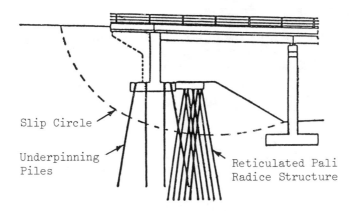

Slip Circle

Underpinning
Piles

Reticulated Pali
Radice Structure

Fig. 1 Typical section through the North Abutment

7.   The Reticulated Structure was extended by 3 m to both
the North and South extremities of the 12 m wide abutment to
cater for any 'side' effects of unstable ground movements.

8.   Where possible, all piles were dry-drilled using a
continuous flight auger to limit the amount of drilling water
injected into the ground.

9.   A total of 223 No. piles were installed within a tight
programme period of eights weeks using three drilling rigs.
The piles under the bridge deck were constructed in a headroom
of 3.5 metres. Movement of the abutment was continuously
monitored throughout the pile installation period and was
shown to be progressively arrested as the intensity of the
piles installed increased.

10.   The affected bridge deck was subsequently jacked back
to it's original position.

SLOPE INSTABILITY EFFECTING HOUSING ESTATE

11. Lethbridge Close Housing Estate at Lewisham consists of
high rise blocks of housing, situated within an old chalk
quarry, the North facing slope of which is an extension of a
natural escarpment of Eocene strata. The surface of this
slope is covered by fill materials up to 4 metres thick.

12. At the top of the slope is a small development of
private houses constructed in the 1930's.

13. Following the removal of diseased elm trees, the slope
showed signs of instability with serious subsidence to the
service road to the garages of the private houses at the top
of the slope, together with minor surface slips on the slope
itself.

14. The condition of the slope was monitored and a full
site investigation instigated.

15. This investigation, together with slope stability
calculations, identified the problem to be mainly in the fill
materials facing the slope. Critical slip surfaces were
identified having factors of safety approaching unity
extending over the whole of the slope. Deep seated failures
affecting the houses at the top of the slope were considered
to be extremely unlikely.

16. The area of the slope that required to be stabilised
consisted of a section approximately 70 metres in length,
varying in height from 10 metres at the Western end to 15
metres at the Eastern end. The slope angles varied
considerably both along the slope and in some areas from top
to bottom of the slope. The steepest parts were at a slope of
about sixty degrees to the horizontal with an average slope
angle of about forty degrees.

17. A Reticulated Pali Radice Structure, constructed by
Fondedile Foundations Limited provided the solution.

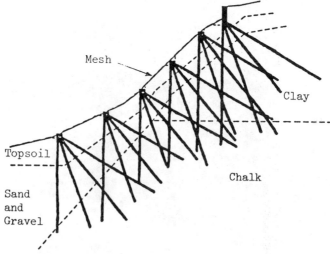

Fig. 2 Typical section showing Reticulated Pali Radice
           Structure

18. The Reticulated Pali Radice Structure is a method of providing an in-situ earth reinforcement, and is based on installing a network of Pali Radice grouted piles in a predetermined network such that any critical slip plane is intersected by a configuration of piles. Thus, the natural resistive forces on the slip plane are improved by the combined shear resistance of the piles together with the soil mass. At the same time, the ability of the general soil mass to withstand tensile and shear forces is greatly improved.

19. Since the resistance of the piles adds to that already naturally existing, a gradual improvement is achieved as the piles are installed and the treatment can be applied to active slips in a very cost effective way.

20. A computer analysis based on the Janbu method was carried out to determine the extent of the intervention required. This showed that an average of nine piles per metre run of slope would strengthen all potential slip planes with a calculated factor of safety greater than 1.3.

21. Piles were constructed as shown in Fig. 2.

22. The 100 mm nominal diameter piles had lengths varying from 6 to 10 metres to found generally in the underlying chalk. All piles were constructed using sand/cement grout and reinforced with a single 25 mm reinforcing bar.

23. The piles were connected on the face of the slope by 500 mm x 400 mm reinforced concete capping beams, generally following the contour lines, to improve load distribution. These beams were landscaped into the slope and covered with top soil.

TN12. The stabilisation of a landslipped area to incorporate a highway by use of a system of bored piles

E. A. SNEDKER, BSc, MICE, Regional Geotechnical Engineer, Department of Transport, West Midlands Region

SYNOPSIS. The purpose of this Note is to record a technique which has been successfully used to build a highway across an area of landslip. Although the technique (referred to as 'Stitching') was designed some 15 years ago, and appeared novel at the time, it now seems to be finding more favour. The system of 'stitching' this hillside has been in position for almost 14 years and has proved to be successful.

THE AREA REQUIRING TREATMENT

1. The hillside which was stabilised is an escarpment face in Staffordshire (UK) where the outcropping strata are Newcastle Beds overlying Etruria Marl. The Newcastle Beds generally consist of grey sandstones and shales whilst the Etruria Marl is mainly mottled red and purple marls and clays. Both beds form part of the Upper Coal Measures series which dip into the hillside at about 6°. For a distance of about 150m downslope of the Newcastle Beds the hillside had been subject to natural land slippage. Considerably evidence of active and past slippage was available and many springs indicated high water tables. Aerial photographs suggested that the area may have been part of a more extensive past movement. A generalised section is shown in Figure 1.

2. The line of a new dual carriageway trunk road had been established part way up this slope, crossing the slipped area transversely. The alignment of the road in the longitudinal direction could not avoid areas of shallow cut and fill although these were minimised. The depth of disturbed material containing slip planes appeared to extend as deep as 7m and generally consisted of weathered marl with detritus from the Newcastle Beds. The residual angle of friction of this material was shown to be 13° both by laboratory testing and back analysis.

3. The design problem therefore was to stabilise the hillside both under any formed embankment and within any cutting slope excavated in the hillside such that any major movement would not affect the highway. It soon became clear that within the boundary constraints of the highway the initial design proposals of relatively flat slopes would prove inadequate. In a number of places they reduced the factors of safety to well below unity on the already slipped escarpment and some additional measures

were obviously needed.

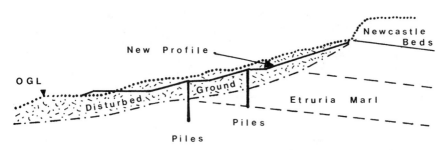

FIG 1

THE DESIGN OF STITCHING

4. A design was developed incorporating two rows of reinforced bored piles along the escarpment running almost parallel to the highway. The topography and varying ground conditions
created the need to divert slightly from this parallel alignment. In effect the piles 'stitched' the foundered mass to the
stable material below, and they were bored such that at least
one third of their length was founded in undisturbed Etruria
Marl.

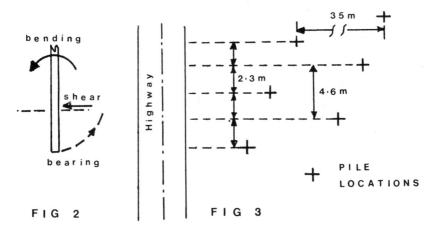

FIG 2

FIG 3

5. An assessment was required of the design considerations
which needed to be taken into account. These can be seen in
diagramatic form in Figure 2. On each of the worst potential
slip surfaces the shear resistance which would be required to a-
chieve a minimum Factor of Safety of 1.2 was calculated. The
shear resistance available from the soil was established and
the deficit thus needed to obtain this factor of safety was
considered to be provided by the piles intersecting the poten-

tial slip surfaces. The bored piles therefore were designed to provide this shear resistance and supplement that provided by the soil.

6. If these shear forces were to be developed in the bored piles then a corresponding bending moment would be developed and consequently the reinforcement was designed to cope with this as well. Having produced a design to cater for the assessed bending and shear it was then necessary to check the bearing of the portion of the pile embedded in undisturbed material, such that it could not fail under the designed loading by rotational failure.

7. The resultant design for this particular slope was for the use of 915mm diameter bored cast in place concrete piles with main reinforcement of 36Nᵒ, 32mm diameter steel bars equally spaced on a pitch circle diameter of 762mm. The piles were designed to be spaced at 4.6m centres parallel to the line of the highway but staggered in two parallel rows to give an effective spacing of 2.3m. The pile lengths varied depending on location from 9.2m to 13.7m. A typical plan of the pile locations is shown in Figure 3.

8. Berms were incorporated in the slope to provide access for the boring rigs and also to provide some small improvement in stability. Counterfort drainage was provided in the new cutting slope to provide some improvement in the upper foundered layers particularly upslope of the piles.

PERFORMANCE

9. The tops of the piles at ground level were monitored for several years after completion of construction and no measureable movement was recorded. The overall slope has remained stable in spite of some very wet conditions.

ACKNOWLEDGEMENT

10. I would like to thank the Director (Transport) West Midlands Regional Office Department of Transport for permission to record this Technical Note.

TN13. Incidence of highway slope stability problems in Lower Lias and Weald Clay

J. PERRY, BSc, MSc, MIMM, FGS, Higher Scientific Officer, Transport and Road Research Laboratory

INTRODUCTON

1. A survey of highway earthworks has been conducted with the aim of identifying the basic factors affecting the stability of the side slopes of cuttings and embankments and quantifying long-term problems. The survey concentrated on lengths of motorway where overconsolidated clays predominate. The first part of the survey, with details of procedures and analyses, has already been presented to the Symposium (see Parsons and Perry. Slope stability problems in ageing highway earthworks). As an extension of this survey a study has been made of earthworks in two further predominantly clay materials, the Lower Lias and Weald Clay on the M5 and M23 motorways respectively.

2. The lengths of motorway slope surveyed were approximately 52 km of Lower Lias and 18 km of Weald Clay. The ages of the earthworks on the M5 and M23 motorways, at the time they were surveyed, were 13 and 9 years respectively, measured from the date the motorways were opened to traffic.

SLOPE FAILURES

3. During the survey, failures were encountered mostly on embankment slopes with only a single slip on a cutting slope of Lower Lias.

4. The depth of the failure plane below the slope surface rarely exceeded 2m. The type of slope failure observed varied from slab to shallow circular but with most slips being a combination of translational and rotational movement. These types of shallow slip have been observed in many other geologies.

5. The observed method of reinstatement of a slope failure was to excavate the failed material and replace with a free-draining granular material such as crushed limestone. Top-soil has been applied in most of the repaired areas.

ANALYSIS OF DATA
Geology

6. The failures in areas of Lower Lias and Weald Clay were, with one exception, on embankment slopes. Embankments of Lower Lias construction exhibited a 1.1% failure rate with a

total length of failed slope of 320m. The failure rate is defined as the length of failed slope, parallel to the centre of the road, expressed as a percentage of the total length of slope with given characteristics. Weald Clay embankments showed a failure rate of 1.6% with a total length of failed slope of 190m. When compared to other geologies studied earlier in the survey these failure rates are low. Gault Clay, for example, exhibited a 9.1% failure rate in rather older embankments of 22 years of age.

7. The one failure in a cutting slope on Lias clay was over a 15m length and represented a 0.1% failure rate.

Geometry of slope

8. Most of the embankment slopes constructed with Lower Lias or Weald Clay in the areas surveyed tended to have a slope gradient of either 1:2 or 1:2 1/2 (vertical:horizontal). However, sufficient variation existed to provide some indication of the effect of slope angle on the occurrence of failures. Results are given in Fig 1 for the two geologies. Only results for geometries with a length of slope in excess of 50m are included. Generally the trend of failure rates is logical for both geologies, with the failure rate increasing with increasing height and angle to a maximum of about 10% for slopes of 1:2, more than 5.0m high. Although earlier studies of different geologies used in embankment construction have shown a trend in many instances for a decrease of failure rate with increased slope angle this trend is not exhibited in the embankments of Lower Lias and Weald Clay.

9. Angles of cutting slopes were not as uniform as those of embankments but the predominant slope for both geologies in the areas surveyed was 1:3. The one failure in the Lower Lias was at a point where the geometry was most severe with a slope of 1:2 1/2, more than 5.0m high. The calculated failure rate is 9.2% corresponding to a 15m length of failed slope at this

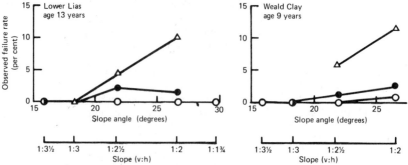

Fig. 1 Relations between failure rate of embankment slopes and geometry of slope

geometry but the sample size is very small, 0.16 km out of a total length of Lower Lias cutting of 24.28 km. Cutting slopes of Weald Clay showed no signs of failure even at the steepest observed slopes of 1:2 1/2, more than 5.0m high.

Age of earthworks

10. Although only one construction age is available for each geology, comparisons can be made between the two over-consolidated clays. It can be seen from Fig 1 that they both have fairly similar failure rates for a given geometry, but the Lower Lias earthworks, at 13 years of age at the time of survey, were older than the Weald Clay earthworks at 9 years of age.

CONCLUSIONS

11. As a result of a survey of earthworks in Lower Lias and Weald Clay the following conclusions have been reached.

1. Lower Lias and Weald Clay embankments 13 and 9 years old respectively, exhibit the same types of shallow slip failure seen in earthworks constructed of other geological strata, but not to such a large extent as other overconsolidated clay embankments.

2. There have been no failures in cuttings in Weald Clay in the areas surveyed. A short length of Lower Lias has failed but only at the most severe geometry.

3. Generally the Lower Lias and Weald Clay embankments show a greater failure rate as slope angle and height increase.

4. Embankments constructed of either material show similar failure rates for the same slope geometry.

ACKNOWLEDGEMENTS

12. The work described in this paper forms part of the programme of the Transport and Road Research Laboratory and the paper is published by permission of the Director.

Any views expressed in this paper are not necessarily those of the Department of Transport.

TN14. Slope failure in low and high plasticity clays

A. J. BATCHELOR, Director, Holequest Ltd, M. C. FORDE, BEng, MSc, PhD, MICE, MIHT, Senior Lecturer, and B. H. V. TOPPING, BSc, PhD, Lecturer, Department of Civil Engineering and Building Science, University of Edinburgh

SYNOPSIS

Effective stress slope stability analyses are discussed with respect to earthworks in both low and high plasticity soils. Attention is drawn to the significance of equalisation of porewater pressure in consolidated undrained triaxial testing.

INTRODUCTION

1. In general slope stability failure is not anticipated by Engineers. Frequently failure occurs as a result of an inadequate site investigation, an inadequate evaluation of the geological history of the site or an inadequate understanding of the interaction between the groundwater regime and the soil parameters. The net result is that the slope of the earthworks, be it cutting or embankment, is too steep.

HISTORY OF EARTHWORKS

2. The recent and geological history of the site are of vital importance. For example the widening of an existing deep cutting in London Clay where a pre-existing slip surface exists would give rise to the need to consider long term residual shear strength parameters for O ´. In earthworks in Glacial Tills, where the Brittleness Index is lower, the problem would be lower. It might even be innapropiate to even consider choosing residual shear strength parameters for first time excavations in these latter instances.

SPECIFIC FACTORS TO CONSIDER

3. The specific geotechnical factors to be considered in the evaluation of the stability of earthworks include:

    (a) Geological History – whether large strains have already occurred or are likely to occur, thus leading to residual shear strength parameters.

    (b) Whether fissuring of the soils may have occurred – e.g. London Clay and the Glacial Tills of Strathclyde Region.

    (c) The phreatic surface in the short term and the long term, including possible variations.

    (d) The soil density profile – both vertically and laterally.

(e) The soil shear strength parameters and drainage environment - i.e. total stress or effective stress shear strength parameters.

(f) The accuracy of the site investigation undertaken in order to establish the parameters.

TYPICAL PARAMETERS

4.   Early investigations into slope stability problems and the design of new works frequently relied upon total stress shear strength parameters.

5.   More recent investigations into slope stability problems have concentrated upon effective stress shear strength parameters. However due to errors in the testing procedure these parameters have frequently been incorrect. The most common source of error has been in the laboratory testing.

6.   The problem arises due to the fact that for reasons of expediency effective stress parameters are obtained in practice from consolidated undrained triaxial tests with pwp and volume change measurement made against a back pressure. This test utilises Terzaghi's principle of effective stress whereby the effective principal stresses are calculated from total stress minus porewater pressure. If the test is undertaken too rapidly and porewater equalisation has not occurred then there is an overestimate of c' and an underestimate of ϕ'.

8.   From the above it is argued by the writers that where c' is significantly in excess of zero, with the possible exception of undisturbed London Clay, then the test has been undertaken too rapidly.

9.   Typical peak effective stress shear strength parameters are given below:

Soil Type	LL/PL	Apparent Cohesion c'	Angle of Shearing Resistance ϕ'
London Clay	75/27	0	21
Cheshire Clay	35/20	0	30
Lothian Till	40/20	0	35
Highland Till	25/15	0	40
Silty Clay	27/18	0	25

Table 1

CASE STUDIES

10.   The first, unamed, case study refers to a major slip on a Trunk Road in Scotland. The highway cutting slipped before the road was opened. The original S.I. involved total stress shear strength parameters. Remedial works involved a further investigation where more total stress parameters were obtained. Remedial works involved physical restraint of the slope without altering the basic slope of 1:1.5. The remedial works have shown ongoing signs of distress. The authors have established that a major stratum of parameters: $c'=0$, $\phi' = 26$ lies in between strata with parameters: $c'=0$, $\phi'=35$. The above parameters were combined with a high phreatic surface seeping out of the slope 1 metre above the toe of the slope. It will thus be appreciated that in both the original geotechnical investigation and the subsequent remedial works that insufficient attention was paid to effective stress shear strength parameters.

11.   The second, also unamed, case study refers to a dam in the South of England. The original design parameters used were (Ref 1):

Analysis	Dam		Dam Foundation		F.O.S.
	c'	ϕ'	c'	ϕ'	
1	9	20	6	20	1.45
2	0	20	6	20	1.36
3	0	20	0	20	0.96

Table 2

12.   It can thus be seen from the above example that to assume $c' > 0$ kN/m**2 is potentially dangerous. In many S.I. reports the authors have seen numerous examples where C' has been quoted as being of the order of 25 kN/m**2.

CONCLUSIONS

13.   These may be summarised as:

(1) Essential to establish soil macro-structure or fabric.

(2) Effective stress shear strength parameters should be investigated (including residual stress parameters as appropriate).

(3) CUD triaxial tests with pwp measurement should be undertaken sufficiently slowly to permit equalisation of pwp.

(4) The phreatic surface must be accurately established.

REFERENCES

1.   TOPPING B.H.V, FORDE M.C. and LEE C.H. Interactive microcomputer aided design of earth dams. Proc CIVIL COMP-83, London, 1983.

TN15. Smooth slip planes in clay fills resulting from soil machine interaction

I. L. WHYTE, BSc(Tech), DipASE, MICE, MIHT, Lecturer, and
I. G. VAKALIS, PhD, Former Research Student, University of
Manchester Institute of Science and Technology

SYNOPSIS.   Smooth slickensided slip planes can develop between the compacted layers of clay fills under particular conditions of soil-machine interaction. A laboratory model study
using a smooth wheel roller reproduced these layers and enabled both a formation mechanism and limiting condition to be
identified. Methods of detecting these surfaces on site are
discussed along with possible remedial measures.

INTRODUCTION

1. Compacting clay soils in layers produces anisotropy of
fill properties. In addition, incipient slip plane discontinuities can result from the interaction of earthmoving plant
and the compacting soil. These smooth surfaces have been reported with most types of equipment (smooth wheel, tamping
roller, pneumatic tyres) on highway embankments and earth dams
such as Bewl Bridge, Ardingley, Pournari and Carsington (refs.
1-3). Detailed research information had not previously been
reported and so a laboratory study was carried out using a
model smooth wheel roller on an intermediate to high plasticity clay.

TEST PROCEDURE

2. The model roller chosen was the front drum of an 8-tonne
roller at 1/3 scale. Preliminary studies showed that maximum
compaction was virtually achieved after 10 passes of the roller,
and compactive effort was varied by adjusting the loose layer
thickness between 20 mm and 70 mm (these correlating with prototype placement practice). Compaction curves were established
and measurements taken of soil movements (displacement and distortion) both at the top surface and at the base interface of a
compacted layer. Shear strengths were found from shear box and
triaxial testing, test details being given in refs.1-3.

DISCUSSION

3. Three forms of interface were found in the model and
termed 'bonded', 'intermediate' and 'polished'.
4. 'Bonded' interfaces occurred with moisture contents dry
of optimum, the term being used since layer separation was
difficult, if not impossible, to achieve. The fill however,
exhibited anisotropy of dry density and void ratio with depth

with discontinuities of these parameters at an interface.

5. 'Polished', or smooth interfaces developed with moisture contents wet of optimum and were dependent upon compactive effort for their occurrence. Layer separation was readily achived and the interface surface exhibited a polished appearance. Fill compaction was more uniform with depth but a shear strength discontinuity existed at the interface.

6. 'Intermediate' interfaces were noted with moisture contents around the optimum value. Layer separation was possible, but the surface texture had not been smoothed or 'polished' by the action of the compacting equipment.

7. Smooth interface development was found to be dependent upon both soil consistency and compactive effort, increasing compaction reducing the moisture content at which polishing occurred. The wetness index (W.I) was found to be a good indicator of the problem.

$$W.I. = \frac{w_L - w}{w_L - w_{opt}} \qquad (1)$$

where w_L = liquid limit, w = moisture content,
w_{opt} = optimum moisture content for effort used.

8. In the model, smooth interface surfaces developed between successively compacted clay layers when W.I.< 0.96 (model). Modelling laws (refs.2,3) predicted that such surfaces could be expected at prototype scale when W.I.< 0.93 (prototype).

9. A mechanism for the development of a smooth interface was obtained from movements of the soil peds during compaction, Fig.1.

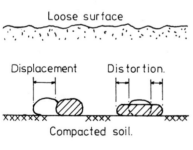

Loose surface

Displacement   Distortion.

XXXXXXX   XXXX   XXXX   XXXX

Compacted soil.

Fig.1 Soil movements.

10. Soil displacements were greatest with the first pass of a roller and were very small with subsequent passes. The displacements, however, did reveal a trend of being greater with roller passage as moisture content increased. For 'bonded' interfaces, distortion of a ped ceased after the initial roller passes whilst for 'smooth' interfaces distortion continued (at a decreasing rate) with each pass of the roller. Distortion measurement was found to indicate the development of polished surfaces in the fill and enabled the limiting wetness index to be identified.

11. Both displacement and deformation of soil peds in the compacting layer would appear to be necessary since soils displaced with normal loading or distorted without displacement on a compacted surface failed to develop polishing. Smoothness developed, however, if both these movements were combined.

12. Fig.2 shows the reduction in unconfined compression strength along an interface relative to a homogeneous uniform sample, and Fig.3 shows the results of drained shear box tests

446

along an interface.

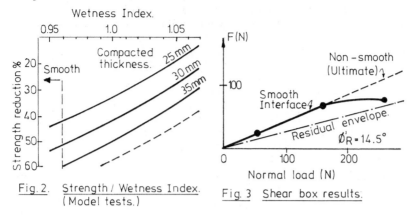

Fig.2. Strength / Wetness Index. (Model tests.)

Fig. 3 Shear box results.

13. In fills with smooth interfaces, the reduction of > 40% on unconfined strength shows that immediate stability could be greatly reduced, whilst Fig.3 shows that residual states can occur under increasing effective vertical stresses. Even at low vertical stress, the brittleness is such that residual states can be approached with relatively small displacements beyond failure.

SUMMARY

14. Incipient slip planes can develop in wet clay fills due to soil displacements induced by earthmoving plant. Such smooth interfaces are not detectable at the fill surface and can only be revealed by careful excavation. The following site conditions can be indicative of problem interface development.
(a) Excessive soil deformations induced by a roller
(b) A compacted surface of even texture with few surface voids
(c) Overcompacting weak fills at trafficability limits,particu-
    larly with a general method specification.
15. Should smooth interfaces develop on site then they can be eliminated by varying compaction procedures (e.g. reducing the number of passes) or by scarification (ref.4).

REFERENCES
1. WHYTE I.L. and VAKALIS I.G. An investigation into the development of polished interfaces in clay fills. Proceedings 8th Eur.Conf. Soil Mech.Found.Eng., Helsinki, 1983, Vol.1, 323-326.
2. VAKALIS I.G. The development of smooth interfaces in clay fills. Ph.D. Thesis, 260 pp. UMIST, 1984.
3. WHYTE I.L. and VAKALIS I.G. Modelling the interface and shear movements between compacted clay layers. To be published.
4. PAVLAKIS G. Discussion. Proc.of the Clay Fills Conference, I.C.E., London, 1978, 266-269.

TN16. Slope drains in an old cutting in London Clay

D. M. FARRAR, MSc, MInstP, Principal Scientific Officer, Ground
Engineering Division, Transport and Road Research Laboratory

SYNOPSIS.   This note describes observations made following
installation of slope drains in an old clay cutting to control
instability. The drains have been effective throughout the
first three years, keeping water tables down to predicted
values except very near to the drain heads. However, the
results indicate that such drains may not always be effective
during the first winter after their installation.

INTRODUCTION
1. Slope ('Counterfort') drains are often used as a
remedial measure to prevent or control instability in cutting
slopes. Hutchinson (1977) gives a method for predicting the
reduction in pore water pressures as a result of a drainage
installation. He also reviews some case histories in natural
slopes – but not cuttings – and concluded that slope drains
may not be effective near the drain head, or in the first
winter after their installation. In addition, the filter
materials may become clogged.
2. In order to test the validity of Hutchinson's method in
cuttings, observations have been made following installation
of slope drains in an old clay cutting on the A12 Trunk Road
at Romford, Essex.

HISTORY OF SITE
3. The cutting for the A12 Trunk Road was originally made
in 1934, through brown London Clay. After a failure in 1969,
the slope was cut back to 10 – 13degs. There was some further
movement, and it was decided to stabilise the slope with slope
drainage. The drains were at 5m centres and 0.6m width (Fig 1).
The trench was lined with a geotextile fabric (Dupont Typar
3407), then filled with 'Type B filter material' (Dept of
Transport 1976). Finally the ends of the fabric were turned
over the top of the trench. The drains were constructed in
Sept 1981. The opportunity was taken by TRRL to install
standpipe piezometers between three pairs of drains.

RESULTS
4. The minimum depth to ground water measured between Sept
1981 and Sept 1984, midway between the drains, are shown in

Table 1. Also shown are the depths predicted using Hutchinson's method.

Table 1. Measured and predicted depths to ground water

Measured minimum depth to water (m)		Predicted minimum depth (m)
Piezometers 1.5m deep	Piezometers 3m deep	
Undrained parts of slope		
0.9, 0.7	0.6, 0.8, 0.9	–
Top of drained slope (less than 1m from drain head)		
1.1	1.2	2.0
Top of drained slope (3 – 4m from drain heads)		
1.2	2.0, 1.9	2.0
Middle of drained slope		
dry	1.9, 2.1, 2.5	2.0
Bottom of drained slope		
1.3	1.3	1.3
	1.6	1.6

5. The drains have been effective during the monitoring period. There is good agreement between measured minimum depths to ground water and predicted depths, except within 3m of the drain heads. There is a seasonal fluctuation in the undrained slope (Fig 2) and a corresponding though much smaller fluctuation in the drained slope, with a markedly delayed response in some of the lower parts of the slope. The piezometers were also used to measure in situ permeability. Values of between 10^{-8} and less than 10^{-10} m/sec were obtained, the lowest permeabilities being measured at the bottom of the slope.

6. The drains at this site were installed at a favourable time of year at the end of summer when water tables were either low or decreasing. The low permeabilities and delayed seasonal fluctuations observed in parts of the slope suggest that the drains would not have immediately lowered a winter water-table. It would therefore seem prudent to assume that this type of drain will not always achieve the desired effect during at least the first winter after its installation.

7. It should also be noted that these observations were made on an old cutting. In a new cutting in over-consolidated clays, long term changes in pore pressures due to stress relief will also be an important consideration.

CONCLUSIONS

8. Slope drains were found to be effective in keeping down the water table in a 50 year old cutting in brown London Clay throughout the first three years after their installation, except very near the drain heads. The observations suggest, however, that drains may not always be effective during the first winter after their installation.

The work described in this paper forms part of the programme of the Transport and Road Research Laboratory and the paper is published by permission of the Director. Thanks are due to Mr F C Holt, Borough Engineer and Surveyor of the London Borough of Havering, for permission to use the site, and Mr R K Bhandari, also of the London Borough of Havering, for his invaluable advice and assistance.

REFERENCES

DEPARTMENT OF TRANSPORT. Specification for road and bridge. HM Stationery Office, London, 1976.

HUTCHINSON, J N. Assessment of the effectiveness of corrective measures in relation to geological conditions and types of slope movement. Bull. Internat. Assocn. of Engng. Geology, 1976, Vol 16 131 – 155.

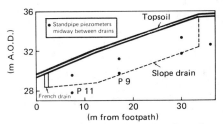

Fig.1 Typical cross-section through slope at Romford

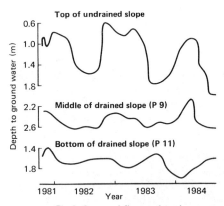

Fig.2 Seasonal fluctuations in water-table at Romford

TN17. Centrifuge model of an embankment failure

M. C. R. DAVIES, BSc(Eng), AKC, MPhil, PhD, FGS, Lecturer in Soil Mechanics, Department of Civil and Structural Engineering, University College, Cardiff

SYNOPSIS. The 'case study' of an embankment failure in a centrifuge model shortly after construction during centrifuge operation is described. Monitoring of deformations and pore water pressures throughout the foundation prior to failure showed the development of a mechanism of progressive failure.

INTRODUCTION
1.   Geotechnical processes may be reproduced realistically at a small model scale in the centrifuge. The ability to repeat experiments, the control of model parameters and the quantity of high quality information available from centrifuge models combined with the relative cheapness of performing a suite of model tests, compared with a single well instrumented proto-type, make centrifuge modelling an attractive method for the study of failure in earthworks.
2.   During a series of centrifuge tests to examine the staged construction of embankments on soft clay foundations (ref. 1) one model embankment failed a short time after construction. The instrumentation was monitored during the period between the end of construction and the delayed failure allowing the development of the embankment slip to be followed.

MODEL TEST
3.   The 1/100 scale foundation layer consisted of a cake of kaolin clay overlain by a thin layer of sand. Miniature pore water pressure transducers were installed in the cake and load cells on the surface of the foundation. Dark clay was sprinkled in a grid on the visible face of the model; breaks in which indicated the presence of any dis-continuities in the found-ation, fig.1. A reflective displacement marker was placed in the centre of each clay grid square; the location of which was measured from photographs taken throughout the test.

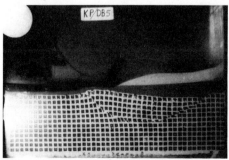

Fig. 1 Failed Embankment

Failures in earthworks. Thomas Telford Ltd, London, 1985

4. Employing the method described in reference 2 the foundation was induced with a typical stress history for a soft alluvial clay; an overconsolidated layer overlying normally consolidated clay. The embankment was constructed during centrifuge operation, by pouring sand from a hopper to a height of 6.0m (prototype scale). About 6 seconds after the end of construction, which corresponds to a prototype time of 17 hours, the embankment failed with a near perfect circular slip, fig. 1. It is clear from the delay in the occurrence of the failure that the factor of safety at the end of construction was very close to unity.

5. Using the in-situ undrained shear strength profile of the foundation measured with a vane during centrifuge operation a total stress stability analysis produced a factor of safety at failure for the actual failure surface of 1.04. Measured values of construction pore water pressures were used in a Bishop type effective stress analysis yielding a factor of safety at failure of 0.90. The lower factor of safety in the effective stress analysis conforms with the observations collated by Parry (ref. 3).

EMBANKMENT FAILURE

6. Measurement from the photographs taken between the end of construction and the failure of the embankment indicated a large rotation around the toe resulting in the development of large shear strains in the region of the eventual failure surface. Immediately after construction, fig. 2(a) high shear strains were recorded in the region adjacent to the potential slip surface. As time progressed there was a spreading of the highly stressed zones, values as high as 29% being recorded prior to failure. In laboratory tests the shear strain at failure for the clay was 8% in compression and 18% in extension. This indicates that immediately after construction the shear strains near the centreline were well in excess of the failure value, but beyond the toe were below the value necessary to initiate failure.

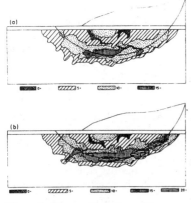 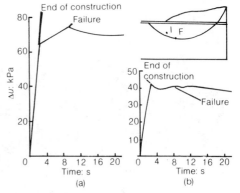

Fig. 2 Shear Strains %
   (a) after construction
   (b) before failure

Fig. 3 Construction pore
   pressures at piezometers
   (a)   F   (b)   I

7. Development of high shear strains along the eventual failure surface was caused by a redistribution of stress in the foundation. Pore pressure response indicates that total stress at F, fig. 3(a), a region which was close to failure, increased from the end of construction until failure. The increased rate in pore pressure generation 1.75 seconds before the embankment failed, indicated the onset of contained local failure. At I, fig. 3(b), a region which initially only developed 10% shear strains, the rate of pore pressure dissipation was greater than the rise in pore pressure due to stress distribution for the first 3 seconds following construction after which pore pressure started to rise, indicating a local increase in total stress.

8. Progressive failure is a phenomenon, found in soil which experiences a reduction in strength below its peak value as the shear strain is increased and is believed to be the major cause of stress redistribution in the centrifuge model. A slip surface will not generally immediately form along its entire length, but will develop initially in the highest stressed zones. Further strains will then cause the ruptured zones to unload, propagation of the failure surface occurring if the 'spare' load capacity of the adjoining unfailed zones is insufficient to compensate for the reduction of load in the previously ruptured zones. Laboratory tests indicate lightly overconsolidated kaolin clay does not display a marked drop in strength following failure. It follows that kaolin has to be subjected to very large strains to cause sufficient drop in strength for progressive failure to propagate.

CONCLUSIONS

9. (i) Centrifuge models provide a convenient method for investigating failure of earthworks. (ii) In the model test of a granular embankment, which failed shortly after construction on a soft clay foundation, zones of high shear strains were observed to propagate in the region of the eventual failure surface resulting in a progressive failure. (iii) Pore pressure measurements indicated the rise in total stress due to redistribution of stress during progressive failure.

REFERENCES

1. DAVIES, M.C.R. Centrifugal modelling of embankments on clay foundations. Ph.D. Thesis, University of Cambridge, 1981.
2. DAVIES, M.C.R. and PARRY, R.H.G. Determining the shear strength of clay cakes in the centrifuge using a vane. Geotechnique, Vol. 32, No. 1, 1982.
3. PARRY, R.H.G. Stability analysis for low embankments on soft clays. Proc. Roscoe Memorial Symp. Stress Strain Behaviour of Soils. Pub. Foulis. 1972.

TN18. Failure of rock slope at Powrie Brae, Dundee

W. M. REID, BSc, MICE, MIStructE, MIHT, Director, Thorburn Associates

SYNOPSIS   This technical note describes the failure during construction of a site slope to a roadway cutting. The failure took place along a bedding plane inclined at 13° to the horizontal and involved an estimated 20,000 cubic metres of glacial till and bedrock. A brief description of the remedial works is also presented.

INTRODUCTION
During 1979 the A929 was being realigned at a location just north of Dundee in the Tayside Region of Scotland. On the 3rd November a large planar rock slip occurred resulting in the lateral translation of an estimated 20,000 cubic metres of glacial till and rock. The failed mass of soil and rock was stabilised on a temporary basis by infilling the area at the front of the slip debris with previously excavated material. Completion of the cutting was delayed while stabilising works to the rock slope were completed.

PLATE 1 - General View of Failure

DISCUSSION OF THE FAILURE

1.   The rock strata in the area of the failure consisted
of micaceous siltstones and sandstones overlain by a 5-6
metre deep layer of glacial till.   On average, the dip of
the strata was about 13° eastwards and bedding planes
existed at this angle over many square metres. A feature of
the strata was the presence of a thin layer of clay on many
of the bedding planes.

2.   At the time of the failure the excavation had
penetrated to a depth of between 11 metres and 15 metres
below rockhead and movement took place from the west face of
the excavation along a clearly defined bedding plane.   At
the north end of the slip the failure plane outcropped at
the pavement of the excavation while at the southern end of
the slip the failure plane outcropped some three metres
above the base of the excavation.   The rock mass apparently
moved as a relatively intact mass and only broke up at the
point when it moved over the face of the excavation.   The
magnitude of forward translation was greater at the southern
end of the failure with minimal forward movement at the
northern end.

3.   Plate 1·shows a general view of the failure taken from
above the east face looking westwards. Plate 2 shows a view
of part of the failed mass of rock and illustrates the
natural joints present within the rock mass.

PLATE 2 - Section of Failed Rock

REASONS FOR THE FAILURE

4.   Tests on the rock strata established that the angle of
friction for rock to rock contact was 24°.   Undrained tests
on the clay taken from the bedding planes also indicated a ϕ

value of $24°$ but with increasing strain the ϕ value reduced to as low as $9°$. Analyses were carried out assuming a friction angle across the failure plane of $24°$ and this indicated that failure was possible if water pressure were allowed to develop within the rock mass. Once initiated, movement would continue even after release of the water pressure, as the friction angle on the sliding surface would reduce to the $9°$ characteristic of the residual value for the clay layer in the bedding plane.

5. The rainfall in the week preceding the failure was relatively high and although no large seepage was noted on the face of the excavation prior to the slip, a relatively strong flow was subsequently observed from the failure debris.

REMEDIAL WORK

6. At the time of failure there remained a requirement for a further 2 metres of excavation to achieve the required carriageway elevations. A redesign exercise was carried out but this failed to achieve major changes to the required depth of the cutting. As further excavation would expose new potential failure planes, it was considered essential that steps were taken to preclude the possibility of a further failure.

7. Two principal alternatives were investigated. The first was to excavate all the rock above the potential failure planes and to re-infill this to create sloping sides to the cutting.

8. The second alternative was to stabilise the potentially unstable mass of rock by the use of anchors or dowels and to install a drainage system within the slope which would reduce the water pressure which could develop along the potential failure planes.

9. After an initial study, it was concluded that the additional cost and the requirement for a very significant increase in land acquisition, made the first alternative an uneconomic proposition.

10. Various alternative forms of anchor systems were considered, including the provision of an anchored concrete wall at the face of the excavation, the use of inclined anchors and the use of shear dowels installed perpendicular to the potential failure planes. The provision of a concrete face wall was unattractive as this required the removal of the stabilising infill prior to the installation of the anchors. The use of inclined anchors was rejected due to the difficulty in creating stable inclined rock faces on which to cast the required anchor blocks.

11. The solution adopted and subsequently installed is illustrated in Figure 1. This involved the installation of 40mm diameter Macaloy dowels in a regular grid pattern across the potential failure planes. Each dowel was first grouted into the rock beneath the lowest potential failure

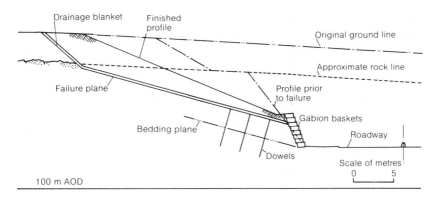

Drainage blanket
Finished profile
Original ground line
Approximate rock line
Failure plane
Profile prior to failure
Gabion baskets
Bedding plane
Roadway
Dowels
Scale of metres
0   5
100 m AOD

FIGURE 1

surface. The dowels were then given a nominal prestress and fully grouted for their complete length. The shear capacity available from each dowel was established by tests which demonstrated that the ultimate shear available per dowel was 20 tonnes.

12.   530 dowels were installed varying in length from 5 metres to 9 metres over a total length of slope of 260 metres.

13.   In addition to the shear dowels, a drainage blanket was installed at the upper surface of the rock prior to infilling above with excavated material to form inclined side slopes to the cutting.

14.   Where rock faces were exposed these were protected from deterioration by the installation of gabions.

CONCLUDING REMARKS

15.   Although the friction angle of the material on the failure surface was 24°, water pressure initiated a failure along a bedding plane inclined at 13° to the horizontal.

16.   The most economical form of remedial work proved to be the installation of shear dowels perpendicular to the potential failure planes.

TN19. Collapse test on a reinforced soil wall

N. W. M. JOHN, BSc, PhD, FGS, Lecturer, Queen Mary College, University of London

SYNOPSIS. An instrumented reinforced soil wall was constructed as part of an investigation into the behaviour of walls with plastic soil reinforcement. The safety factor in part of the structure was progressively reduced over two months until an adherence failure was eventually induced. As the structure approached failure interesting changes in the load distribution were detected.

THE STRUCTURE

1. This was rectangular in plan and covered an area of 20m x 5m. Each of the structure's four faces consisted of a $2\frac{1}{2}$m high reinforced soil wall using plastic strip soil reinforcement. Single size 20mm gravel aggregate was used as fill both within and behind the reinforced soil walls. The soil reinforcement consisted of 50mm wide Paraweb strips, with an ultimate strength of 400 kg, placed at a spacing of 300mm vertically and 600mm horizontally.

2. During construction the reinforcement supporting one of the short end walls was inserted through horizontal slotted small plastic pipes passing through the structure. These permitted a cutting device to be inserted which reduced the reinforcement length in stages during the collapse test. Load transducers were used as connectors between the reinforcement

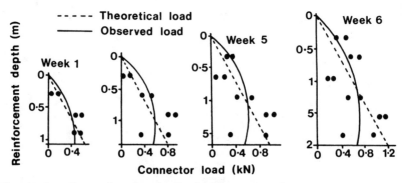

Figure 1. Connector loads during construction

Failures in earthworks. Thomas Telford Ltd, London, 1985

and the facing panels. In addition foil resistance strain gauges were used to monitor the strain distribution along some of the reinforcement strips.

RESULTS

3. Figure 1 shows the increase in load on the connectors between the reinforcement and facing panels during construction. The theoretical load indicated in this figure is based on a simple tie-back wedge analysis. It was observed that the lateral pressures at the base of the structure were less than the theoretical analysis predicted, while those at the top were greater than predicted. As compaction was deliberately avoided in this structure this lateral pressure distribution is probably the result of the mode of wall deflection (ref.1).

4. There was little change in the lateral pressure distribution (fig. 2) throughout the 14 months prior to the collapse test or during the first 7 weeks of the collapse test. However, during the last stages of this test there was a very noticable change in the load distribution at the facing panel (fig. 3).

5. The strain distribution along one of the reinforcement strips placed at a depth of 2m below the top of the finished wall is shown in fig. 4. This indicates that during construction the strain increased as expected when the fill was placed. Readings for the same reinforcement strip show the transfer of load towards the face of the structure as the reinforcement becomes shorter (fig. 5). Eventually the friction bond on the remaining intact reinforcement is so reduced that the load on the facing then decreases.

CONCLUSIONS

6. It appears that the tie-back wedge method of analysis under-estimates the true adherence safety factor. In this particular case the error in the calculated value was in the region of 70%, even allowing for side wall and base friction. As reinforced soil walls are flexible and tend to rotate about their upper edge (ref. 1) the true failure zone should be smaller than the simple Coulomb wedge which is assumed in

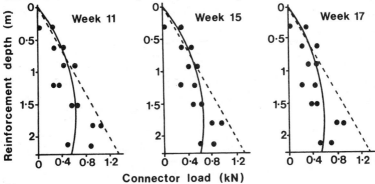

Figure 2. Connector loads prior to collapse test

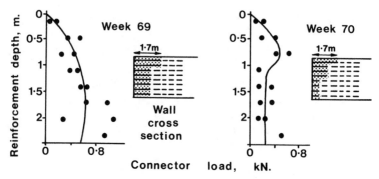

Figure 3. Connector loads during collapse test

the tie-back wedge analysis. However, the other extreme of
ignoring the failure zone and considering the full length of
intact reinforcement to contribute to adherence stability is
invalid as this consierably over-estimates, the true adherence
safety factor.

7. The observations from this test demonstrate the ability
of reinforced soil structures to redistribute internal loads,
suggesting that a local reinforcement failure does not inevitably lead to collapse.

REFERENCE

1. JOHN N.W.M. Behaviour of Fabric Reinforced Soil Walls,
Ph.D. thesis, Portsmouth Polytechnic, 1983.

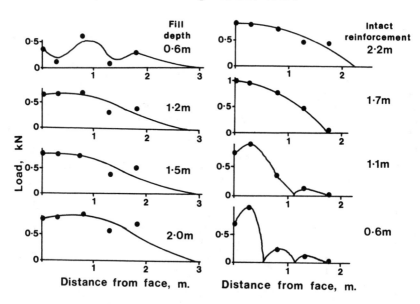

Fig.4 Strain distribution
      during construction

Fig.5 Strain distribution
      during collapse test

TN 20. Selecting design strength of granular fills

M. D. BOLTON, MA, MSc, MICE, Lecturer, Cambridge University
Engineering Department

SYNOPSIS. Slip displacements of the order of 5 to 10 particle
diameters cause the angle of shearing resistance of dense sand
to drop from about 50° to a critical state value of about 33°.
A rationale is offered for the selection of the design
strength of fills in relation to their particle size.

ANGLES OF SHEARING RESISTANCE

1. A large body of triaxial and plane strain data exists
which suggests that the effective shearing resistance of any
soil fill is most conveniently thought of in terms of a secant
angle ϕ', using $c' = 0$ on a τ/σ' diagram. The maximum angle
angle ϕ'_{max} in such uniform strain tests is a function of the
soil's density and the mean effective stress level.
Ultimately, the angle of shearing drops to the critical state
value ϕ'_{crit}.

2. ϕ'_{crit} can be identified with shearing at constant
volume. It is found to be between 32° and 34° for the
majority of quartz soils, but it has been found to be about 5°
higher in felspar soils.

3. The extra angle of shearing ($\phi'_{max} - \phi'_{crit}$) of
initially dense soils is directly proportional to the rate of
dilatancy, and can only be mobilized if the volume of the soil
is expanding while it is sheared. Maximum dilatancy
components of 20° in plane strain and about 12° in triaxial
strain have been observed in all soils when fully compacted
and at a mean effective stress p' of about 300kN/m². Any
increase in p', or reduction in relative density, is observed
to cause a reduction in the dilatant contribution.

LOCALISATION OF SHEAR

4. Angles of shearing measured using a 60mm square shear
box are habitually found to be less than those measured in
uniformly strained plane tests. One reason for this maybe
that the horizontal shear plane need not coincide with the
planes of maximum stress obliquity. Another reason may be
that strains localise, permitting softening in shear bands
prior to the mobilization of maximum shear force. Such shear
bands have been observed to propagate from the ends in a

specially constructed long shear box (ref.1). It was observed that the thickness of such bands was of the order of 5 to 10 particle diameters over a particle size range of d_{50} = 0.1 to 0.9mm, though the box displacements required to attain maximum shear force were not so strongly influenced by particle size.
5.   Complex conditions must exist inside the box in order to achieve compatibility between the dilatant and softened zones. There is, of course, no guarantee that the kinematic restraint offered by a shear box is similar to that offered by a body of soil in the field, engaging in some earthwork failure.

PREVIOUS RESEARCH ON SMALL-SCALE MODELS

6.   In passive retaining wall tests (ref.2) with H/d_{50} = 370 a boundary translation δ_b/d_{50} = 12 at maximum mobilization caused negligible softening from peak plane strain strengths.
7.   In active retaining wall tests with H/d_{50} = 6500 a boundary translation δ_b/d_{50} = 12 at maximum mobilization was sufficient to cause an average 38% softening from peak to critical.   In passive tests with H/d_{50} = 2000 a boundary translation δ_b/d_{50} = 80 at maximum mobilization was sufficient to cause 72% softening (ref.3).
8.   In two series of passive wall tests of identical geometry (ref.4), at H/d_{50} =2300 a boundary translation δ_b/d_{50} = 115   at maximum mobilization was sufficient to cause 60% softening, while at H/d_{50} = 23000 a boundary translation δ_b/d_{50} = 692 at maximum mobilization was sufficient to cause 95% softening.
9.   These studies suggest
   (a) that each problem, eg active or passive, has its own characteristics.
   (b) that in passive translation of a retaining wall, the movement for maximum mobilization may not be a strong function of H/d_{50} whereas δ_b/H = 0.04 offers a reasonable fit.
   (c) that the amount of softening prior to maximum mobilization is a strong function of H/d_{50}, being of the order of

   $\quad$ 0% for H/d_{50} =   100
   $\quad$ 50% for H/d_{50} =   1000
   $\quad$ 100% for H/d_{50} = 10000

RECENT RESEARCH ON MODELS

10.   Centrifuge tests were conducted on mass walls retaining dry sand subjected to a strip footing under increasing load (ref.5). Although some progressive failure was observed at H/d_{50} = 164, rather more was observed at H/d_{50} = 658 using a finer soil.
11.   Centrifuge tests were conducted on mass walls retaining dry sand and subjected to lateral base shaking (ref.6). Progressive failure, leading to total softening to a critical state, was observed to require a wall movement δ_b/H = 0.06 for both H/d_{50}= 100 and for H/d_{50} = 400. Apparently, full peak

stengths were initially mobilized in each of these active failures.

RECOMMENDATIONS

12. If the degree of localisation of shear strains in a particular failure scenario is unknown, fully softened critical state strengths should be used in design.

13. In static passive earth pressure situations an extrapolation of model experience suggests that the use of crushed rock to achieve $H/d_{50} \ll 1000$ might permit the mobilization of a significant proportion of the dilatant peak. This would be consistent with practice for rock fill dams, in which localisation is much less likely, of course.

14. In active earth pressure situations, the tendency to localise strains is very great, but the displacements necessary for maximum mobilization are very small. The use of larger particles to achieve $H/d_{50} \ll 1000$ might similarly be sufficient to ensure a significant dilatancy contribution.

15. More model tests and field trials are necessary to establish these relationships with greater certainty. Any reliance on the dilatant peak strength component confers a vulnerability to catastrophic failure following unforseen deformations, such as those generated by earthquakes.

REFERENCES

1. Scarpelli G. and Wood D.M. (1982) Experimental observations of shear band patterns in direct shear tests. IUTAM Conf. on Deformation and Failure of Granular Materials, Delft, 473-483.

2. Roscoe K.H. (1970) The influence of strains in soil mechanics. Geotechnique, vol. 20, No.2, 129-170

3. Rowe P.W. (1969) Progressive failure and strength of a sand mass. Proc. 7th Int. Conf. on Soil Mechanics and Foundation Eng., Mexico, vol. 1, 341-349.

4. Davis A.G. and Auger D. (1979) La butée des sables: essais en vraie grandeur. Annales de L'Institut Technique du Batiment et des Travaux Publics, vol. 375, Sols et Fondations No. 166, September, 69-92.

5. Mak K.W. (1984) Modelling the effects of a strip load behind rigid retaining walls. PhD Thesis, Cambridge University.

5. Bolton M.D. and Steedman R.S. (1985) Modelling the seismic resistance of retaining structures. To be published at 11th Int. Conf. on Soil Mechanics and Foundation Engineering, San Francisco, August.

TN21. Failures in buried pipes

L. ADESTAM, Civil Engineer, Swedish Geotechnical Institute

SYNOPSIS. Since 1980 the Swedish Geotechnical Institute has been using the hose settlement gauge to measure the elevation of buried pipes, e.g. sewers. Many results indicate that un-expected heaves and settlements have occured since backfil-ling, even in areas with firm soils and rock. Investigations show that most of the failures are caused by incorrect const-ruction as a result of errors in workmanship or recommenda-tions.

THE HOSE SETTLEMENT GAUGE. The equipment used for measuring the elevations of buried pipes is shown i Figure 1, together with the result obtained.

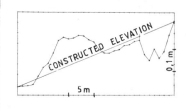

Figure 1. The hose settlement gauge. Elevation in a 225 mm dia sewer.

The tube is filled with fluid and the pressure (relative level) in the transducer can be read off the instrument. The accuracy in data obtained is closer than ±5 mm. The field level mea-surement capacity is about 400 m a day. A computer is used to calculate and draw up the actual elevation (see the example in Figure 1).

Failures in earthworks. Thomas Telford Ltd, London, 1985

FAILURES. During the past four years the elevations of more than 40 000 m of pipes have been measured, generally to check contract work. In many cases, unacceptable changes in level, mainly settlements, have been found. For example the elevation shown i Figure 2 was measured in a two-year old sewer founded on rock, firm clay and firm silt.

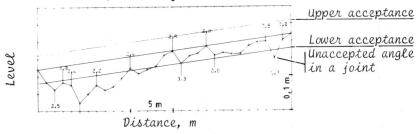

Figure 2. Measured elevation in a 150 mm dia sewer.

Due to the risk of sedimentation at low points and leakage at joints this pipe had to be reinstalled. The example is just one of a number of pipes under similar conditions with similar problems.

Investigations show that many of these failures are related to incorrect construction of the pipe bed, mostly as a result of bad workmanship. One type that is often seen is shown in Figure 3.

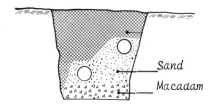

Figure 3. Incorrect pipe bed

Figure 4. Unsealed sub-
surface

The bottom of the trench is filled with a coarse material (macadam). The pipe bed is then prepared with a fine grained material (sand), which is also used as fill around the pipe. The construction performs satisfactorily until water enters the pipe bed and the sand falls into the voids in the coarse

465

material.

A similar situation has been found in a trench of blasted rock, where the pipe bed of sand was prepared without sealing the subsurface (se Figure 4).

Problems with liquefaction and "piping" are frequent in pipe trenches in fine grained soils below the groundwater level. Figure 5 shows how a pipe placed on a softened silt sank through approximately half its diameter. Uneven settlements caused high loads in the stiff grouted, joints, and the pipe cracked.

Figure 5.   Sunken pipe in   Crack near joint.
            silt.

Settlements in pipes have also been found when the subsoil is disturbed by excavation work, blasting etc.

Heaves are often explained as a result of faulty levelling. However, one example showed that improper filling/compaction around a pipe can cause heaves together with a lateral movement.

CONCLUSION.   The hose settlement gauge has shown to be a reliable and useful equipment to check the elevations in buried pipes. Some of the observed failures that have been examined are results of bad workmanship.

One way to improve construction work is to carry out continous inspection and supervision. It has also been found that education and information about incorrect constructions and their consequences are important.

TN22. Serviceability failure of a new railway embankment

C. L. LAIRD, BSc, MICE, Project Engineer, Thorburn Associates

SYNOPSIS. High excess porewater pressures and associated slow rates of settlement were recorded during construction of a railway embankment for the Selby Diversion, despite pretreatment with vertical drains. In order to meet programme requirements, berms were constructed to allow increased construction rates and surcharging. However as it became apparent that the final settlement performance criteria would not be met, the embankment was removed and reconstructed on a piled foundation.

DESIGN

1. The Temple Hirst embankment was constructed over an area of recent alluvium for a length of 60 metres at the southern end of the Selby Diversion (Ref. 1 and 2). A very important design consideration for the embankment was that the specified settlement limitations required that not more than 25 mm occurred in the first month of operation and not more than 75 mm would occur in the first year.

2. Recognising the settlement performance criteria and also to ensure stability of the embankment during construction, the original design incorporated the installation of vertical drains at 2.5 metres spacings beneath the 4.5 metres high rockfill embankment. Settlement at design stage was predicted to be up to a maximum of 900 mm. Two full sections of the embankment were instrumented with piezometers, settlement gauges and inclinometers to monitor the performance of the embankment which was immediately adjacent to the existing east coast main line embankment.

CONSTRUCTION

3. The construction programme dictated that the embankment be completed approximately 25 months after installation of the vertical drains at 2.5 metres centres down to the underlying granular deposits. The first 1.8 metres of embankment was placed in December 1980 without any appreciable build up of porewater pressure. Further upfilling was restricted by adjacent bridge fabrication works until 900 mm of rockfill was placed in February 1981 with 225 mm layers being added in October and November 1981, and February 1982 amounting to a

total of 3.375 metres of rock fill. At this time it was noted
that excess porewater pressures were dissipating relatively
slowly and that lateral movements of 120 mm had been recorded
in the inclinometer at the west toe of the embankment with a
maximum centreline settlement of 575 mm (Fig. 1).

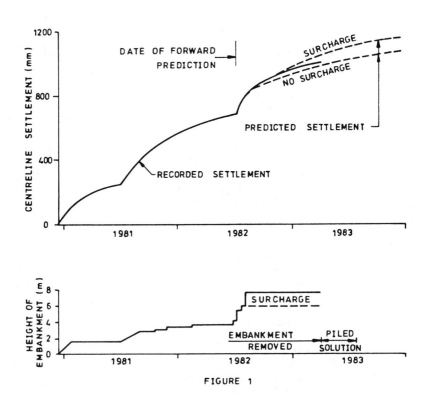

FIGURE 1

4. A review of embankment design, stability and projected
performance was initiated at this time. In order to be able
to make a reasonable prediction of future subsoil response to
further embankment loading, a consolidation analysis was
carried out using the program SETTLE (Ref. 3). A two-layer
soil model was adopted, with the upper 2.5 metres representing
the soft to firm brown and grey silty CLAY and the lower 5.5
metres representing the soft to firm grey/brown peaty CLAY
(with layers of peat in places) which existed beneath the
embankment. Two-way vertical drainage was used in view of the
persistent granular deposits beneath the peaty alluvium. A
back analysis was carried out initially to calibrate the
consolidation parameters. A best fit of the model was found
with the following coefficients:

Upper silty CLAY - C_v and C_h = 30 m^2/year
Lower peaty CLAY - C_v and C_h - 5 m^2/year

5. The model was then used to predict soil/embankment response to further loading and enabled assessments of stability to be made. A berm was subsequently formed along the toe of the embankment to ensure stability while the remainder of upfill was placed rapidly. In addition a surcharge loading of 32 kN/m^2 was provided by track ballast. Predictions of settlement made at this time indicated a revised total settlement of 1200 mm and that it would be unlikely that the settlement criteria would be met within the contract programme.(Fig.1)

6. An alternative design was prepared which involved removing the majority of the embankment before reconstructing the problematic 60 metres length on universal bearing piles using the BASP method (ref. 4). This alternative was only to be adopted if monitoring records during the period September 1982 to March 1983 indicated that the settlement criteria would not be achieved. In the event, by March 1983, substantial excess porewater pressures of up to 6 metres above hydrostatic level remained in the peaty clays with a fairly uniform monthly settlement rate of 15-20 mm being recorded. Even with the ballast surcharge removed it was considered that the settlement performance would be well outside the permissible limits, and hence the decision was taken to carry out the alternative piled design. This work was completed in a 4 month period April-July 1983 and allowed completion of the tracklaying without delay to the overall completion of the project by early October 1983.

REFERENCES
1. DAVIS P.B, FENWICK T.H. and BASTIN R.D. Selby diversion of the East Coast Main Line. Proceedings of Institution of Civil Engineers, Part 1, 1983, Vol.74, November, 719-747.
2. COLLINGWOOD R.W. and FENWICK T.H. Selby diversion of the East Coast Main Line; construction. Proceedings of Institution of Civil Engineers,Part 1,1985,Vol.77, February, 49-84.
3. MURRAY R.T. Two dimensional analysis of settlement by computer program. Transport and Road Research Laboratory, Report No. 538.
4. REID W.M. and BUCHANAN N.W. Bridge approach support piling. Proceedings of the International Conference on Advances in Piling and Ground Treatment for Foundations, ICE, London, 1983, 267-274.

TN23. Treatment of shallow underground fires

D. J. AYRES, FSE, FGS, Soil Mechanics Engineer, British Railways Board

TREATMENT OF SHALLOW UNDERGROUND FIRES

Most organic and many inorganic materials in nature react chemically with their surroundings, usually with the production of heat. This is spontaneous combustion. These exothermic reactions are rarely noticeable as the small amounts of heat produced are lost to the surroundings. When the reaction is such that there is a large exotherm and the heat cannot dissipate quickly there is a rise in the temperature which can reach flashpoint, at which stage fire can occur. This general concept is often applied to embankment or shallow underground fires in describing them as "spontaneous" but, in the writer's experience, this is uncommon, as the prime agent of fire is man.

A fire can only continue if the three essential prerequisites exist, namely combustible material, a supply of air and an ignition temperature, obtain. Special chemical fires are not considered here. All fire fighting deals with the removal or control of one or more of these factors.

In the century up to 1960 large amounts of ash from steam locomotives were available and this material was used as fill in many places: colliery shale waste has also been used in mining areas. Such materials were not compacted to modern specifications and sparks from locomotives in the past or the lighting of bonfires has been sufficient to cause a fire. This has also happened where a contractor was removing ash fill from a disused embankment and the angle of the exposed faces in relation to the prevailing winds was sufficient either to take spontaneous combustion to flash conditions or to allow a cigarette stub to ignite it. Power stations take account of the problem by stacking coal at limiting slopes.

When fires start early investigation and treatment is required. Vegetation may become more lush locally and will wither at hot spots. The smell is typical and at this point the Fire Brigade is called. Most senior fire officers appreciate that a specialist approach is required but some will apply large quantities of water in an attempt to dowse the fire. The effect of water is to make a large number of channels through the combustible material such that, when the water is turned off, they dry swiftly and air flows through the warm paths at a high rate with a consequent return of the fire.

Investigation

The surface effects, smoke, vegetation change, may swiftly be observed by eye and this can be supplemented using an infra red camera which can produce instantly black and white or coloured thermal images. Boreholes are of limited use, as a large number would be required to locate the main fire and they could not be capped while boring. Probing using driven steel tubes set out to a specified grid is appropriate as they can be used for checking the depth of non-combustible material, for temperature measurements and for grouting. All tubes must be capped. For a very large area or to check heat proximity to sensitive locations such as fuel tanks, infra red cameras mounted in helicopters may be used. B.R. has developed a portable thermocouple system such that the hot junction can be lowered up to 15m down a 12mm steel or plastics tube to measure to within $1^{\circ}C$. As the centre of heat moves with time its location is plotted in three dimensions so the nature of air flow may be inferred and treatment applied accordingly.

Treatments

According to site conditions, if the combustible material is not too thick, say 4m or 5m, limiting cut-off trenches may be placed; otherwise the fire may be allowed to burn itself out or, in small sites, the combustibles may be dug out. Excavation must be swift and the backfill should be of crusher-run rock (tailings) with a size distribution down to silt size. Crushed limestone may be used, which expands under sufficient heat and gives off carbon dioxide. Exposed slopes allow air entry. Blankets to cover them should have a high uniformity coefficient to allow the wind to remove only the top few mm, below which level the dust content can still control gas permeability.

Permeation grouting is a well-established technique using suspensions of cement, pulverized fuel ash or limestone flour. Injection should be near the centre of fire, applied on the side towards which it appears to be moving. At temperatures much over $150^{\circ}C$ thinner suspensions should be used to prevent blockage of the tube and allow travel of grout to block air passages within the requirements of the grid of points.

The fire may also be checked by densifying the ground in its vicinity.

A fire occurred in a 20m high embankment in S.Wales, extending from the cone, where the double track ran on to a viaduct, back some 170m. The top 12m of the embankment was composed of mine waste and ashes, which had been tipped over a buried timber viaduct. The slope was steep and subject to high wind gusting; fires moved about in the combustible waste with frequent temperatures of 700°C. Thin grout suspensions were injected to keep the fire away from the sleepered track to permit trains to pass safely at low speed. The line was closed over several weekends to allow sections of a new viaduct to be installed and the access was used to apply deep compaction techniques. The top of the embankment was from 16m to 20m and three rows of compaction points (prints) were set 6m apart to run along the central 12m and to avoid slip failure in the slope. A weight of 15 tonnes was used, dropping from increasing heights, with up to 30 drops per print as the ground was compacted. Dirty ballast taken from other sites was imported to keep down air permeability and placed to fill the prints. Drop heights were up to 20m followed by levelling the ballast with a blade. Further drops of 10m constituted the "ironing" pass.

Penetration tests had been carried out initially to assess the energy of drop; post treatment tests indicated general densification but all probe areas were subsequently grouted in view of the chimneys being produced in the fill. A total of 2000T of ballast was used in the prints, which extended only to 25m from the edge of the viaduct to avoid damaging it.

The work was carried out over three weekends and was checked by a helicopter I.R. scan 8 weeks later, in April, 1979. This produced isotherm pictures which implied surface temperatures generally of 8°C, except for a warmer area of 11°C towards the base of the cone of the embankment. This area was blanketed with a crushed stone sand and no further problems have been encountered.

Experience of various sites indicates that it may take one or two years before cooling of the body of the embankment brings the internal temperature much below 35°C.

Acknowledgements

The author is grateful to Mr. M. C. Purbrick, Director of Civil Engineering, British Railways Board, for permission to publish this paper.